MOLECULAR COLLISION THEORY

A Series of Monographs

THEORETICAL CHEMISTRY

Consulting Editors

MOLECULAR COLLISION THEORY

M. S. CHILD
Fellow of St Edmund Hall and Lecturer in Theoretical Chemistry,
The University of Oxford

1974

ACADEMIC PRESS
London and New York
A subsidiary of Harcourt Brace Jovanovich, Publishers

ACADEMIC PRESS INC. (LONDON) LTD.
24/28 Oval Road,
London NW1

United States Edition published by
ACADEMIC PRESS INC.
111 Fifth Avenue
New York, New York 10003

Library of Congress Catalog Card Number: 74 5663
ISBN: 0 12 172650 9

Set in 'Monophoto' 569 Mathematical Times and printed offset litho in Great Britain by
Page Bros (Norwich) Ltd, Norwich

PREFACE

This book is intended as an introduction to the theory required for interpretation of an increasingly sophisticated range of molecular scattering experiments. The contents have been chosen to emphasize the quantum mechanical and semi-classical nature of these events, with little attention being given to purely classical behaviour. A further selection has been made in giving an essentially analytical treatment, with no attempt at comprehensive coverage either of the formal or of the computational aspects of the subject.

Some knowledge of the quantum mechanics of bound states has been assumed. The reader seeking a first introduction may, without loss of continuity, bypass certain inessential or more difficult sections marked by an asterisk, and postpone considerations of mathematical detail until turning to the Appendix.

While the weaknesses of the book remain my own, my gratitude is due to Dr. J. N. L. Connor and Mr. A. G. Ritchie for critical comments on parts of the manuscript, and to Mrs. M. Long and Mrs. J. Nordman for secretarial assistance.

M. S. CHILD

Oxford
January 1974

CONTENTS

* Sections marked by an asterisk may be omitted without loss of continuity in the argument.

CHAPTER 1

Introduction

1.1 Objectives

This book is concerned with the basic theory of molecular collisions; that is with the relation between the intermolecular interaction potential and the simplest experimentally observable events. The scope of the theory is therefore circumscribed by the nature of possible experiments, and by the behaviour of matter on a molecular scale.

The first problem is that, unlike an impact between billiard balls for example, no single molecular collision is ever observed in isolation. Furthermore any information obtained is limited to an overall change in the state of the system before and after the event. The theoretical quantity for comparison with experiment is therefore not the single particle trajectory, even when this can be defined, but the cross-section, $\sigma_{ij}(E)$, or notional area, through which a representative incident particle must pass at a given energy to achieve a given change $i \rightarrow j$ in the system.

Any breakdown of classical mechanics must also be taken into account. This applies not only to the familiar quantization of internal states, but also to the relative translational motion of molecules in the system. Quantum mechanical considerations may lead, on one hand, to interference effects, but they may also weaken the very concept of a collision. Suppose for simplicity that an elastic collision is specified by marking a point on the incident path and noting either the deflection from this path, or equivalently the momentum change due to the collision. This requires that uncertainties in the marker point and in the momentum change are small compared with their absolute values. Thus as the deflection angle, and momentum change, decrease to zero, it may become impossible by virture of the uncertainty principle to specify the marker point with any accuracy. The concept of a collision then loses any meaning in a classical sense. Such breakdowns are however rare on the tolerably massive molecular scale. Thus the appropriate theory must be quantum-mechanically based but also loosely tied to classical mechanics, in order to benefit from classical experience when the de Broglie wavelength becomes short compared with the range of the relevant interaction. The result is best described as semi-classical, and one purpose of this book is to offer a background to current rapid developments in this field. The other is to provide an introduction to available analytical methods for the interpretation of an increasingly sophisticated range of possible experiments.

Three broad classes of such experiment may be recognized. These are termed elastic in cases of simple momentum transfer, inelastic if there is also a change in internal energy, and reactive when accompanied by a change in chemical composition. The first give information about the mean intermolecular potential function, averaged over internal states, the second about the dependence of this function on the relevant internal coordinate, and the third about the nature of the potential energy surface in the neighbourhood of the "reaction coordinate".

The material in subsequent chapters follows this natural division. Thus elastic scattering is covered by Chapters 2–5, inelastic by Chapters 6–9 and reactive by Chapter 10, with the balance of material being governed partly by the current development of the theory, and partly by the extent to which, for example, the reactive discussion may be taken to depend on the previous treatment of inelastic events. While, as indicated above, broadly semi-classical in intention, each of the three parts of the book is firmly based on quantum mechanics. This sets the semi-classical analysis of Section 4.2 and Chapters 5, 7, 8 and 9 in context, and also justifies the use of non semi-classical methods such as those employed in Born and Distorted Wave approximations, where these have proved valuable in practice. Emphasis in selecting material has been given to obtaining analytical results of practical value in the interpretation of experimental data. Thus the book is intended to complement wider and more formal approaches to molecular rate phenomena such as that given by Levine (1969).

1.2 Laboratory and Centre of Mass Coordinates

Subsequent calculations are simplified by transforming from the laboratory to the centre of mass coordinate systems. This involves replacing the positions $\mathbf{r}_1$, $\mathbf{r}_2$ and velocities $\mathbf{v}_1$, $\mathbf{v}_2$ of the collision partners, with assumed masses m_1 and m_2 respectively, by their centre of mass ($\mathbf{R}$, $\mathbf{V}$) and relative ($\mathbf{r}$, $\mathbf{v}$) counterparts;

$$\mathbf{R} = \left(\frac{m_1}{M}\right)\mathbf{r}_1 + \left(\frac{m_2}{M}\right)\mathbf{r}_2$$

$$\mathbf{r} = \mathbf{r}_1 - \mathbf{r}_2 \tag{1.1}$$

and

$$\mathbf{V} = \frac{m_1}{M}\mathbf{v}_1 + \frac{m_2}{M}\mathbf{v}_2$$

$$\mathbf{v} = \mathbf{v}_1 - \mathbf{v}_2 \qquad (1.2)$$

in terms of which the classical kinetic energy T and angular momentum L may be written

$$T = \tfrac{1}{2}m_1 v_1^2 + \tfrac{1}{2}m_2 v_2^2 = \tfrac{1}{2}MV^2 + \tfrac{1}{2}mv^2$$

$$L = m_1(\mathbf{r}_1 \wedge \mathbf{v}_1) + m_2(\mathbf{r}_2 \wedge \mathbf{v}_2) = m\mathbf{r} \wedge \mathbf{v}, \qquad (1.3)$$

where

$$M = m_1 + m_2, \quad m = m_1 m_2/(m_1 + m_2). \qquad (1.4)$$

This means since the potential energy is necessarily independent of $\mathbf{R}$, and of the absolute orientation of the system, that the centre of mass motion may be factored out. The remaining problem, to which the subsequent theory is confined, is therefore equivalent to that of a single (possibly structured) particle with reduced mass m, in a given initial internal state, moving with initial translational energy, $\frac{1}{2}mv^2$, and (orbital) angular momentum $\mathbf{L}$ about a scattering centre at $r = 0$.

The relation between this motion in the centre of mass frame and laboratory measurements is conveniently expressed by means of the Newton diagram in Fig. 1.1, which gives a pictorial representation of Eqns (1.2) or the equivalent forms

$$\mathbf{v}_1 = \mathbf{V} + \left(\frac{m_2}{M}\right)\mathbf{v}$$

$$\mathbf{v}_2 = \mathbf{V} - \left(\frac{m_1}{M}\right)\mathbf{v}. \qquad (1.5)$$

The important points are that the magnitude of the relative velocity, $\mathbf{v}$, is a direct measure of the available translational energy, and that the centre of mass point, G, divides $\mathbf{v}$ in inverse ratio to the masses of the collision partners. Overall kinematic changes in the system for different types of collision therefore lead to different changes in the diagram. In the discussion below, dashed circles in Figs 1.2–1.4 are used to indicate the constraints imposed on $\mathbf{v}$, and primes to denote the final state of the system. In all cases the centre of mass velocity $\mathbf{V}$ remains of course invariant.

Thus in a purely elastic collision, with both the magnitude and the point of division of $\mathbf{v}$ conserved, the momentum transfer

$$m_1(\mathbf{v}_1' - \mathbf{v}_1) = -m_2(\mathbf{v}_2' - \mathbf{v}_2), \qquad (1.6)$$

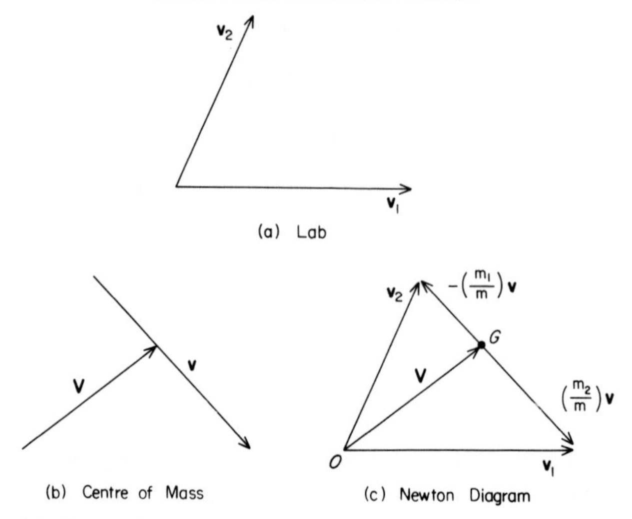

FIG. 1.1. Centre of mass and laboratory coordinates, and the Newton diagram.

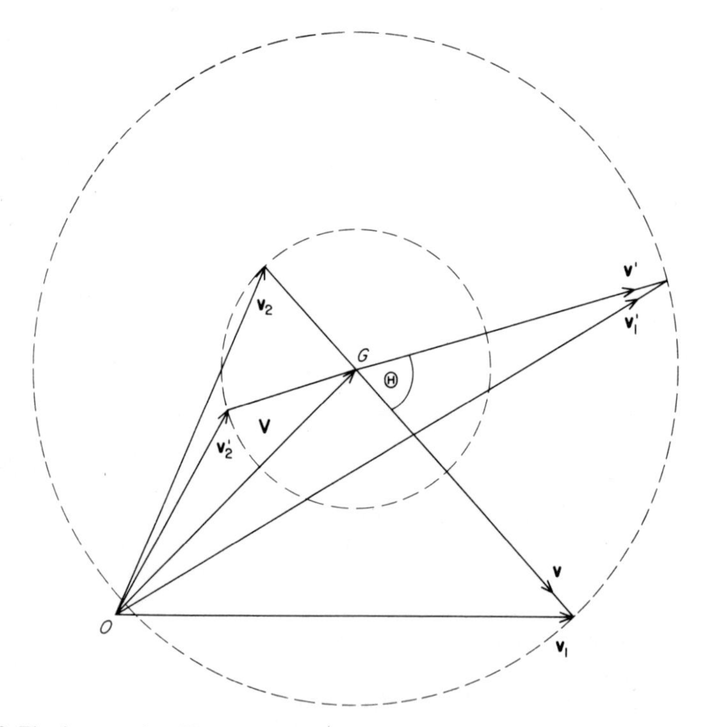

FIG. 1.2. Elastic scattering. The centre of mass deflection angle Θ lies between the initial and final relative velocity vectors $\mathbf{v}$ and $\mathbf{v}'$ respectively.

measured in the laboratory appears in Fig. 1.2 simply as a rotation of **v**. The deflection angle Θ depends however not only on the linear momentum and kinetic energy terms illustrated in the figure, but also on the available angular momentum and the potential energy function. Hence the diagram alone cannot determine Θ; it can only assert that the ends of the vector **v**′ should lie on the circles shown. Similarly the possibility of energy transfer between internal and translational motion, in an inelastic collision, is represented in Fig. 1.3 by allowing a change in both magnitude and direction of **v**.

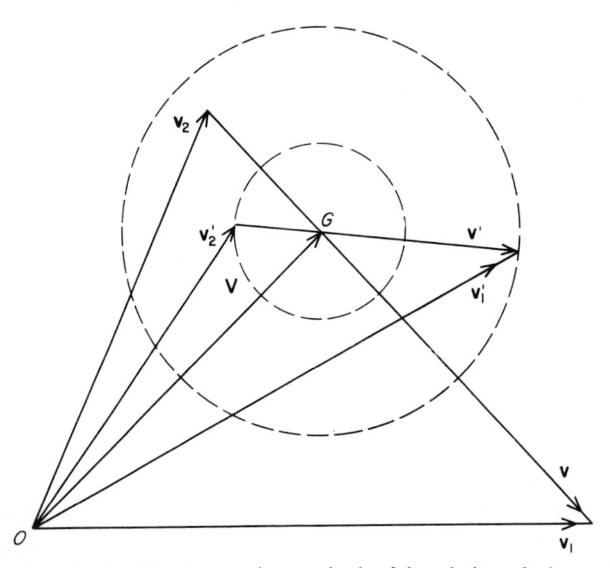

FIG. 1.3. Inelastic scattering. The decrease in magnitude of the relative velocity vector indicates a case of internal excitation, leading to a loss of translational energy.

Finally in the reactive case the initial relative velocity **v** refers to the reactants, and **v**′ to the products. Hence neither the magnitude nor the direction of **v** is conserved, and the divisions of **v** and **v**′ by the centre of mass point, *G*, depend on the reactant and product mass ratios respectively. It is clear however from Fig. 1.4 that if the reaction products have widely differing masses, the velocity of the heavier product must lie close to **V**, and that the possible angular spread of the product velocity vectors (given by the locus of $\mathbf{v}'_d$ for example) about the centre of mass velocity **V**, is a measure of the relative magnitudes of **v**′ and **V**.

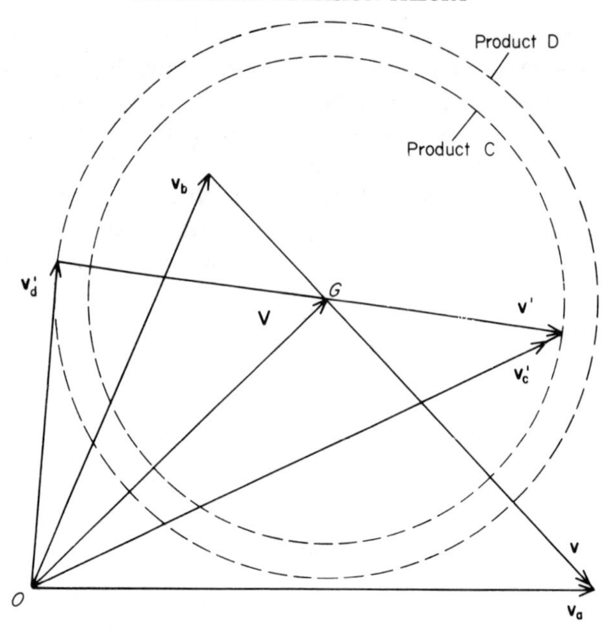

FIG. 1.4. Reactive scattering. The division of the final relative velocity vector indicates reaction products of almost equal mass, with $m_C \gtrsim m_D$.

1.3 Experimental Observables

The simplest type of molecular collision experiment involves two beams with defined velocities $\mathbf{v}_1$ and $\mathbf{v}_2$ and a detector tuned to observe a given change $i \to j$ in the internal states or chemical composition of the system (see Ross, 1966). Scattering occurs in all directions and the resulting intensity distribution in the centre of mass frame gives the differential cross-section defined by the ratio

$$\frac{\mathrm{d}\sigma_{ij}}{\mathrm{d}\Omega} = I_{ij}(\theta,\phi) = \frac{\text{scattered current† per unit solid angle}}{\text{incident current per unit area}}, \tag{1.7}$$

This differential cross-section is the fundamental observable quantity to which all others may be related. The second is the analogous total cross-section

$$\sigma_{ij}(E) = \int_0^{2\pi}\int_0^{\pi} I_{ij}(\theta,\phi)\sin\theta\,\mathrm{d}\theta\,\mathrm{d}\phi, \tag{1.8}$$

dependent on the initial kinetic energy of relative motion, $E = \frac{1}{2}mv^2$. The theory given below is restricted to these two quantities.

† Number of particles per unit time.

We must recognize however that many interesting systems at present defy investigation by this elegant beam technique and recourse must be had to a bulk phase experiment. The measurable quantity is then typically a rate constant which is related to the cross-section in the form

$$k(T) = \langle k_{ij}(T) \rangle = \left\langle \int_0^\infty v\, \sigma_{ij}(\tfrac{1}{2}mv^2)\, P(v)\, v^2\, \mathrm{d}v \right\rangle \tag{1.9}$$

where $P(v)$ denotes the (usually Maxwellian) relative velocity distribution, and the brackets $\langle k_{ij}(T) \rangle$ indicate an average over initial internal states. Experiments then differ according to the states j accepted by the detector. Thus in the case of a conventional primary chemical reaction step, i and j would cover all reactant and all product states respectively, while in more refined experiments, such as those described by Carrington and Polyani (1972), the detector might be tuned to a particular product vibrational level. Similarly vibrational relaxation rates measured by ultra-sonic and shock-wave techniques (see Herzfeld and Litovitz, 1959; Burnett and North, 1969) yield a rate constant for a given vibrational transition, thermally averaged over the initial rotational distribution and summed over final rotational states. The measurable in a gas transport process, which is the bulk phase manifestation of elastic scattering, is marginally different in that the coefficients of viscosity and thermal conductivity depend not on a rate constant but on the collision integral (see Present, 1958),

$$\Omega_\eta = \left(\frac{2kT}{\pi m}\right)^{\frac{1}{2}} \int_0^\infty \left(\frac{mv^2}{2kT}\right)^{\frac{7}{2}} \sigma_\eta(v) \exp\left(-\frac{mv^2}{2kT}\right) \mathrm{d}v, \tag{1.10}$$

where $\sigma_\eta(v)$ is the weighted cross-section

$$\sigma_\eta(v) = \int_0^{2\pi} \int_0^\pi I(\theta, \phi)(1 - \cos^2\theta) \sin\theta\, \mathrm{d}\theta\, \mathrm{d}\phi, \tag{1.11}$$

rather than the simple form defined by (1.8). Again knowledge of the differential cross-section, $I(\theta, \phi)$, as a function of collision velocity, v, is sufficient to determine the viscosity or thermal conductivity coefficient.

The problem in using such bulk phase data to elucidate the form of the interaction potential is of course that the necessary averaging process may destroy detailed information contained in the differential or total cross-section, leaving only strongly marked features in the velocity dependence of σ_{ij} to be reflected in rate constant. Thus if $\sigma_{ij}(E)$ vanishes below a threshold E^0_{ij}, $k_{ij}(T)$ takes the familiar Arrhenius form

$$k_{ij}(T) = A_{ij}(T) \exp(-E^0_{ij}/kT), \tag{1.12}$$

where $A_{ij}(T)$ is relatively insensitive to T. Similarly a cross-section of the form

$$\sigma_{ij}(\tfrac{1}{2}mv^2) \simeq \sigma_{ij}^0 \exp(-v_0/v), \tag{1.13}$$

is readily shown (see Herzfeld and Litovitz (1959)) to yield a rate constant which behaves as

$$k_{ij}(T) = B_{ij}(T) \exp\left[-\frac{3}{2}\left(\frac{mv_0^2}{kT}\right)^{\frac{1}{3}}\right] \tag{1.14}$$

where again $B_{ij}(T)$ is weakly dependent on T.† This form is of importance in vibrational relaxation theory, with v_0 taken to depend on the range of the intermolecular potential (see 7.2). Again however there are only two significant observables: v_0 and the magnitude $B_{ij}(T)$.

The most important role of molecular collision theory is therefore to identify the dominant characteristics of a given model potential function and to extract from them all possible observable consequences. To this we may add that in the special field of elastic scattering there is a unique direct inversion procedure to pass from the differential cross-section to the form of the interaction potential. This is described in Section 5.6.

Valuable general references may be found under Ross (1966), Hartmann (1968), Schlier (1970), Levine (1973) and Polyani and Schreiber (1973).

† The argument relies on approximating an integral of the form

$$k(T) = \int_0^\infty b(v) \exp\{-[(v_0/v) + (\tfrac{1}{2}mv^2/kT)]\} \, dv \tag{1.15}$$

by the method of steepest descents (Jeffreys and Jeffreys, 1956). This involves a quadratic approximation to the exponent about the stationary point, v^*, given by

$$\frac{d}{dv}\left[\left(\frac{v_0}{v}\right) + \left(\frac{mv^2}{2kT}\right)\right] = -\frac{v_0}{v^2} + \frac{mv}{kT} = 0, \tag{1.16}$$

so that

$$v^* = (v_0 kT/m)^{\frac{1}{3}}. \tag{1.17}$$

Hence

$$\left(\frac{v_0}{v}\right) + \left(\frac{mv^2}{2kT}\right) \simeq \frac{3}{2}\left(\frac{mv_0^2}{kT}\right)^{\frac{1}{3}} + \left(\frac{3m}{2kT}\right)(v - v^*)^2, \tag{1.18}$$

with the result, when $b(v)$ is replaced by its value at v^*, as given by (1.14).

CHAPTER 2

Classical Scattering by a Central Force

The purpose of this chapter is to give substance to the central ideas and terminology of collision theory by setting them in the familiar classical context. The assumption of a central force implies an interaction potential, $V(r)$, dependent only on the particle separation, and restricts the results of the collision to elastic scattering effects. The discussion leads from the classical trajectory to the collision cross-sections, and concludes with some comments on the validity of the classical approximation.

2.1 The Classical Trajectory

The equations of motion for the classical trajectory, in the centre of mass coordinate system, follow from conservation of the initial energy E and angular momentum $\mathbf{L}$ given by (1.3) and (1.4). The conditions of a given event are therefore conveniently defined by the incident relative velocity $\mathbf{v}$ and impact parameter b shown in Fig. 2.1. Since the direction of $\mathbf{L}$ is conserved, the trajectory must lie in a plane, and the equations of motion may be written, in polar coordinates:

$$L = mvb = mr^2\dot{\phi} \tag{2.1}$$

$$\begin{aligned} E = \tfrac{1}{2}mv^2 &= \tfrac{1}{2}m\dot{r}^2 + \tfrac{1}{2}mr^2\dot{\phi}^2 + V(r) \\ &= \tfrac{1}{2}m\dot{r}^2 + \frac{L^2}{2mr^2} + V(r). \end{aligned} \tag{2.2}$$

It follows by elimination of the time derivatives that

$$\frac{\mathrm{d}\phi}{\mathrm{d}r} = \left(\frac{\dot{\phi}}{\dot{r}}\right) = \pm\frac{b}{r^2[1 - b^2/r^2 - V(r)/E]^{\frac{1}{2}}}, \tag{2.3}$$

the sign being chosen according to the sign of the radial velocity, positive for outward and negative for inward motion respectively. The closest distance of approach, a in Fig. 2.1, at which $\dot{r} = 0$, is termed the *classical turning point*; a is therefore the largest root of the equation

$$1 - b^2/r^2 - V(r)/E = 0$$

or

$$E = V(r) + L^2/2mr^2 \tag{2.4}$$

The significance of multiple and repeated roots of (2.4) is discussed below.

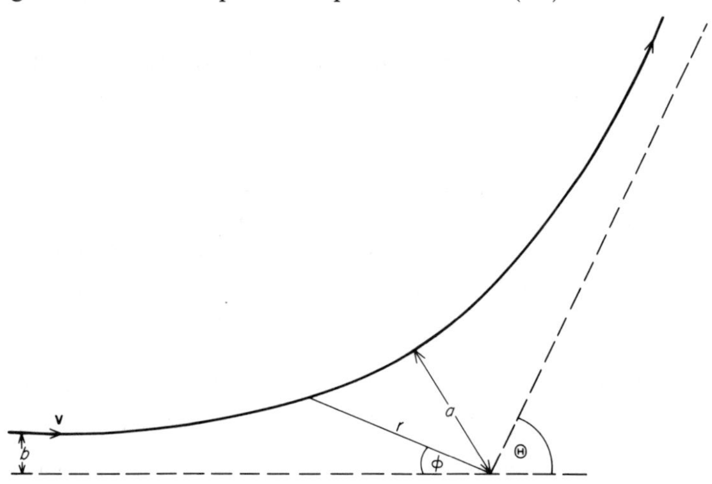

FIG. 2.1. A classical trajectory at velocity **v** and impact parameter b, with deflection angle Θ. a is the classical turning point.

The complete classical trajectory may now be determined by integrating (2.3) from infinity to a and out again. It is however unnecessary to follow the trajectory in detail if, as in the molecular case, the effects of the collision are observed only after the event. All that is necessary is the overall deflection Θ in Fig. 2.1;†

$$\Theta(E, L) = \pi - 2b \int_a^\infty \frac{\mathrm{d}r}{r^2[1 - b^2/r^2 - V(r)/E]^{\frac{1}{2}}} \tag{2.5}$$

Two other quantities of interest in later chapters may also be derived from (2.1) and (2.2). The first and most important is the radial component of the classical collision action† (Smith, 1965).

$$\Delta(E, L) = 2\int_a^\infty p(r)\,\mathrm{d}r - 2\int_b^\infty p^0(r)\,\mathrm{d}r$$

$$= 2mv\left\{\int_a^\infty [1 - b^2/r^2 - V(r)/E]^{\frac{1}{2}}\,\mathrm{d}r - \int_b^\infty [1 - b^2/r^2]^{\frac{1}{2}}\,\mathrm{d}r\right\}, \tag{2.6}$$

† See footnote on p. 11.

where $p(r)$ and $p^0(r)$ are the radial momenta in the presence and absence of the potential $V(r)$ respectively. The second is the collision delay time, or difference between the actual collision time, $2\int_a^\infty \mathrm{d}r/\dot{r}$, and its value in the absence of the potential†

$$\tau(E, L) = \frac{2}{v}\left\{\int_a^\infty \frac{\mathrm{d}r}{[1 - b^2/r^2 - V(r)/E]^{\frac{1}{2}}} - \int_b^\infty \frac{\mathrm{d}r}{[1 - b^2/r^2]^{\frac{1}{2}}}\right\}. \tag{2.7}$$

$\Theta(E, L)$ and $\tau(E, L)$ are actually partial derivatives of $\Delta(E, L)$

$$\begin{aligned}\Theta(E, L) &= (\partial\Delta/\partial L)_E \\ \tau(E, L) &= (\partial\Delta/\partial E)_L\end{aligned} \tag{2.8}$$

from which

$$\mathrm{d}\Delta = \Theta(E, L)\,\mathrm{d}L + \tau(E, L)\,\mathrm{d}E \tag{2.9}$$

At present, however, att ntion largely focuses on the deflection function $\Theta(E, L)$, typical forms for which are illustrated in Figs 2.3 (a), (b) and (c), for repulsive, attractive and non-monotonic intermolecular potentials respectively. The general appearance of these diagrams may be interpreted with the help of Fig. 2.2, which shows for the non-monotonic potential, the variation with L of an *effective* or *centrifugally* corrected potential

$$V_L(r) = V(r) + L^2/2mr^2, \tag{2.10}$$

the second term of which may be interpreted as the rotational energy required by angular momentum conservation; hence the derivative of this term with respect to r gives the familiar centrifugal force. It is seen from Fig. 2.2 that as the impact parameter, and hence according to (2.1) as L increases, the classical turning point a_L, which represents the limit of the accessible part of the potential, moves outwards. It follows that the deflection angles for high impact parameters are determined by the long-range part of the potential, being positive or negative for repulsive or attractive potentials respectively,

† For a two parameter potential, with strength ε and range r_0, $V(r) = \varepsilon W(\rho)$ where $\rho = r/r_0$, say, the quantities

$$\Theta = \pi - 2\beta\int_\alpha^\infty \rho^{-2}[1 - W(\rho)/K - \beta^2/\rho^2]^{-\frac{1}{2}}\,\mathrm{d}\rho,$$

$$\Delta^* = \Delta/(2mvr_0) = \int_\alpha^\infty [1 - W(\rho)/K - \beta^2/\rho^2]^{\frac{1}{2}}\,\mathrm{d}\rho - \int_\beta^\infty [1 - \beta^2/\rho^2]^{\frac{1}{2}}\,\mathrm{d}\rho$$

$$\tau^* = \tau v/2r_0$$

depend only on the reduced energy $K = E/\varepsilon$ and reduced impact parameter $\beta = b/r_0$. Numerical techniques for the evaluation of Θ, Δ^* and τ^* are discussed by Smith (1964), Kennedy and Smith (1967) and Smith (1965). Tables of Θ for the Lennard–Jones and exp-6 forms of potential are given by Hirschfelder, Curtis and Bird (1954) and Mason (1957) respectively.

as shown in Fig. 2.3(c). Low impact parameter collisions on the other hand experience the full energetically accessible range of the potential, but, since the angle ϕ in (2.4) and Fig. 2.1 is small over the attractive part of the potential, the tendency to a negative deflection is small and the trajectory is dominated by any repulsive core of the potential. Consequently the limiting trajectory as $b \rightarrow 0$ shows no deflection for a purely attractive potential (Fig. 2.3b), and strong backward scattering ($\Theta = \pi$) in the presence of a repulsive core (Fig. 2.3a, c). Furthermore since for a potential with an attractive part $\Theta = 0$ or π for $b = 0$ and $\Theta < 0$ for large b, the deflection function at a given energy

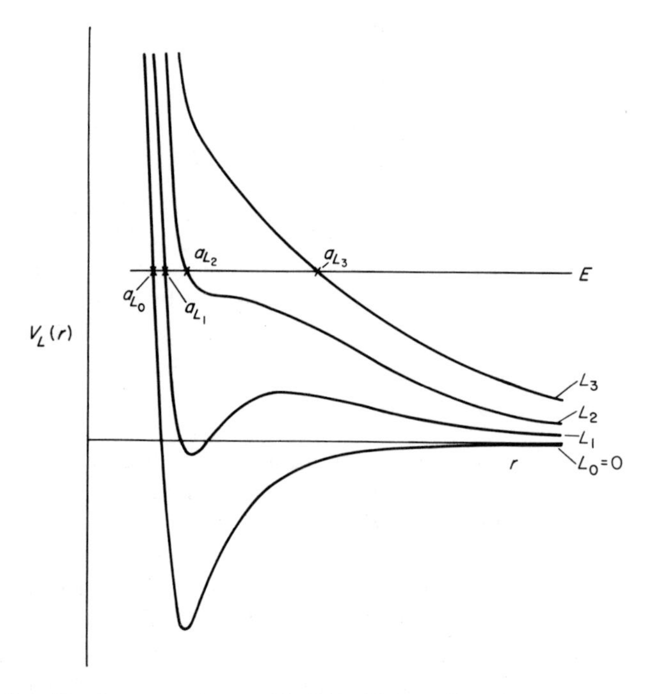

FIG. 2.2. Centrifugally corrected potentials $V_L(r)$ with $L_1 < L_2 < L_3$. a_0, a_{L_1}, a_{L_2} and a_{L_3} are the classical turning points at collision energy E.

must pass through a minimum, so called *rainbow* value, which has an important effect on the scattering cross-sections discussed below. A further peculiarity of attractive forces is shown by the appearance of *orbiting* singularities in Figs 2.3b and 2.3c, which correspond physically to the mutual capture of the colliding particles. The classical orbiting condition is actually that Eqn (2.4) should have repeated roots leading to a divergent integrals in (2.5) and (2.7); in other words the energy should coincide with a maximum in $V_L(r)$ in Fig. 2.2. This situation can arise for any attractive potential with long range

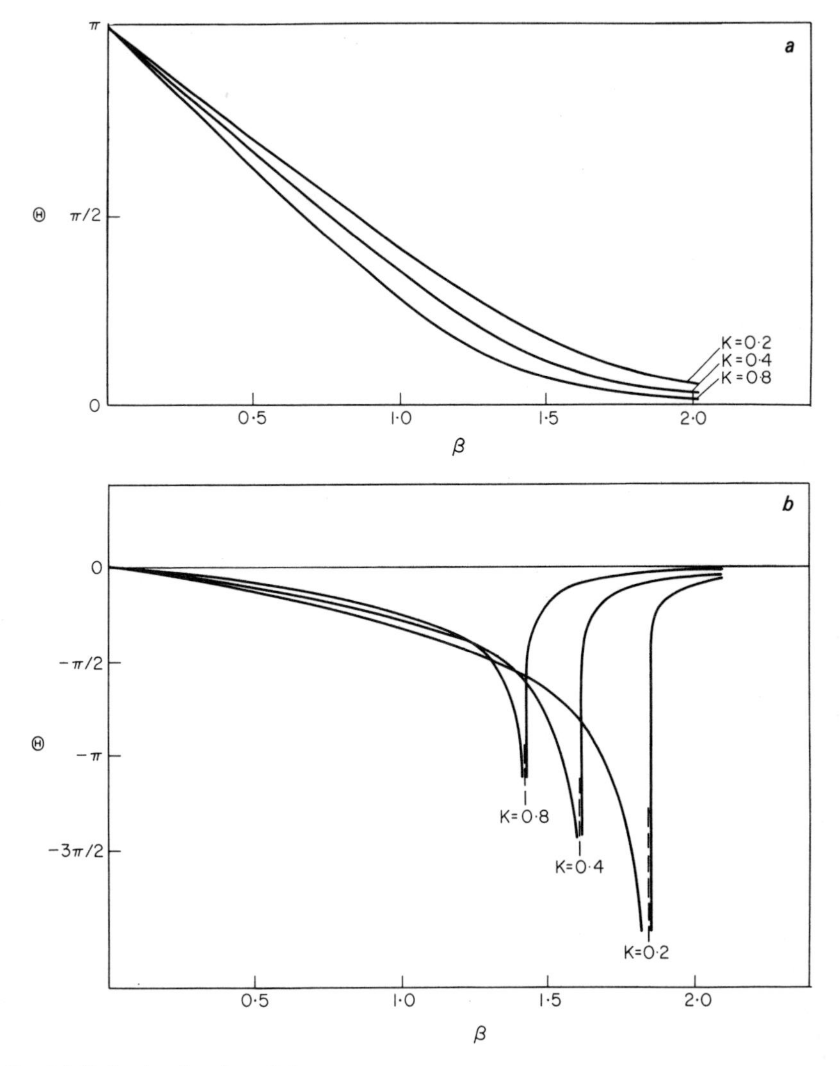

FIG. 2.3. Deflection functions Θ for

(a) repulsive $\left[V(r) = \varepsilon\left(\frac{r_0}{r}\right)^6\right]$,

(b) attractive $\left(V(r) = \begin{cases} -\varepsilon & r \leqslant r_0 \\ -\varepsilon(r_0/r)^6 & r > r_0 \end{cases}\right)$

interatomic potentials. β is the reduced impact parameter (b/r_0) and K the reduced collision energy (E/ε). Fig. 2.3(c) is shown overleaf.

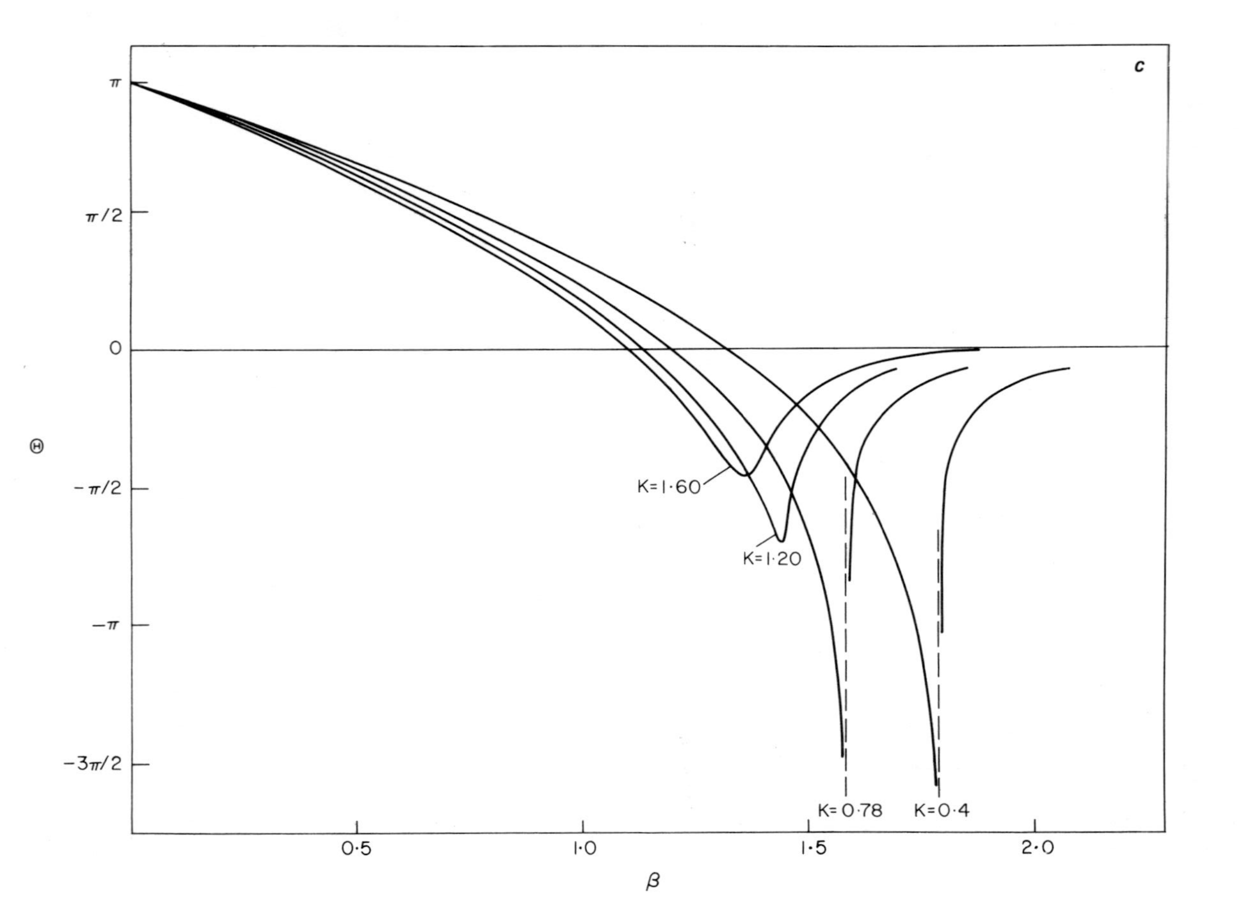

FIG. 2.3(c) Deflection functions for a non-motionic potential
$V(r) = \varepsilon[\exp\{12[1 - (r/r_0)]\} - 2(r_0/r)^6]$.

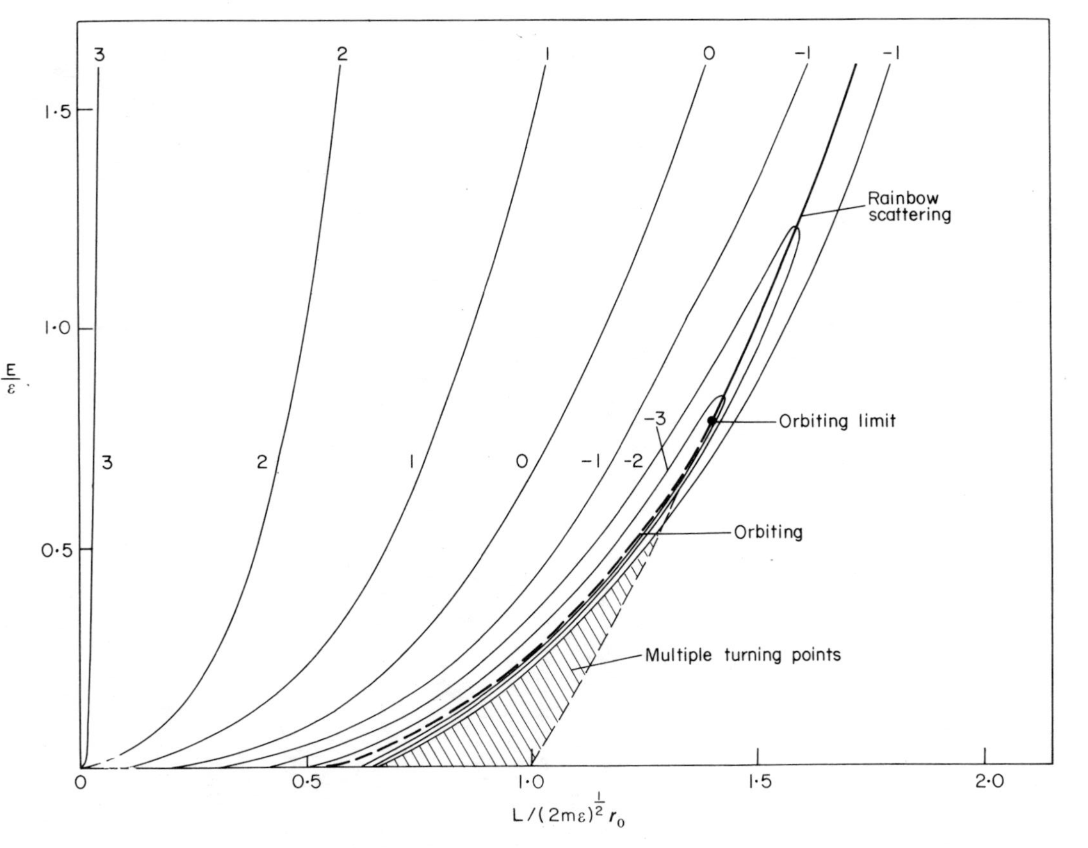

FIG. 2.4. Contours of the deflection function (values in radians), for the exp-6 potential. K is the reduced energy, E/ε, and L^* the reduced angular momentum. Quantum mechanical tunnelling may lead to resonance effects in the shaded region (compare Fig. 5.8).

forms $C^{(s)}r^{-s}$, $s > 2$, provided that the energy is sufficiently low that the maximum in $V_L(r)$ is not replaced by a point of inflection.

The nature of this orbiting singularity is readily obtained by expanding the effective potential about the maximum point, $r = r_{max}$ say

$$V_L(r) = V_L^{max} - \tfrac{1}{2}\kappa r_{max}^4 \left(\frac{1}{r} - \frac{1}{r_{max}}\right)^2 \tag{2.11}$$

so that $\kappa = (\partial^2 V_L/\partial r^2)_{r=r_{max}}$. It then follows from (2.5) that

$$\Theta = \Theta_1 + \frac{b}{r_{max}^2}\left[\frac{2E}{\kappa}\right]^{\frac{1}{2}} \ln\left(\frac{b - b^*}{b^*}\right) \text{ for } b > b^* \tag{2.12a}$$

$$= \Theta_2 + \frac{2b}{r_{max}^2}\left[\frac{2E}{\kappa}\right]^{\frac{1}{2}} \ln\left(\frac{b^* - b}{b^*}\right) \text{ for } b < b^* \tag{2.12b}$$

where b^* is the critical impact parameter at which $V_L^{max} = E$, and the factor 2 appears in (2.12a) because the integration range includes both sides of r_{max} when $b < b^*$. The values of Θ_1 and Θ_2 depend on the form of $V(r)$ far from the maximum point.

These general characteristics of the classical elastic deflection function $\Theta(E, L)$ are conveniently summarized by the contour diagram in Fig. 2.4, calculated for the exp-6 potential function

$$V(r) = \varepsilon\left[\exp\{12[1 - (r/r_0)]\} - 2(r_0/r)^6\right]. \tag{2.13}$$

Thus typical sections through the surface $\Theta(E, L)$ at constant E, take the forms illustrated in Fig. 2.3c. Note particularly the possibility of orbiting only in the low energy region, and the general decrease in the rainbow angle contour as the energy increases.

2.2 Collision Cross-sections

The discussion up to this point has centred on the trajectory for a single event, specified by an initial energy and impact parameter. From a practical point of view however, although the incident energy may be well-defined in for example a molecular beam experiment, it is impossible on a molecular scale to single out a particular impact parameter. Experimental results are therefore reported in terms of the *collision cross-section*, which is a notional cross-sectional area normal to the incident direction containing all impact parameters leading to the appropriate result. In the case of elastic scattering for example, we may distinguish a total cross-section $\sigma(E)$ which includes all elastic scattering events and a differential cross-section $d\sigma/d\Omega$ which refers to scattering at a given angle, while in the inelastic case there are total

$\sigma_{ij}(E)$, and differential $d\sigma_{ij}/d\Omega$, cross-sections for scattering from the ith to the jth internal state.

The elastic *differential cross-section*, which is most closely related to the classical trajectory, is defined in terms of the *intensity* of scattering per unit solid angle at a given measured angle θ. Since it is impossible to distinguish positive and negative deflections experimentally, θ is the magnitude of the deflection angle Θ of Eqn (2.5), modulo π,

$$\theta = |\Theta|, \qquad 0 < \theta < \pi. \tag{2.14}$$

θ defined in this way will be termed the *scattering* angle. It is clear from Fig. 2.3 that apart from discontinuities at $\theta = 0$, and possible orbiting singularities, θ varies smoothly with b, and hence trajectories within an impact parameter range b to $b + db$, defining (since the scattering is cylindrically symmetrical) an area $2\pi b\, db$, are deflected into an appropriate solid angle $d\Omega = 2\pi \sin\theta\, d\theta$. The differential cross-section is therefore

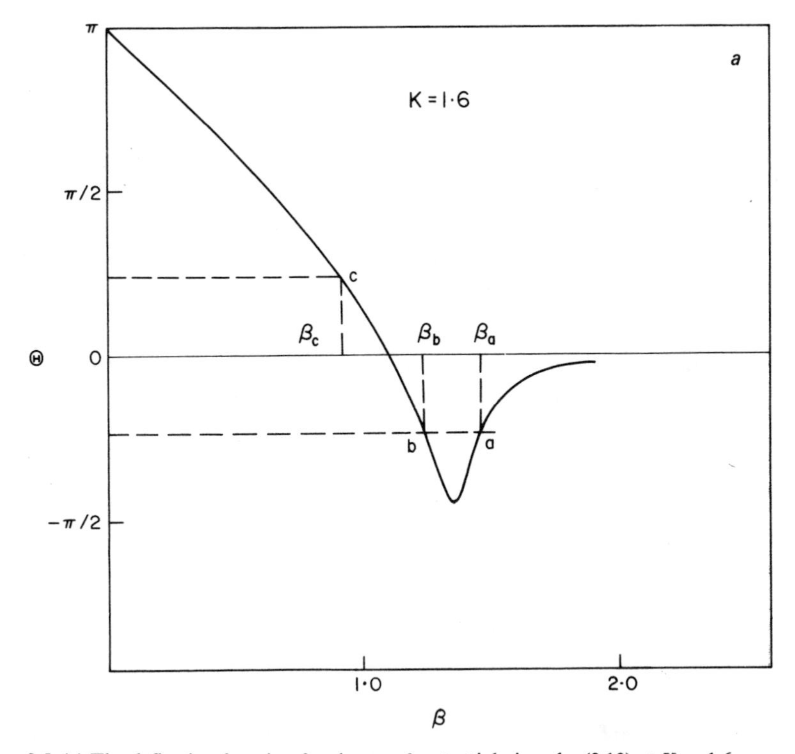

FIG. 2.5. (a) The deflection function for the exp-6 potential given by (2.13) at $K = 1{\cdot}6$.

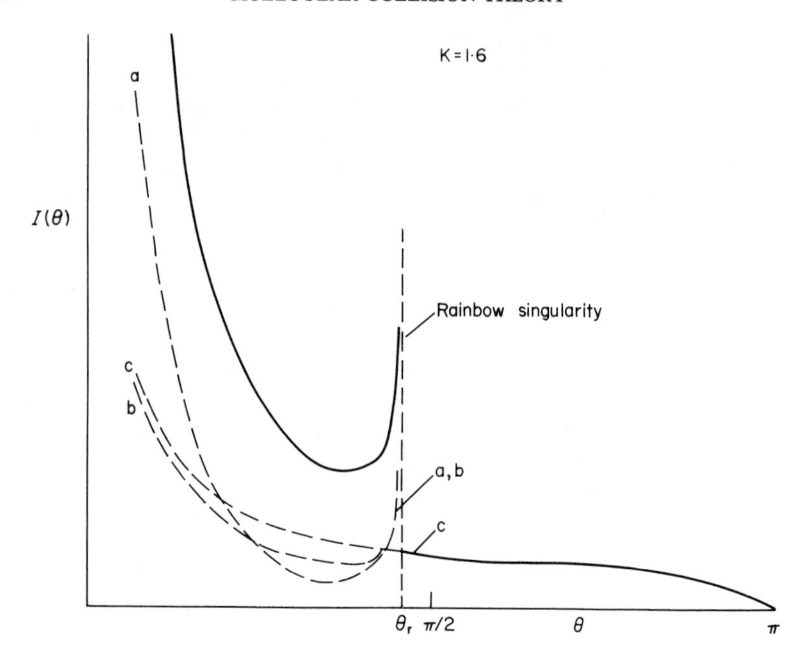

FIG. 2.5 (b) The differential cross-section $I(\theta)$ derived from the deflection function shown in Fig. 2.5(a). Dashed lines indicate contributions from the three branches indicated in Fig. 2.5(a). Note the increasing dominance of the a branch contribution as $\theta \to 0$.

$$\frac{d\sigma}{d\Omega} = I(\theta, E) = \left|\frac{2\pi b \, db}{2\pi \sin\theta \, d\theta}\right| = \left|\frac{b}{\sin\theta \, (d\theta/db)}\right|. \tag{2.15}$$

In cases where more than one value of b leads to the same angle θ, the classical intensity is correspondingly increased and (2.12) is replaced by an appropriate sum:

$$\frac{d\sigma}{d\Omega} = I(\theta, E) = \sum \left|\frac{b}{\sin\theta \, (d\theta/db)}\right|. \tag{2.16}$$

(Three such branches may be identified in the low angle region of Fig. 2.5a). The *total elastic cross-section* is then defined as the integral of the differential form;

$$\sigma(E) = \int \frac{d\sigma}{d\Omega} d\Omega = 2\pi \int_0^\pi I(\theta, E) \sin\theta \, d\theta \tag{2.17}$$

Note that the definition of $d\sigma/d\Omega$ as an intensity measurement leads to the possibility of two types of singularity, the so-called *glory* and *rainbow* effects in the classical differential cross-section as shown in Fig. 2.5. The *glory* effect, which may be forward ($\theta = 0$) or backward ($\theta = \pi$), arises from the disappearance of the $\sin\theta$ term in Eqn (2.15), because the element of solid angle becomes vanishingly small at these points. The observation of a backward glory in intermolecular scattering is however limited to low energies for which the deflection angle given by (2.5) takes values $\Theta = -\pi$, $-3\pi\ldots$, because the apparent singularity at $\Theta = \pi$ can be seen from Fig. 2.3c to be balanced by the simultaneous disappearance of b in the numerator in (2.15). The *rainbow* singularity, on the other hand corresponds with a turning point in the deflection function, at which $(d\theta/db) = 0$. Hence the rainbow angle corresponds physically to a point at which the density of classical trajectories rises to infinity.

As its title suggests, this latter singularity has a close connection with the familiar optical rainbow. Figure 2.6a shows the path of a typical light ray through a spherical droplet, the angles α and β being related by the refractive index, n, of the medium in the form $\sin\alpha = n\sin\beta$. The result is that the scattering angle $\phi = 4\beta - 2\alpha$ expressed as a function of α, passes through a maximum at the point $\cos\alpha = [(n^2 - 1)/3]^{\frac{1}{2}}$, as shown in Fig. 2.6b. It is the presence of this maximum, corresponding to infinite scattering intensity at ϕ_r, which when corrected for the wave nature of light, leads to the optical rainbow effect.

The presence of an orbiting singularity in the deflection function also results in a characteristic contribution to the differential cross-section, as may be seen by substituting for $(d\theta/db)$ in (2.15), using the form implied by (2.12)

$$\begin{aligned}\frac{d\Theta}{db} &= \frac{\beta}{b - b^*} = \frac{\beta}{b^*}\exp\left[-(\Theta - \Theta_1)/\beta\right] \text{ for } b > b^* \\ &= \frac{2\beta}{b^* - b} = \frac{2\beta}{b^*}\exp\left[-(\Theta - \Theta_2)/2\beta\right] \text{ for } b < b^*\end{aligned} \tag{2.18}$$

where

$$\beta = \frac{b}{r_{\max}{}^2}\left[\frac{2E}{\kappa}\right]^{\frac{1}{2}}.$$

Note that contributions to the scattering at a measured angle θ may arise from deflections $\Theta = -\theta,\ -\theta - \pi,\ -\theta - 2\pi\ldots$, but that according to (2.18), both contributions to $(d\Theta/db)$ increase along this series. Hence the dominant orbiting contribution to the differential cross-section may be written,

$$\frac{d\sigma_{orb}}{d\Omega} \simeq \frac{b^{*2}}{\beta \sin\theta}\{\exp[-(\theta-\theta_1)/\beta] + \tfrac{1}{2}\exp[-(\theta-\theta_2)/2\beta] + \ldots\}. \quad (2.19)$$

Important quantum mechanical corrections to these singularities are discussed in Chapter 5.

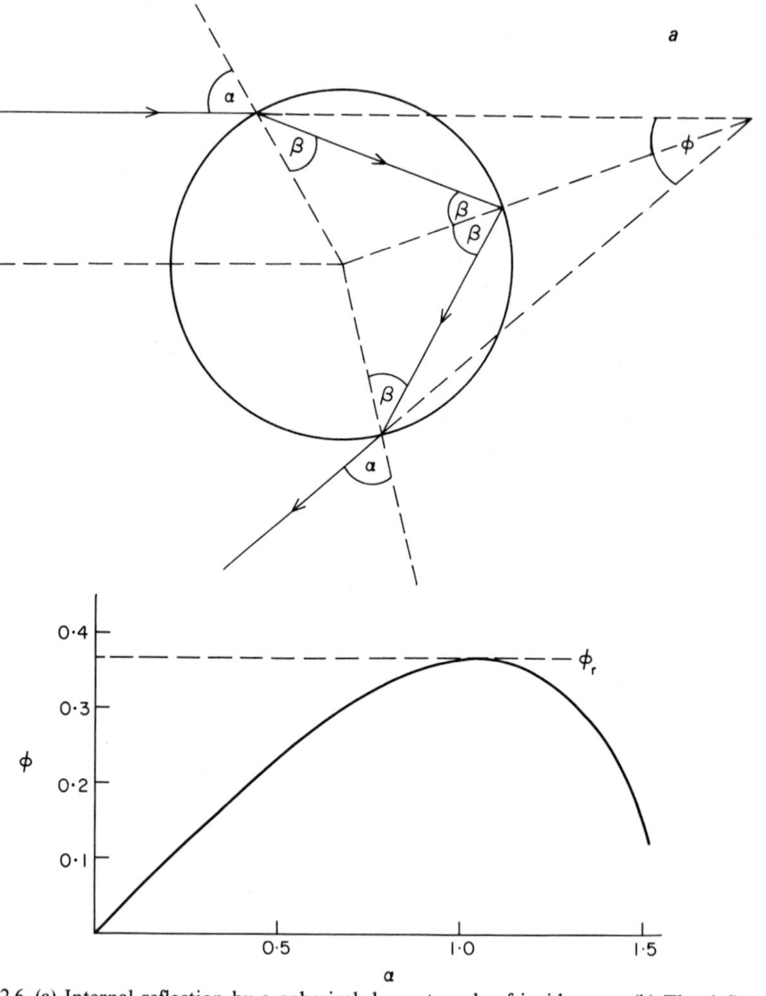

FIG. 2.6. (a) Internal reflection by a spherical drop at angle of incidence α. (b) The deflection angle, ϕ as a function of the incident angle, α.

2.3. Hard Sphere, Rutherford and Low Angle Scattering

The use of Eqns (2.5), (2.15) and (2.17) in deducing the appearance of the cross-section from the form of the potential $V(r)$ will first be illustrated for the models of hard-sphere and Rutherford scattering, which can be treated analytically. Some general features of elastic intermolecular scattering will then be discussed.

In the *hard sphere* model

$$\begin{aligned} V(r) &= 0 \qquad r \geqslant d \\ &= \infty \qquad r < d, \end{aligned} \tag{2.20}$$

and the classical turning point is given by

$$\begin{aligned} a &= d \text{ for } b \leqslant d \\ &= b \text{ for } b > d. \end{aligned} \tag{2.21}$$

Since $V(r) = 0$ over the accessible region, we find on substituting $z = 1/r$ in Eqn (2.5) that

$$\begin{aligned} \theta = \Theta &= \pi - 2b \int_0^{1/a} \frac{\mathrm{d}z}{[1 - b^2 z^2]^{\frac{1}{2}}} \\ &= \pi - 2 \sin^{-1} (b/d) \text{ for } b \leqslant d \\ &= 0 \qquad\qquad \text{for } b > d. \end{aligned} \tag{2.22}$$

It follows for the scattered ($\theta \neq 0$) trajectories with $b \leqslant d$, that

$$b = d \sin(\pi/2 - \theta/2) = d \cos\theta/2, \tag{2.23}$$

and hence on using Eqns (2.15) and (2.17), that

$$\frac{\mathrm{d}\sigma}{\mathrm{d}\Omega} = \frac{b}{\sin\theta \, |\mathrm{d}\theta/\mathrm{d}b|} = \tfrac{1}{4} d^2 \tag{2.24}$$

and

$$\sigma(E) = \int \frac{\mathrm{d}\sigma}{\mathrm{d}\Omega} \mathrm{d}\Omega = 2\pi \int_0^{\pi} \left(\frac{\mathrm{d}\sigma}{\mathrm{d}\Omega}\right) \sin\theta \, \mathrm{d}\theta = \pi d^2. \tag{2.25}$$

In other words the scattering intensity is uniform in all directions and independent of energy, and the total cross-section is simply the cross-sectional area presented by the sphere. These results are of course readily obtained by more elementary methods.

The case of *Rutherford scattering* refers to a repulsive Coulomb potential

$$V(r) = B/r, \qquad B > 0, \tag{2.26}$$

for which direct integration of (2.5) yields the result

$$\theta = |\Theta(E, b)| = 2 \operatorname{cosec}^{-1} [1 + (2bE/B)^2]^{\frac{1}{2}}. \tag{2.27}$$

It follows by inversion of (2.27) that

$$b = (B/2E) \cot\theta/2, \tag{2.28}$$

and hence, using (2.15), that

$$\frac{d\sigma}{d\Omega} = (B/4E)^2 \operatorname{cosec}^4 (\theta/2). \tag{2.29}$$

The total cross-section therefore diverges in this case, a result which is confirmed by quantum mechanical analysis, this being a rare case in which the quantum differential cross-section is identical with the classical form. The validity of (2.29) for high angle α particle scattering by atoms, corresponding according to (2.28) to low impact parameters, $b \sim 10^{-12}$ cm, is of historical interest in providing the first experimental evidence for the nuclear structure of the atom.

For most forms of intermolecular potential the above analysis must be replaced by numerical methods (Mason, 1957). Two important pointers for the interpretation of differential cross-section data may however be recognized. First, although the appearance of the rainbow singularity is modified by quantum mechanical corrections, it still forms a prominent feature in the scattering pattern, and for potentials of a given long range form the position of the rainbow angle θ_r is found to be largely determined by the ratio of the collision energy E to the well depth ε, the value of this ratio differing by only 10% for various assumed forms of potential (Bernstein and Muckerman, 1967).

Secondly an analytical approximation is available for the (dominant) high impact parameter contribution to the low angle scattering, on the assumption that only the long range part of the potential, with assumed form $V(r) \sim -C/r^s$ ($s \geqslant 2$) need be considered. The argument rests on the observation that the inequality $b^2/r^2 \gg |V(r)/E|$ must hold for sufficiently large impact parameters, over the whole of the (classically accessible) integration region $a < r < \infty$ in Eqn (2.5). Hence an expansion for $\Theta(E, L)$ in powers of (C/E) may be derived by the substitution

$$b^2/r^2 - (C/E)r^{-s} = \sin^2 \xi. \tag{2.30}$$

which may itself be rearranged to give an expansion for (b/r). To zeroth order for example, $(b/r) \simeq \sin \xi$, giving the first order equation

$$\begin{aligned}(b/r) &\simeq [\sin^2 \xi + (C/Eb^s) \sin^s \xi]^{\frac{1}{2}} \\ &\simeq \sin \xi + (C/2Eb^s) \sin^{s-1} \xi,\end{aligned} \tag{2.31}$$

and higher terms may be obtained with increasing labour by repeated iteration. A more powerful approach to this general problem is described by F. T. Smith (1966). Restricting the present discussion to first order, and combining (2.5), (2.30) and (2.31), the deflection function becomes

$$\theta = |\Theta(E, L)| = \frac{(s-1)C}{Eb^s} \int_0^{\pi/2} \sin^{s-2} \xi \, d\xi$$
$$= (s-1)Cf(s)/Eb^s, \tag{2.32}$$

where (Dwight,1961)

$$f(s) = \pi^{\frac{1}{2}}\Gamma(\tfrac{1}{2}s - \tfrac{1}{2})/2\Gamma(\tfrac{1}{2}s). \tag{2.33}$$

In the important case $s = 6$, $f(s) = 3/16$. With this expression for, the dominant contribution to the scattering intensity, (2.15) becomes

$$\frac{d\sigma}{d\Omega} = g(s)\,(C/E\theta)^{2/s}/\theta \sin \theta, \tag{2.34}$$

where

$$g(s) = [(s-1)f(s)]^{2/s}/s. \tag{2.35}$$

A log–log plot of $\sin \theta(d\sigma/d\Omega)$ against E may therefore be used to determine the coefficient C and the exponent s in the long range form of the potential. A further feature of (2.34), which F. T. Smith, (1966) has shown to apply to more general forms of potential is that to a first approximation, the reduced differential cross-section

$$\rho(\theta, E) = \theta \sin \theta(d\sigma/d\Omega) \tag{2.36}$$

depends only on the product $\tau = E\theta$. Indeed, in order to make optimum use of experimental data, Smith has suggested that it should be reported in this form.

2.4 Limitations of the Classical Theory

The fundamental defect in the classical theory is of course that no particle trajectory can ever be precisely defined, because of the inherent uncertainties associated with it. Possible consequent limitations to the theory are most simply illustrated by reference to the elastic scattering problem discussed above.

The most obvious constraint is that the impact parameter can be prescribed within an uncertainty Δb only by accepting an uncertainty $\Delta p_{\perp} \simeq \hbar/\Delta b$ in the transverse component of momentum. It is assumed above, however, that $p_{\perp} = 0$. Hence we have an uncertainty

$$\Delta\theta = \Delta p_{\perp}/mv = \hbar/mv\,\Delta b \tag{2.37}$$

in the incident direction of motion, leading to an equal uncertainty in the final scattering angle. Taken to the limit this means that for an exact description of the scattering at a given angle one must accept complete uncertainty in the impact parameter; this is the basis of the partial wave formulation in Section 3.1. Alternatively, on rearranging (2.37) to obtain

$$mv\,\Delta b\Delta\theta = \Delta L\Delta\theta \simeq \hbar, \tag{2.38}$$

it appears that an exact specification of the angular momentum implies complete uncertainty in the scattering angle; this is accepted for each partial wave in Eqn (3.8).

It is, however, unnecessary to go to these extremes; the real question is whether Δb (or ΔL) and $\Delta\theta$ may be chosen small compared with b (or L) and θ respectively. The true constraint is therefore, according to (2.37)

$$\theta \gg \Delta\theta \simeq \hbar/mv\,\Delta b \gg \hbar/mvb \simeq \hbar/L. \tag{2.39}$$

Particular care is clearly required in the small angle region, and here we may take the long range form of the potential to be dominant. The relation (2.32)

$$\theta = Kb^{-s}, \tag{2.40}$$

between θ and b for a potential with asymptotic form C/r^{-s} $(s > 2)$, shows according to (2.39) that the classical approximation must break down at angles outside the range

$$\theta \gg (\hbar/mv)^{s/(s-1)}K^{-1/(s-1)}, \tag{2.41}$$

or equivalently for impact parameters not covered by

$$b \ll (mvK/\hbar)^{1/(s-1)}. \tag{2.42}$$

This implies the existence of a (velocity dependent) critical angle θ^* below which the classical concept of a collision breaks down. The quantum theory developed in Chapter 5 shows that deflections within the cone defined by θ^* contribute negligibly to the scattering, and that in contrast with the classical prediction, the total cross-section is finite. For another view of this limitation we may note that (2.42) also represents a constraint on the radial collision action $\Delta(E, L)$, since with θ given by (2.40),

$$\Delta(E, L) = \int_{\infty}^{L} \Theta(E, L)\,\mathrm{d}L = mv\int_{\infty}^{b} \theta(E, b)\,\mathrm{d}b = mvK/(s-1)\,b^{s-1}. \tag{2.43}$$

It follows from (2.42) that the condition for a valid classical description may also be stated in the form

$$\Delta(E, L) \gg \hbar. \tag{2.44}$$

This is in fact one example of a much wider result (see Landau and Lifshitz (1965) Section 6).

The problem of Rutherford scattering must be treated separately because according to (2.27), (2.40) must be replaced by

$$\theta/2 < \tan \theta/2 = B/2Eb. \tag{2.45}$$

Hence by comparison with (2.39) the classical description is valid at all scattering angles provided that

$$B \gg E\hbar/mv. \tag{2.46}$$

The proper quantum mechanical treatment given in Section 3.2. shows that even this condition may be relaxed. Hence the divergent total cross-section implied by (2.29) is upheld in the quantum theory. This may be ascribed to the long range of the potential B/r which ensures that $\Delta(E, L)$ is infinite for all L.

The uncertainty principle also implies modifications to the classical rainbow phenomenon, association above with a prescribed angular momentum L_r and a prescribed deflection function Θ_r. The difficulty here arises not from uncertainty in the incident direction, since L_r and Θ_r normally satisfy (2.39), but from the presence of a minimum in the deflection function at the rainbow point, around which

$$\Theta = \Theta_r + Q(L - L_r)^2 \tag{2.47}$$

say. This implies the existence of two classical trajectories with the same angular deflection $\Delta\Theta$ from Θ_r, and hence an associated uncertainty in angular momentum given according to (2.47) by

$$\Delta L \simeq [(\Theta - \Theta_r)/Q]^{\frac{1}{2}} = [\Delta\Theta/Q]^{\frac{1}{2}} \tag{2.48}$$

This means by the uncertainty relation $\Delta L \Delta\Theta \simeq \hbar$, that no classical description can be valid unless

$$\Delta\Theta > [\hbar^2 Q]^{\frac{1}{3}}, \tag{2.49}$$

The result, as confirmed in Section 5.2, is that the rainbow singularity is smoothed out over the range given by (2.49).

This catastrophic failure of classical mechanics at the rainbow point is also accompanied by a weaker breakdown at neighbouring angles served by more than one classical trajectory, because the classical theory of additive scattering intensity must be replaced by one based on the amplitude of the wavefunction.

Interference between different trajectories is therefore expected to lead to oscillations in the differential cross-section. The general significance of such oscillations is discussed in Chapter 9. We note here merely that the phase change associated with a given trajectory may be obtained from the de Broglie formula $\lambda(r) = h/p(r)$, generalized to allow for slow variation in the momentum, which implies a phase change of the form

$$\phi = \int \frac{2\pi \, dr}{\lambda(r)} = \frac{1}{\hbar} \int p(r) \, dr. \tag{2.50}$$

The appropriate phase is therefore directly related to the classical action $\int p(r) \, dr$.

A final type of quantum correction applies to the orbiting phenomenon associated above with the presence of a maximum in the effective potential function $V_L(r)$ shown in Fig. 2.2. The possibility of tunnelling through this barrier not only shades the importance of the maximum point to the extent that the classical singularity in the deflection function disappears, it also makes accessible the region of the central dip in Fig. 2.2. This results in orbiting peculiarities in the quantum theory, which arise from resonances associated with quasi-bound states within this dip, rather than from an energy coincidence with the barrier maximum.

CHAPTER 3

Quantum Scattering by a Central Force

3.1 Introduction

Quantum mechanical theory employs a wavefunction and its associated interference pattern in place of the limiting deflection for a well defined classical trajectory. This wavefunction is taken to include the whole system and to be independent of time, the object being to describe the average behaviour of the system over a long period.

In the simple case of scattering by a central force, the interaction potential $V(r)$ depends only on the interparticle separation r, and the wavefunction is therefore a function only of position $\mathbf{r}$.

It is defined to be regular at the origin, $\Psi(0) = 0$, and to behave in the asymptotic region, where physical measurements are made, in the form†

$$\Psi(\mathbf{r}) \overset{r \to \infty}{\sim} \tilde{\Psi}(r) + f(\theta)\, \mathrm{e}^{ikr}/r. \tag{3.1}$$

Here $\tilde{\Psi}(r)$, which represents the state of the system in the absence of the scattering potential, is determined by the physical conditions of the experiment, normally in the form of a plane wave, e^{ikz}, to include all impact parameter collisions with given momentum $k\hbar$ in the z direction. Hence $\tilde{\Psi}(r)$ is an eigenfunction of $\hat{p}_z$;

$$\hat{p}_z \tilde{\Psi}(r) = -i\hbar \frac{\mathrm{d}}{\mathrm{d}z} \mathrm{e}^{ikz} = k\hbar\, \mathrm{e}^{ikz} \tag{3.2}$$

Similarly the factor e^{ikr} in the second (scattered) term in (3.1) ensures that it is outgoing in character, again with momentum $k\hbar$, while the scattering amplitude, $f(\theta)$, governs the angular distribution of the scattering.

The precise connection between $f(\theta)$ and the differential cross-section is obtained by comparing the incident flux (number of particles per unit area

† This expression must be symmetrized or antisymmetrized when applied to collisions of like particles (Mott and Massey, 1965, p. 294). $f(\theta)$ would be replaced by $f(\theta, \phi)$ in the case of scattering by an anisotropic potential (see Section 6.4).

per unit time) with the rate of scattering into a given solid angle $\mathrm{d}\Omega$. Since $\tilde{\Psi}(r)$ represents a wave with unit density ($|\tilde{\Psi}(r)|^2 = 1$), moving with velocity $v = k\hbar/m$, the incident flux is simply $k\hbar/m$, as may be confirmed by evaluating the z component of the current density

$$\mathbf{j} = -\frac{i\hbar}{2m}(\tilde{\Psi}^*\nabla\tilde{\Psi} - \tilde{\Psi}\nabla\tilde{\Psi}^*). \tag{3.3}$$

Similarly the rate of scattering through an element of area $\mathrm{d}S = 2\pi r^2 \sin\theta \mathrm{d}\theta$ is $v|f(\theta)/r|^2\,\mathrm{d}S$. Hence by the definition of the scattering intensity (compare (2.15)),

$$2\pi\frac{\mathrm{d}\sigma}{\mathrm{d}\Omega}\sin\theta\,\mathrm{d}\theta = |f(\theta)/r|^2\,\mathrm{d}S = 2\pi|f(\theta)|^2\sin\theta\,\mathrm{d}\theta$$

or

$$\frac{\mathrm{d}\sigma}{\mathrm{d}\Omega} = I(\theta) = |f(\theta)|^2. \tag{3.4}$$

The total cross-section is then obtained, as before, by direct integration;

$$\sigma = \int\frac{\mathrm{d}\sigma}{\mathrm{d}\Omega}\mathrm{d}\Omega = 2\pi\int_0^\pi |f(\theta)|^2\sin\theta\,\mathrm{d}\theta. \tag{3.5}$$

The problem is therefore to determine the form of the scattering amplitude, $f(\theta)$, for a given function $V(r)$. To do this it is necessary to turn from the asymptotic region to the range where the interaction potential is significant. This implies solving the appropriate Schrödinger equation.

3.2 Partial Wave Analysis

In the case of elastic scattering under a potential $V(r)$ the Schrödinger equation

$$\left[-\frac{\hbar^2}{2m}\nabla^2 + V(r)\right]\Psi(\mathbf{r}) = E\Psi(\mathbf{r}) \tag{3.6}$$

may conveniently be written in the reduced form

$$[\nabla^2 + k^2 - U(r)]\,\Psi(\mathbf{r}) = 0, \tag{3.7}$$

where

$$k^2 = 2mE/\hbar^2,\ U(r) = 2mV(r)/\hbar^2.$$

Since the system has spherical symmetry and the incident motion is assumed to be in the z direction, angular momentum is conserved, and only waves with a zero component about the z axis need be taken into account. $\Psi(r)$ is therefore expanded in the form

$$\Psi(\mathbf{r}) = \frac{1}{r}\sum_{l=0}^{\infty} A_l\,\psi_l(r)P_l(\cos\theta), \tag{3.8}$$

where the Legendre functions $P_l(\cos\theta)$ (Abramowitz and Stegun (1965)) are the appropriate angular momentum states. The coefficients A_l will be determined later by the condition that $\Psi(\mathbf{r})$ should behave asymptotically as (3.1). The immediate problem is to describe the radial functions $\psi_l(r)$ which may be verified to satisfy equations of one dimensional form,†

$$\left[\frac{d^2}{dr^2} + k^2 - U_l(r)\right]\psi_l(r) = 0, \tag{3.9}$$

where the effective potential

$$U_l(r) = U(r) + l(l+1)/r^2 \tag{3.10}$$

is closely related to the centrifugally corrected potential $V_L(r)$ of (2.10), because

$$b \simeq L/mv \simeq l/k, \tag{3.11}$$

the reduced incident momentum or *wavenumber* of the collision, being given according to (3.7) by $k = mv/\hbar$.

It is shown in Appendix A that the acceptable (everywhere finite) continuum solution of (3.10) may be taken to be real, and that if $U_l(r)$ vanishes sufficiently rapidly ($rU(r) \to 0$ as $r \to \infty$ Mott and Massey (1965), p. 23) this solution has asymptotic form

$$\psi_l(r) \sim \sin(kr + \delta_l), \tag{3.12}$$

where the phase term δ_l depends on the function $U_l(r)$. Written another way, each partial wave

$$\psi_l(r) \sim \frac{1}{2i}[\exp(ikr + i\delta_l) - \exp(-ikr - i\delta_l)], \tag{3.13}$$

is seen to contain an incoming, e^{-ikr}, and an outgoing, e^{ikr}, part, the amplitudes of which are equal since δ_l is real.‡ Elastic scattering therefore simply affects the relative phases of the incoming and outgoing parts of ψ_l, and it is

† The factor r^{-1} in (3.8) accounts for the disappearance of the first derivative term contained in ∇^2.

‡ In the presence of inelastic or reactive processes which may lead to a reduced amplitude in the elastically scattered wave, it will be found that δ_l may be complex with a negative imaginary part. (See Eqn (6.64)–(6.68)).

these phase changes which determine the interference pattern represented by the scattering amplitude, $f(\theta)$.

In order to emphasize the contribution of the scattering potential $U(r)$ itself to the phase term δ_l, it is conventional to discount the contribution due to the centrifugal term $l(l + 1)/r^2$ in (3.10) by defining a *phase shift*,

$$\eta_l = \delta_l - \tilde{\delta}_l, \tag{3.14}$$

where $\tilde{\delta}_l$ which arises from the term $l(l + 1)/r^2$ alone may be determined by setting $U(r) = 0$ in (3.9), to obtain the equation

$$\left[\frac{\mathrm{d}^2}{\mathrm{d}r^2} + k^2 - \frac{l(l + 1)}{r^2}\right] \tilde{\psi}_l(r) = 0, \tag{3.15}$$

the bounded solution of which may be expressed in terms of a spherical Bessel function (Abramowitz and Stegun 1965),

$$\tilde{\psi}_l(r) = kr\,j_l(kr), \tag{3.16}$$

with asymptotic form

$$\tilde{\psi}_l(r) \sim \sin(kr - l\pi/2). \tag{3.17}$$

It follows that $\tilde{\delta}_l = -l\pi/2$, and hence in (3.14), that

$$\eta_l = \delta_l + l\pi/2. \tag{3.18}$$

The solutions of (3.10) are therefore taken to have general asymptotic form

$$\psi_l(r) \sim \sin(kr - l\pi/2 + \eta_l). \tag{3.19}$$

The case of scattering by a Coulomb potential $U(r) = \beta/r$ poses special problems because the condition for the validity of (3.12) breaks down. It is found, however, by direct integration of (3.4) in terms of confluent hypergeometric functions (Mott and Massey (1965), p. 60), that the regular solution has asymptotic form

$$\psi_l(r) \sim \sin(kr - \alpha \ln 2kr - l\pi/2 + \mu_l). \tag{3.20}$$

where

$$\alpha = B/\hbar v,\ B = \beta\hbar^2/2m,$$

and

$$\mu_l = \arg \Gamma(l + 1 + i\alpha). \tag{3.21}$$

This means that the phase shift η_l of (3.19) is a divergent function of r. If the field departs from the form $V(r) = B/r$ only at short range, μ_l in (3.11) is replaced by

$$\mu_l = \arg\Gamma(l + 1 + i\alpha) + \sigma_l, \tag{3.22}$$

where σ_l depends on the difference potential $V(r) - B/r$.

The exact determination of η_l in (3.19) or σ_l in (3.22) normally requires numerical solution of (3.9), (see Bernstein (1960)) but the Born and WKB approximations discussed in sections 4.1 and 4.2 are adequate for most

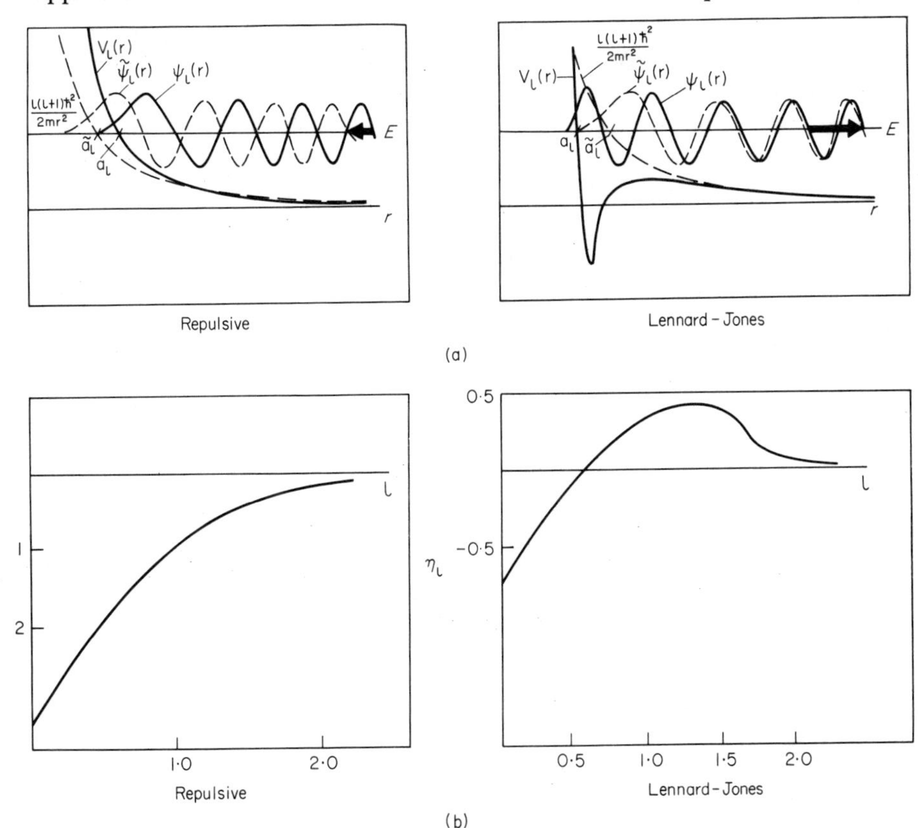

FIG. 3.1 (a) Wavefunctions and (b) phase shift variations with l for repulsive and Lennard–Jones interaction potentials. Heavy arrows indicate displacements required to superimpose $\psi_l(r)$ and the reference function $\tilde{\psi}_l(r)$ in the asymptotic region.

molecular problems. Certain important qualitative features of η_l as a function of k and l may be understood with the help of Fig. 3.1. First since η_l is the difference between the asymptotic phases of $\psi_l(r)$ and $\tilde{\psi}_l(r)$ it will be affected by a difference in the origins of the two oscillatory patterns (i.e. between the classical turning points a_l and $\tilde{a}_l$) and also by any difference in the oscillation frequency. A repulsive potential, which increases both the classical turning point, $a_l > \tilde{a}_l$, and the de Broglie wavelength therefore results in a negative phase shift, while a purely attractive potential would have the opposite effect. For the Lennard–Jones case illustrated in Fig. 3.1(a), the reference wave $\tilde{\psi}_l(r)$ oscillates over the attractive region of the potential, and hence the phase shift is positive; but at lower l values, the repulsive branch of the potential becomes dominant and the phase shift is negative as illustrated in Fig. 3.1(b). In every case an increase in l must cut down the region of $U(r)$ sampled by the oscillatory part of $\psi_l(r)$, so that if $U(r) \sim C/r^s$ with $s > 2$, $\eta_l \to 0$ as $1 \to \infty$. Furthermore, provided that $r^2U(r) \to 0$ as $r \to \infty$, an increase in k reduces the proportional contribution of the potential $U(r)$ towards the de Broglie wavelength and hence $\eta_l \to 0$ as $k \to \infty$. This condition may be applied to remove an ambiguity of 2π in the value of η_l defined by (3.19).

It now remains to demonstrate the relation between the scattering amplitude and the phase shifts. This involves determination of the coefficients A_l in (3.8), using the asymptotic boundary condition (3.1). The first step is to decompose the incident wave, $\tilde{\Psi}(r)$, which is assumed to be a plane wave, e^{ikz}, into partial waves; by analogy with (3.8),

$$\tilde{\Psi}(\mathbf{r}) = e^{ikz} = \frac{1}{r}\sum_{l=0}^{\infty} \tilde{A}_l\, \tilde{\psi}_l(r)\, P_l(\cos\theta), \tag{3.23}$$

where the $\tilde{\psi}_l(r)$ are given by (3.16), and the coefficients $\tilde{A}_l$ may be shown to follow from the orthogonality between Legendre functions; with $\cos\theta = x$,

$$\int_{-1}^{1} P_l(x)P_{l'}(x)\,dx = [2/(2l+1)]\delta_{ll'}. \tag{3.24}$$

Thus, on multiplying (3.23) by $P_l(\cos\theta)$ and integrating by parts,

$$\begin{aligned} r^{-1}\tilde{A}_l\tilde{\psi}_l(r) &= \left[\int_{-1}^{1} e^{ikrx}\, P_l(x)\,dx\right] \Big/ \left[\int_{-1}^{1} P_l(x)^2\,dx\right] \\ &= \frac{(2l+1)}{2ikr}\left\{[e^{ikrx}P_l(x)]_{-1}^{1} - \frac{1}{ikr}\int_{-1}^{1}\frac{dP_l}{dx}\,d\,e^{ikrx}\right\} \\ &\sim \frac{(2l+1)}{kr}\, i^l \sin(kr - l\pi/2), \end{aligned} \tag{3.25}$$

because $P_l(\pm 1) = (\pm 1)^l$, and terms of order $(kr)^{-2}$ become negligibly small in the asymptotic region. Consequently, by comparison between (3.17) and (3.23)

$$\tilde{A}_l = (2l+1)i^l/k. \tag{3.26}$$

Now the outgoing boundary condition (3.1) requires that incoming terms in the difference function

$$\Psi(\mathbf{r}) - \tilde{\Psi}(\mathbf{r}) \sim \frac{1}{r}\sum_{l=0}^{\infty} [A_l \sin(kr - l\pi/2 + \eta_l) - \tilde{A}_l \sin(kr - l\pi/2)]\, P_l(\cos\theta) \tag{3.27}$$

must vanish, from which it is readily verified by expressing the oscillatory terms in exponential form that

$$A_l = \tilde{A}_l\, e^{i\eta_l} = i^l(2l+1)\, e^{i\eta_l}/k,$$

and hence that the coefficient of e^{ikr}/r in (3.1) is

$$f(\theta) = \frac{1}{2ik}\sum_{l=0}^{\infty}(2l+1)(e^{2i\eta_l} - 1)\, P_l(\cos\theta). \tag{3.28}$$

This expression, which provides the essential link between the potential $U(r)$ and the observed cross-section, clearly demonstrates the wave nature of the quantum mechanical description. The angular dependence due to any given partial wave, determined by $P_l(\cos\theta)$, oscillates with θ but shows no strong preference for any given direction, as may be emphasized by the convenient asymptotic approximation (Abramowitz and Stegun 1965)

$$P_l(\cos\theta) \simeq \left[\frac{2}{\pi l \sin\theta}\right]^{\frac{1}{2}} \sin[(l + \tfrac{1}{2})\theta + \pi/4], \qquad l\theta \gg 1.$$

It is the *interference* between neighbouring waves which determines the angular variation of the scattering pattern. Of the two terms in the central bracket in (3.28) the first, since it involves the phase shifts, is directly due to the potential $U(r)$; the second which is required to remove any contribution from the unscattered plane wave may be omitted for $\theta \neq 0$ because by definition the plane wave moves only in the forward direction; mathematically

$$\sum_{l=0}^{\infty}(2l+1)\, P_l(\cos\theta) = 0 \qquad \text{for} \quad \theta \neq 0. \tag{3.29}$$

Equation (3.29) may therefore be written

$$\left.\begin{aligned} f(\theta) &= \frac{1}{2ik}\sum_{l=0}^{\infty}(2l+1)\, e^{2i\eta_l}\, P_l(\cos\theta) \qquad &\text{for} \quad \theta \neq 0 \\ &= \frac{1}{k}\sum_{l=0}^{\infty}(2l+1)\sin\eta_l\, e^{i\eta_l} \qquad &\text{for} \quad \theta = 0 \end{aligned}\right\} \tag{3.30}$$

where the second line follows from the identity $P_l(1) = 1$.

Given the above forms for $f(\theta)$ the total elastic cross-section is readily derived from (3.5),

$$\sigma = 2\pi \int_0^\pi |f(\theta)|^2 \sin\theta \, d\theta$$

$$= \frac{\pi}{2k^2} \sum_{ll'} (2l+1)(2l'+1)|(e^{2i\eta_l} - 1)(e^{2i\eta_{l'}} - 1)| \int_{-1}^{1} P_l(x)P_{l'}(x)\,dx,$$

$$= \frac{4\pi}{k^2} \sum_{l=0}^{\infty} (2l+1)\sin^2\eta_l. \tag{3.31}$$

There is also a direct connection between the total cross-section and the forward scattering amplitude,

$$\sigma = \frac{4\pi}{k}\,\mathrm{Im}\,f(0). \tag{3.32}$$

This result, which is the simplest example of what is known as the optical theorem, originates in the wave nature of the solution and in the conservation of flux during an elastic scattering experiment.

Suppose that the radial part of a typical partial wave in (3.27) is decomposed in the general form

$$A_l\psi_l(r) = \tilde{A}_l\tilde{\psi}_l(r) + c_l\,e^{ikr}$$
$$= k^{-1}i^l(2l+1)\sin(kr - l\pi/2) + c_l\,e^{ikr}, \tag{3.33}$$

the first term being the plane wave component defined by (3.17) and (3.26), and the second the outgoing scattered term. The following more compact expressions for $f(\theta)$ and σ may then be derived in place of (3.30) and (3.31)

$$f(\theta) = \sum_{l=0}^{\infty} c_l\,P_l(\cos\theta)$$

$$\sigma = 4\pi \sum_{l=0}^{\infty} (2l+1)^{-1}|c_l|^2$$

Hence the optical theorem (3.32) will be satisfied if

$$k|c_l|^2 = (2l+1)\,\mathrm{Im}\,c_l \tag{3.34}$$

These equations are simply restatements of more complicated results derived above. The new feature is that Eqn (3.34) may be seen to require conservation of flux for the radial motion, because the radial current density (which must vanish) is given according to (3.33) by

$$j_r = -\frac{i\hbar}{2m}\left\{[A_l\psi_l(r)]^* \frac{\mathrm{d}}{\mathrm{d}r}[A_l\psi_l(r)] - [A_l\psi_l(r)]\frac{\mathrm{d}}{\mathrm{d}r}[A_l\psi_l(r)]^*\right\}$$

$$= \frac{\hbar}{m}\{k|c_l|^2 - (2l+1)\,\mathrm{Im}\,c_l\}, \tag{3.35}$$

where the first term of (3.35) arises from the scattered part of (3.33) while the second originates from *interference* with the plane wave component.

Hence in interpreting Eqn (3.32), the total cross-section is naturally explained as a sum of scattered terms $|c_l|^2$, but the origin of the term in $\mathrm{Im}\, f(0)$ is more subtle. It arises from interference between the two components of (3.33) required, by flux conservation, to account for depletion of the incident plane wave term; the significance of the forward direction, $\theta = 0$, is of course that it is the only direction in which such interference can occur. The extension of (3.32) to cover inelastic events is discussed in sections 6.3 and 6.4.

3.3 Model Problems: Coulomb and Hard Sphere Scattering

The theory requires some modification in the presence of a Coulomb field because of the divergent nature of η_l (see (3.20)–(3.21)). According to Mott and Massey (1965) Section 2.2 the range of the potential is actually sufficiently great to distort the incident wave and (3.23) must be replaced by

$$\tilde{\Psi}(\mathbf{r}) = \left[1 + \frac{\alpha^2}{ik(r-z)}\right]\exp[ikz + i\alpha \ln k(r-z)]. \tag{3.36}$$

Similarly since the divergent part of η_l is independent of l it may be regarded as a distortion of the scattered wave. Thus

$$\Psi(\mathbf{r}) = \tilde{\Psi}(\mathbf{r}) + f(\theta)\exp(ikr - i\alpha \ln 2kr)/r, \tag{3.37}$$

where

$$f(\theta) = \frac{1}{2ik}\sum_{l=0}^{\infty}(2l+1)\frac{\Gamma(l+1+i\alpha)}{\Gamma(l+1-i\alpha)}P_l(\cos\theta)$$

$$= \frac{\alpha}{2k}\operatorname{cosec}^2\theta/2\exp[-i\alpha\ln\tfrac{1}{2}(1-\cos\theta) + i\pi + 2i\mu_0], \tag{3.38}$$

and

$$\exp 2i\mu_0 = \Gamma(1+i\alpha)/\Gamma(1-i\alpha).$$

The differential cross-section given by (3.4) therefore becomes

$$\frac{\mathrm{d}\sigma}{\mathrm{d}\Omega} = |f(\theta)|^2 = \left(\frac{\alpha}{2k}\right)^2\operatorname{cosec}^4\theta/2 = \left(\frac{B}{4E}\right)^2\operatorname{cosec}^4\theta/2 \tag{3.39}$$

in agreement with the classical result (2.29), as anticipated on general grounds in Section 2.4.

The hard sphere model on the other hand illustrates important differences between the quantum mechanical theory and the classical results given by (2.24) and (2.25),

$$\left.\begin{aligned}\frac{d\sigma_{cl}}{d\Omega} &= \tfrac{1}{4}d^2\\ \sigma_{cl} &= \pi d^2\end{aligned}\right\} \tag{3.40}$$

The main features of the discussion, first given by Massey and Mohr (1933) may be summarized with the help of Fig. 3.2. In the first place partial waves

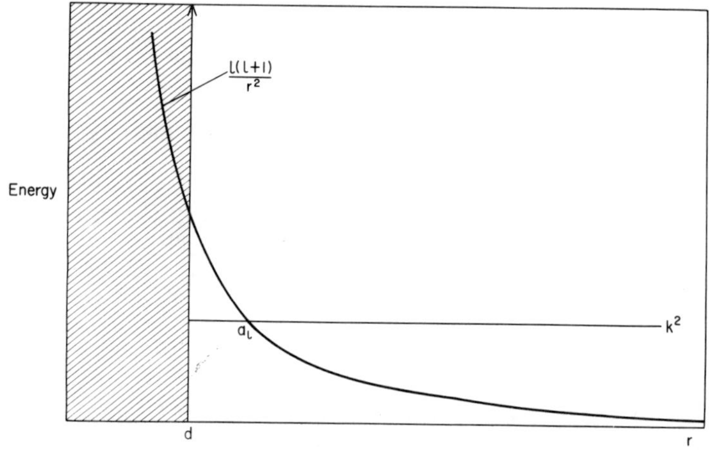

FIG. 3.2. Centrifugal terms and the hard sphere potential. The classical turning point lies at $a_1 = [l(l+1)]^{\frac{1}{2}}/k$ for $l(l+1) \geqslant k^2d^2$ and at d for $l(1+1) < k^2d^2$.

with $1 > kd$ make a negligible contribution to the scattering amplitude in (3.20), because their phase shifts, being governed by the probability of tunnelling from the classical turning point,

$$a_l = [l(l+1)]^{\frac{1}{2}}/k \tag{3.41}$$

are negligibly small. The phase shifts for $1 < kd$ are, however, generally large for large kd, and strongly dependent on 1; it is found by exact solution of (3.9) (Massey and Mohr, 1933) that

$$\eta_l = -kd + l\pi/2 \qquad \text{for} \qquad l \ll kd. \tag{3.42}$$

This means that at very low energies only the zero-order ($l = 0$) partial wave

$$\psi_0 = \sin(kr - kd), \tag{3.43}$$

with

$$\eta_0 = -kd,$$

contributes significantly to the cross-section. Equations (3.29) and (3.34) therefore reduce, for $kd \ll 1$, to

$$\left.\begin{aligned} \frac{d\sigma}{d\Omega} &= d^2 \\ \sigma &= 4\pi d^2 \end{aligned}\right\}. \tag{3.44}$$

The scattering is therefore isotropic, as in the classical theory, but the cross-sections have four times the classical values in (3.40). This may be attributed to diffraction by a sphere whose radius is small compared with the de Broglie wavelength.

At high energies on the other hand the number of partial waves with $l < kd$ is large, but since the phases are strongly dependent on l, the terms $\sin^2 \eta_l$ in (3.34) may be replaced by their average value of one half, so that

$$\sigma \simeq \frac{2\pi}{k^2} \sum_{l=0}^{kd} (2l + 1) \simeq 2\pi d^2, \tag{3.45}$$

the error in this random phase approximation being of order $(kd)^{-\frac{2}{3}}$ (Wergeland, 1945). The total cross-section therefore falls to twice its classical value.

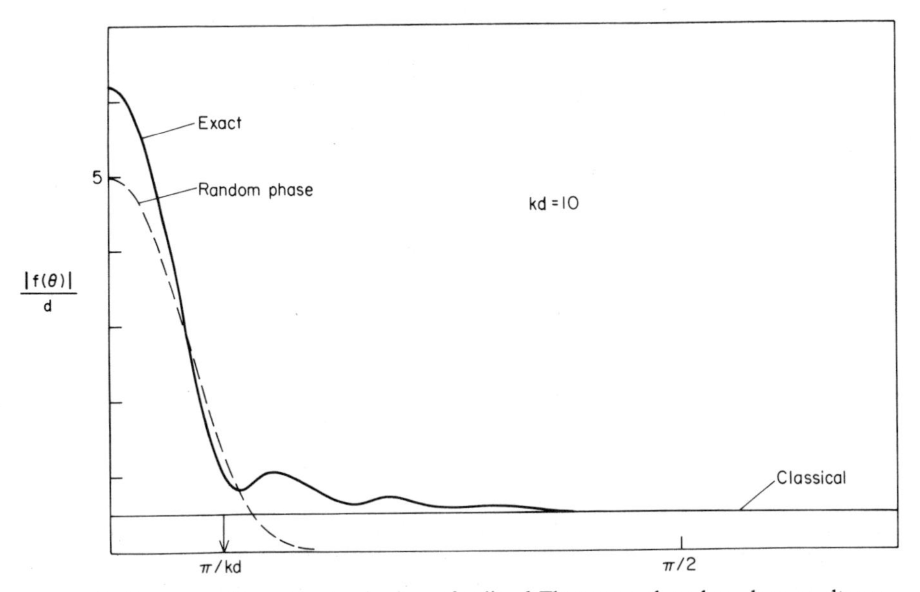

FIG. 3.3. Shadow scattering by a hard sphere of radius d. The exact and random phase results are given by Eqns (3.30) and (3.47) respectively.

Turning to the differential cross-section, according to the arguments of Section 2.3, the quantum corrections responsible for this factor of two are expected to arise from scattering at impact parameters b such that the angle θ is comparable with the uncertainty given by (2.39),

$$\Delta\theta \simeq \hbar/mvb = 1/kb. \tag{3.46}$$

Since $b \simeq d$ for forward scattering, small angle, *shadow scattering* anomalies are therefore expected in the range $0 < \theta < 1/kd$. This general pattern, and the accuracy of (3.48) derived below as an approximation to the forward scattering peak are illustrated in Fig. 3.3.

A quantitative measure of this shadow-scattering anomaly may be obtained by applying the random phase approximation to (3.30) in the small angle region. Thus

$$\operatorname{Re} f(\theta) = \frac{1}{2k}\sum_{l=0}^{kd}(2l+1)\sin 2\eta_l\, P_l(\cos\theta) \simeq 0$$

$$\begin{aligned}\operatorname{Im} f(\theta) &= \frac{1}{k}\sum_{l=0}^{kd}(2l+1)\sin^2\eta_l\, P_l(\cos\theta)\\ &\simeq \frac{1}{2k}\sum_{l=0}^{kd}(2l+1)P_l(\cos\theta)\\ &\simeq \frac{(L+1)}{4k\sin^2\frac{1}{2}\theta}\{P_L(\cos\theta) - P_{L+1}(\cos\theta)\}\end{aligned} \tag{3.47}$$

where $L = kd$. Hence, on expanding the Legendre functions in powers of $\sin^2\frac{1}{2}\theta$ (Abramowitz and Stegun (1965)).

$$P_L(\cos\theta) = 1 + \sum_{r=1}^{\infty}\frac{(L-r+1)(L-r+2)\ldots(L+r)}{(r!)^2}\sin^{2r}\tfrac{1}{2}\theta, \tag{3.48}$$

the small angle scattering intensity for large kd becomes

$$\begin{aligned}\frac{d\sigma}{d\Omega} = I(\theta) = |f(\theta)|^2 &= \frac{k^2d^4}{4}[1 - (kd\sin\tfrac{1}{2}\theta)^2 + \tfrac{5}{12}(kd\sin\tfrac{1}{2}\theta)^4 + \ldots]\\ &\simeq \frac{k^2d^4}{4}\exp[-(kd\sin\tfrac{1}{2}\theta)^2] \simeq \left(\frac{k\sigma}{4\pi}\right)^2\exp\left(-\frac{k^2\sigma\theta^2}{8\pi}\right).\end{aligned} \tag{3.49}$$

The forward scattering intensity, $I(0)$ (which clearly satisfies the optical theorem (3.32)) therefore increases without limit, while the width of the forward peak, being of order $1/kd$, decreases as k increases. In more precise terms this forward peak may be seen to be bounded by an angle $\theta^* \simeq \pi/kd$ corresponding to the first zero of the highest order Legendre function in (3.47), all terms in the sum being positive within this range. This means within the validity of (3.49) over this range that the shadow or diffraction contribution to the total cross-section is given by

$$\sigma_{\text{shadow}} \simeq 2\pi \int_0^{\theta^*} I(\theta) \sin\theta \, d\theta \simeq 2\pi d^2 \int_0^{\pi/2} x\, e^{-x^2} \, dx \simeq 0{\cdot}92\, \pi d^2 \tag{3.50}$$

where $x = \frac{1}{2} kd\, \theta$.

Hence the shadow scattering accounts for the two-fold discrepancy between the high energy classical (3.40) and quantum mechanical (3.45), cross-sections; scattering outside this region may be interpreted by classical methods.

CHAPTER 4

Elastic Scattering Phase Shifts

This chapter is concerned with the determination of phase shifts for a given potential, $V(r)$, the problem, according to (3.9) and (3.19), being to calculate the quantity η_l in the asymptotic form

$$\psi_l(r) \sim \sin(kr - l\pi/2 + \eta_l). \tag{4.1}$$

of the bounded solution ($\psi_l(0) = 0$) of the equation

$$\left[\frac{\mathrm{d}^2}{\mathrm{d}r^2} + k^2 - U(r) - l(l+1)/r^2\right]\psi_l(r) = 0, \tag{4.2}$$

where

$$k^2 = 2mE/\hbar^2, \qquad U(r) = 2mV(r)/\hbar^2 \tag{4.3}$$

This is of course readily achieved by numerical integration of (4.1), and comparison with the Bessel function in (3.16) (Bernstein, 1960). Another approach, extensively discussed by Calogero (1967), involves the solution of equations for the instantaneous phase $S(r)$, with limiting values $S_1(0) = 0$ and $S_1(\infty) = \eta_l$.

Our purpose below is to give a number of useful analytical approximations. The first is the Born approximation which offers an example of continuum perturbation theory. This is followed by sections on the semi-classical phase shift, on resonance effects and on the zero energy phase shift, all of which play a part in the analysis of elastic scattering cross-sections described in Chapter 5.

4.1 The Born Approximation

The following derivation of the Born approximation may be taken as a first example of continuum perturbation theory, and of the use of the Green's function discussed in Appendix B. The aim is to obtain an expansion for the wavefunction in powers of $U(r)$ by a rearrangement of (4.2) to give

$$\left[\frac{d^2}{dr^2} + k^2 - \frac{l(l+1)}{r^2}\right]\psi_l(r) = U(r)\psi_l(r). \tag{4.4}$$

The boundary conditions are that $\psi_l(r)$ should contain a plane wave part $\tilde{\psi}_l(r)$ and an outgoing term. Hence an appropriate formal solution of (4.4) may be expressed in terms of the outgoing Green's function, $\mathscr{G}^{(+)}(r, r')$, (see (B.22) and (B.25)),

$$\psi_l(r) = \tilde{\psi}_l^0(r) + \int_0^\infty \mathscr{G}^{(+)}(r, r')U(r')\psi_l(r')\,dr', \tag{4.5}$$

This Green's function, given by

$$\begin{aligned}\mathscr{G}_l^{(+)}(r, r') &= -k^{-\prime}\,\tilde{\psi}_l^{(+)}(r)\tilde{\psi}_l^0(r') \qquad r > r'\\ &= -k^{-\prime}\,\tilde{\psi}_l^0(r)\tilde{\psi}_l^{(+)}(r') \qquad r < r',\end{aligned} \tag{4.6}$$

contains only the known (unperturbed) solutions of the homogeneous counterpart of (4.4); in this case the Bessel functions (Abramowitz and Stegun, 1964).

$$\begin{aligned}\tilde{\psi}_l^0(r) &= kr\,j_l(kr) \overset{r\to\infty}{\sim} \sin(kr - l\pi/2)\\ \tilde{\psi}_l^{(+)}(r) &= ikr\,h_l^{(2)}(kr) \overset{r\to\infty}{\sim} \exp[i(kr - l\pi/2)].\end{aligned} \tag{4.7}$$

Equation (4.5) is termed an integral equation, because the unknown function $\psi_l(r)$ appears on both sides of the equation. An iterative solution, obtained by successive substitutions for $\psi_l(r')$ in the integrand, yields the Born series,

$$\begin{aligned}\psi_l(r) = \tilde{\psi}_l^0(r) &+ \int_0^\infty \mathscr{G}_l^{(+)}(r, r')\,U(r')\,\tilde{\psi}_l^0(r')\,dr'\\ &+ \int_0^\infty\int_0^\infty \mathscr{G}_l^{(+)}(r, r')\,U(r')\,\mathscr{G}_l^{(+)}(r', r'')\,U(r'')\,\tilde{\psi}_l^0(r'')\,dr'\,dr'' \ldots\end{aligned} \tag{4.8}$$

or in the formal notation of (B.2)

$$\psi_l = (1 + G_l^{(+)}V + G_l^{(+)}VG_l^{(+)}V + \ldots)\,\tilde{\psi}_l^0. \tag{4.9}$$

The problem of the convergence of this series is of considerable general importance (Mott and Massey, 1965; V Section 3), but we shall restrict attention to situations where the effect of $U(r)$ is sufficiently small that terms other than $\tilde{\psi}_l^0(r)$, and first order correction may be neglected. This solution

$$\begin{aligned}\psi_l(r) &\simeq \tilde{\psi}_l^0(r) + \int_0^\infty \mathscr{G}_l^{(+)}(r, r')\,U(r')\,\tilde{\psi}_l^0(r')\,dr'\\ &\overset{r\to\infty}{\sim} \sin(kr - \tfrac{1}{2}l\pi) - k^{-1}\exp(ikr - \tfrac{1}{2}il\pi)\int_0^\infty U(r)\,[\tilde{\psi}_l^0(r)]^2\,dr,\end{aligned} \tag{4.10}$$

gives the first Born approximation, a necessary (but not sufficient) condition for which is that the magnitude of the correction term should be small compared to unity, in which case (4.10) may be written

$$\psi_l(r) \simeq e^{i\eta_l} \sin(kr - \tfrac{1}{2}l\pi + \eta_l), \tag{4.11}$$

where the quantity

$$\eta_l = -k^{-1} \int_0^\infty U(r)\,[\tilde{\psi}_l^0(r)]^2\,dr = -k \int_0^\infty U(r)\,[j_l(kr)]^2\,r^2\,dr, \tag{4.12}$$

is by comparison with (4.1) the Born approximation to the phase shift. If (4.12) is valid for all l, the scattering amplitude (3.28) may be contracted by use of the identity (Abramowitz and Stegun, 1965)

$$\sum_{l=0}^{\infty} (2l + 1)\,[j_l(kr)]^2\,P_l(\cos\theta) = (\sin Kr/Kr), \tag{4.13}$$

where

$$K = 2k \sin \tfrac{1}{2}\theta$$

to the form

$$f(\theta) = -\int_0^\infty \frac{\sin Kr}{Kr}\,U(r)\,r^2\,dr. \tag{4.14}$$

In considering the practical validity of (4.12) and (4.14) we may note that the wavefunction $\tilde{\psi}_l^0(r) = kr\,j_l(kr)$, in (4.12) decreases rapidly to zero within the classically inaccessible region, $r < l/k$, and oscillates within an envelope of order unity for $r > l/k$. A necessary condition on the validity of (4.12) is therefore

$$\eta_l \simeq -\frac{1}{2k} \int_{l/k}^\infty U(r)\,dr \ll 1, \tag{4.15}$$

where, as a crude approximation, the oscillatory term in (4.12) has been replaced by $\frac{1}{2}$. This means, in view of the term k^{-1}, that at a given impact parameter $b \simeq l/k$, there is always a threshold energy

$$E^* = \hbar^2 k^{*2}/2m \simeq \frac{50m}{\hbar^2}\left[\int_b^\infty V(r)\,dr\right]^2, \tag{4.16}$$

above which the Born approximation may be expected to be valid. Secondly, at fixed energy, the approximation improves with increasing l as a result of the increasingly restricted integration range and the long range fall off in the potential.

An analytical expression for the high angular momentum phase shift is available for the important case of a potential with long range form $V(r) \sim -C^{(s)}/r^s$. This is based on use of approximation (Massey and Smith, 1933),

$$\langle[\tilde{\psi}_l^0(r)]^2\rangle = \langle k^2r^2[j_l(kr)]^2\rangle = \tfrac{1}{2}[1 - l^2/k^2r^2]^{-\frac{1}{2}} \tag{4.17}$$

in Eqn (4.12) rather than the crude factor of $\frac{1}{2}$ in (4.15); the phase shift then becomes

$$\begin{aligned}\eta_l &= \frac{mC^{(s)}}{k\hbar^2}\int_{l/k}^{\infty} r^{-s}(1 - l^2/k^2r^2)^{-\frac{1}{2}}\,dr \\ &= f(s)mC^{(s)}k^{s-2}/\hbar^2 l^{s-1} = a_s\, l^{1-s},\end{aligned} \tag{4.18}$$

where $f(s)$ is the same standard integral (Dwight, 1961), as that which occurs in (2.33);

$$f(s) = \int_0^1 z^{s-2}(1 - z^2)^{-\frac{1}{2}}\,dz = \pi^{\frac{1}{2}}\Gamma(\tfrac{1}{2}s - \tfrac{1}{2})/2\Gamma(\tfrac{1}{2}s). \tag{4.19}$$

Hence in the physically important case, $s = 6$, $f(6) = 3/16$. The more accurate formula

$$\begin{aligned}\eta_l &= -\frac{2C^{(s)}mk}{\hbar^2}\int_0^{\infty} r^{2-s}[j_l(kr)]^2\,dr \\ &= \frac{\pi m C^{(s)}}{k\hbar^2}\left(\frac{k}{2}\right)^{s-1}\frac{\Gamma(s-1)\Gamma(l - \frac{1}{2}s + \frac{3}{2})}{[\Gamma(\frac{1}{2}s)]^2\Gamma(l + \frac{1}{2}s - \frac{1}{2})}\end{aligned} \tag{4.20}$$

reduces at large l with the help of Stirling's approximation and the duplication formula (Abramowitz and Stegun, 1965)

$$\Gamma(s - 1) = (2\pi)^{-\frac{1}{2}}\,2^{s-\frac{3}{2}}\,\Gamma(\tfrac{1}{2}s - \tfrac{1}{2})\,\Gamma(\tfrac{1}{2}s), \tag{4.21}$$

to the same result. This so-called Jeffreys–Born approximation is employed in the theory of low-angle scattering in Section 5.3.

4.2 The Semi-classical Phase Shift

The phase shifts for both *simple* and *orbiting* collisions (governed by the effective potential curves illustrated in Figs 4.1) may be obtained by semi-classical methods, based on the mathematical arguments developed in Appendix C.

Here, before considering the *simple* case, we merely recall the simple physical structure of the JWKB wavefunction to the right of an isolated left-hand turning point, such as a_1 in Fig 4.1(a). The general form given by

$$\psi(r) = [k(r)]^{-\frac{1}{2}}\sin\left(\int_a^r k(r)\,dr + \pi/4\right), \qquad r \gg a, \tag{4.22}$$

shows that the phase of the wavefunction is that appropriate to the generalized de Broglie wavelength, $\lambda(r) = h/p(r) = 2\pi/k(r)$, integrated out from the classical turning point, with an additional term $\pi/4$ to cover the non-classical region; the factor $[k_l(r)]^{-\frac{1}{2}}$ ensures that the envelope of $|\psi_l(r)|^2$ depends inversely on the velocity as required by flux conservation.

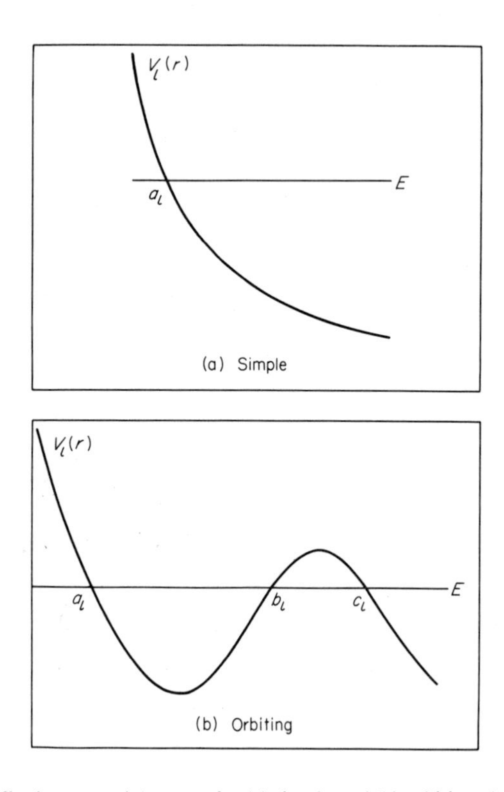

FIG. 4.1. Effective potential curves for (a) simple and (b) orbiting elastic collisions.

It would appear at first sight that, since (4.22) purports to describe the asymptotic behaviour of $\psi(r)$, it would alone suffice to determine the phase shift, provided the required momentum $p_l(r)$ was derived from the appropriate centrifugally corrected potential (see (3.10)).

$$U_l(r) = U(r) + l(l + 1)/r^2 \tag{4.23}$$

However, this ignores a complication foreseen by Langer (1937), namely that the singularity of $U_l(r)$ at the origin may not be sufficiently well separated from the turning point a_l, that the latter can be treated in isolation. Langer (1937) therefore suggests the substitutions

$$r = e^x$$

$$\psi_l(r) = e^{x/2}\phi_l(x) \tag{4.24}$$

in order to remove the singularity to $x = -\infty$, with the result that (4.2) is replaced by

$$\left[\frac{d^2}{dx^2} + \kappa_l^2(x)\right]\phi_l(x) = 0 \tag{4.25}$$

where

$$\kappa_l^2(x) = e^{2x}(k^2 - U(e^x) - (l + \tfrac{1}{2})^2 e^{-2x}). \tag{4.26}$$

The result on transforming back to the original independent variable by inversion of (4.24), is that the semi-classical solution of (4.25) analogous to (4.22) may be written

$$\begin{aligned} \psi_l(r) &= e^{x/2}\phi_l(x) \\ &\simeq e^{x/2}\,[\kappa_l(x)]^{-\frac{1}{2}} \sin\left[\int_{\alpha_l}^{x} \kappa_l(x)\,dx + \pi/4\right] \\ &\simeq [k_l(r)]^{-\frac{1}{2}} \sin\left[\int_{a_l}^{r} k_l(r)\,dr + \pi/4\right], \end{aligned} \tag{4.27}$$

where

$$k_l^2(r) = k^2 - U(r) - (l + \tfrac{1}{2})^2/r^2.$$

The nett effect of the transformation (4.24) is therefore to introduce an apparent term $(l + \frac{1}{2})^2$ rather than $l(l + 1)$ in the effective potential $U_l(r)$. Bernstein (1960) finds by comparison with exact numerical values that the resulting Langer correction to the JWKB phase shift (given by (4.30) below) is physically significant for small l values. Furthermore in the case of free motion ($U(r) = 0$) it leads to the same asymptotic form

$$\begin{aligned} \tilde{\psi}_l(r) &\overset{r\to\infty}{\sim} k^{-\frac{1}{2}} \sin\left[\int_{(l+\frac{1}{2})/k}^{r} [k^2 - (l + \tfrac{1}{2})^2/r^2]^{\frac{1}{2}}\,dr + \pi/4\right] \\ &\sim k^{-\frac{1}{2}} \sin(kr - l\pi/2) \end{aligned} \tag{4.29}$$

as the exact solution (3.17).

The actual formula for the JWKB phase shift is now obtained by direct comparison between (4.27) and (4.1).

$$\eta_l = \lim_{r\to\infty} \left\{ \int_{a_l}^{r} k_l(r)\, dr - kr + (l + \tfrac{1}{2})\pi/2 \right\}$$

$$= \int_{a_l}^{\infty} k_l(r)\, dr - \int_{\tilde{a}_l}^{\infty} \tilde{k}_l(r)\, dr, \tag{4.30†}$$

where $\tilde{k}_l(r) = k^2 - (l + \frac{1}{2})^2/r^2$.
In contrast with the Born approximation, the general application of this result is unaffected by the magnitude of η_l, and the phase shifts are found to be accurate to within about 0·1% (Bernstein (1960)) under realistic molecular scattering conditions.

An important special case arises when the angular momentum is sufficiently large that the centrifugal term $(l + \frac{1}{2})^2/r^2$ dominates the potential $U(r)$ over the classically accessible region, $r > a_l$. The turning points a_l and $\tilde{a}_l$ in (4.30) then coincide, and η_l reduces to the same form

$$\eta_l \simeq -\frac{1}{2k} \int_{(l+\frac{1}{2})/k}^{\infty} U(r)[1 - (l + \tfrac{1}{2})^2/k^2r^2]^{-\frac{1}{2}}\, dr \tag{4.31}$$

as that obtained by the Born approximation (4.18), in the high angular momentum limit.

A final point, before turning to the orbiting problem, concerns the validity of (4.27) in the presence of a Coulomb field, $V(r) = e^2/r$ ($U(r) = 2\alpha k/r$, where $\alpha = e^2/\hbar v$). It is found by direct integration that (4.27) reduces to the form

$$\psi_l(r) \sim \sin[kr - \alpha \ln 2kr - l\pi/2 + \tilde{\mu}_l], \tag{4.32}$$

where

$$\tilde{\mu}_l = \tfrac{1}{2}\alpha \ln[\alpha^2 + (l + \tfrac{1}{2})^2] - \alpha + (l + \tfrac{1}{2}) \arctan[\alpha/(l + \tfrac{1}{2})], \tag{4.33}$$

which is comparable with the exact result, (3.20),

$$\psi_l(r) \sim \sin[kr - \alpha \ln 2kr - l\pi/2 + \arg\Gamma(l + 1 + i\alpha)], \tag{4.34}$$

within the accuracy of Stirling's approximation (Abramowitz and Stegun (1965)),

$$\arg\Gamma(l + 1 + i\alpha) \simeq \mathrm{Im}[(l + \tfrac{1}{2} + i\alpha)\ln(l + 1 + i\alpha) - (l + 1 + i\alpha) - \tfrac{1}{2}\ln 2\pi]$$

$$= \tfrac{1}{2}\alpha \ln[\alpha^2 + (l + 1)^2] - \alpha + (l + \tfrac{1}{2}) \arctan[\alpha/(l + 1)], \tag{4.35}$$

† Note that the semi-classical phase shift given by (4.30) is actually the radial component of the classical collision action, $\Delta(E, L)$ defined by (2.6), divided by $\hbar$. Following the footnote to (2.6) for a two parameter form of potential, $V(r) = \varepsilon W(\rho)$, where $\rho = (r/r_0)$, η_l is conveniently written $\eta_l = (mvr_0/\hbar)\eta_l^*$, where η_l^* depends only on the reduced collision energy $K = E/\varepsilon$ and the reduced equivalent impact parameter, $\beta = (l\hbar/mvr_0)$,

$$\eta_l^* = \int_{\alpha}^{\infty} [1 - W(\rho)/K - \beta^2/\rho^2]^{\frac{1}{2}}\, d\rho - \int_{\beta}^{\infty} [1 - \beta^2/\rho^2]^{\frac{1}{2}}\, d\rho.$$

valid for $|l + 1 + i\alpha| \gg 1$. The conditions therefore are that either α (compare equation (2.46)) or the angular momentum 1 must be large, the difference between $(1 + \frac{1}{2})$ in (4.33) and $(l + 1)$ in (4.35) being of negligible importance in either case. This means, since α is large for a typical molecular collision, ($\alpha \simeq 150$ for reduced mass 10 amu and collision energy 10eV), that all scattering phase shifts for not only a Coulomb potential but also for any screened

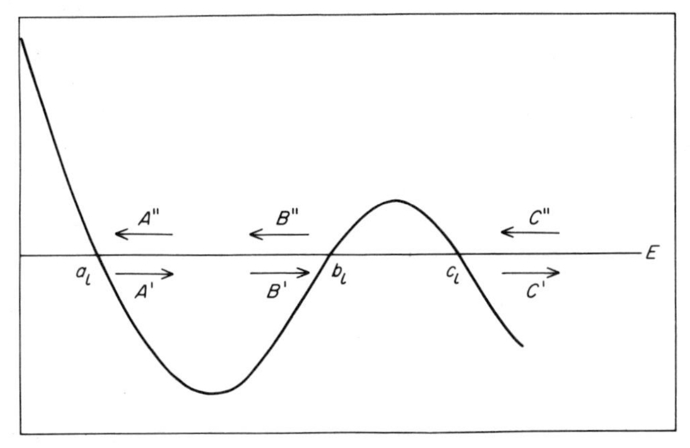

FIG. 4.2. Amplitude coefficients employed in (4.36)–(4.40).

Coulomb potential may be derived from (4.27). The semi-classical approximation breaks down however for electron collisions unless the angular momentum is large (1 > 10).

We turn now to the quantum mechanical orbiting phenomenon which is associated with the possibility of temporary capture in a quasi-bound level between the turning points a_l and b_l in Figs 4.1 and 4.2. Equation (4.27) again describes the wavefunction in this region but it is necessary, with the help of the parabolic connection formulae in section C.3, to continue this solution around the turning points b_l and c_l in order to determine the phase shift.

The procedure, due to Connor (1969), is first to express $\psi_l(r)$ in the general semi-classical form

$$\psi_l(r) = [k_l(r)]^{-\frac{1}{2}}\left[A' \exp\left(i\int_{a_l}^{r} k_l(r)\,\mathrm{d}r\right) + A'' \exp\left(-i\int_{a_l}^{r} k_l(r)\,\mathrm{d}r\right)\right], \quad r \gg a_l \tag{4.36a}$$

$$= [k_l(r)]^{-\frac{1}{2}}\left[B' \exp\left(i\int_{b_l}^{r} k_l(r)\,\mathrm{d}r\right) + B'' \exp\left(-i\int_{b_l}^{r} k_l(r)\,\mathrm{d}r\right)\right], \quad r \ll b_l \tag{4.36b}$$

$$= [k_l(r)]^{-\frac{1}{2}}\left[C' \exp\left(i\int_{c_l}^{r} k_l(r)\,\mathrm{d}r\right) + C'' \exp\left(-i\int_{c_l}^{r} k_l(r)\,\mathrm{d}r\right)\right],$$

$$r \gg c_l, \qquad (4.36c)$$

where the coefficients A, B and C have the physical meanings illustrated in Fig. 4.2.

Secondly the coefficients A' and A'' are necessarily related in the form

$$A' = A\,\mathrm{e}^{-i\pi/4}, \qquad A'' = A\,\mathrm{e}^{i\pi/4}, \qquad (4.37)$$

by means of (4.27) and (C.30). The values of the remaining coefficients are then followed through the turning points b_l and c_l to determine the ratio C' and C'' and hence the phase shift. Thus, since (4.36a) and (4.36b) both cover the region $a_l \ll r_l \ll b_l$, the difference between (A, A'') and (B', B'') lies merely in the choice of phase reference; hence

$$\begin{pmatrix} B' \\ B'' \end{pmatrix} = \begin{pmatrix} \mathrm{e}^{i\alpha_l}, & 0 \\ 0, & \mathrm{e}^{-i\alpha_l} \end{pmatrix} \begin{pmatrix} A' \\ A'' \end{pmatrix}, \qquad (4.38)$$

where

$$\alpha_l = \int_{a_l}^{b_l} k_l(r)\,\mathrm{d}r. \qquad (4.39)$$

(B', B'') and (C', C'') are then related by the parabolic connection formula obtained by inverting (C.49)

$$\begin{pmatrix} C' \\ C'' \end{pmatrix} = \begin{pmatrix} (1 + \mathrm{e}^{-2\pi\varepsilon_l})^{\frac{1}{2}}\,\mathrm{e}^{-i\phi_l}, & -i\,\mathrm{e}^{-\pi\varepsilon_l} \\ i\,\mathrm{e}^{-\pi\varepsilon_l}, & (1 + \mathrm{e}^{-2\pi\varepsilon_l})^{\frac{1}{2}}\,\mathrm{e}^{i\phi_l} \end{pmatrix} \begin{pmatrix} B' \\ B'' \end{pmatrix}, \qquad (4.40)$$

where

$$\varepsilon_l = (E - V_{\max}^l)/\hbar\omega_l^*, \qquad (4.41)$$

$$\omega_l^* = (\kappa_l/m)^{\frac{1}{2}}/2\pi,$$

$$\kappa_l = -(\partial^2 V^l/\partial r^2),$$

$$\phi_l = \arg\Gamma(i\varepsilon_l + \tfrac{1}{2}) - \varepsilon_l \ln|\varepsilon_l| + \varepsilon_l. \qquad (4.42)$$

Overall therefore, on combining (4.37), (4.38) and (4.40),

$$C' = A\exp(-i\phi_l/2 - i\pi/4)\,[(1 + \mathrm{e}^{-2\pi\varepsilon_l})^{\frac{1}{2}}\,\mathrm{e}^{i\delta_l} + \mathrm{e}^{-\pi\varepsilon_l}\,\mathrm{e}^{-i\delta_l}]$$

$$= \tfrac{1}{2}\exp[i(\eta_l^{(r)} - \tfrac{1}{2}\phi_l - \pi/4)], \qquad (4.43)$$

with

$$\delta_l = \alpha_l - \tfrac{1}{2}\phi_l \qquad (4.44)$$

$$\eta_l^{(r)} = \arctan\left\{\left[\frac{(e^{2\pi\varepsilon_l}+1)^{\frac{1}{2}}-1}{(e^{2\pi\varepsilon_l}+1)^{\frac{1}{2}}+1}\right]\tan(\alpha_l - \tfrac{1}{2}\phi_l)\right\} \tag{4.45}$$

$$A = \tfrac{1}{2}[1 + 2e^{-2\pi\varepsilon_l} + 2\,e^{-\pi\varepsilon_l}(1 + e^{-2\pi\varepsilon_l})^{\frac{1}{2}}\cos 2\delta_l]^{-\frac{1}{2}}; \tag{4.46}$$

C'' is the complex conjugate of C'. This means on substituting for C' and C'' in (4.36c) that

$$\psi_l(r) \overset{r\to\infty}{\sim} [k_l(r)]^{-\frac{1}{2}} \sin(kr - l\pi/2 + \eta_l), \tag{4.47}$$

with the phase shift η_l given by

$$\eta_l = \lim_{r\to\infty}\left\{\int_{c_l}^{r} k_l(r)\,dr - kr + (l+\tfrac{1}{2})\pi/2 + \eta_l^{(r)} - \tfrac{1}{2}\phi_l\right\}. \tag{4.48}$$

The essential difference between this orbiting phase shift and the simple form in (4.30) is seen to lie in the two final terms, of which $\eta_l^{(r)}$ is of overriding importance. Two types of limiting behaviour may be recognized.

At energies well above the maximum in Fig. 4.2 (i.e. above the classical orbiting energy), $e^{2\pi\varepsilon_l} \gg 1$ and (4.45) reduces to

$$\eta_l^{(r)} = \alpha_l - \tfrac{1}{2}\phi_l, \tag{4.49}$$

or combining (4.39), (4.48) and (4.49),

$$\eta_l = \lim_{r\to\infty}\left\{\int_{a_l}^{r} k_l(r)\,dr - kr + (l+\tfrac{1}{2})\pi/2 - \phi_l\right\}. \tag{4.50}$$

This formula is identical, apart from the small term ϕ_l ($\phi_l \sim 1/(24\varepsilon_l)$ for $\varepsilon_l \gg 1$ according to (C.40)), with the simple phase shift of (4.30). It may also be noted, according to (4.46) that A takes the constant value $\frac{1}{2}$ under these conditions.

At the opposite, negative energy, limit, $e^{2\pi\varepsilon_l} \ll 1$ and the correction term,

$$\eta_l^{(r)} \simeq \arctan\{\tfrac{1}{2}e^{-2\pi|\varepsilon_l|}\tan(\alpha_l - \tfrac{1}{2}\phi_l)\} \tag{4.51}$$

shows a characteristic resonance pattern. In view of the term $e^{-2\pi|\varepsilon_l|}$ it is generally small, but as the energy passes through coincidence with a quasi-bound level supported by the dip in Fig. 4.1(b), the Bohr quantization condition for which may be written

$$\delta_l = \alpha_l - \tfrac{1}{2}\phi_l = \int_{a_l}^{b_l} k_l(r)\,dr - \tfrac{1}{2}\phi_l = (n+\tfrac{1}{2})\pi, \tag{4.52}$$

the tangent term in (4.51) diverges, and η_l increases rapidly by π as shown in Fig. 4.3. Parallel resonance changes may also be observed in the coefficient A in (4.46), which determines (according to (4.47) and (4.36a)) the ratio of the amplitude of $\psi_l(r)$ in the inner region $a_l < r < b_l$ to its value at infinity. Since $e^{-\pi\varepsilon_l}$ is large, A is of order $e^{\pi\varepsilon_l} \ll 1$ for most energy values. At the resonant point however, $\delta_l = (n + \frac{1}{2})\pi$ and (4.46) reduces to

$$A = \tfrac{1}{2}[(1 + e^{-2\pi\varepsilon_l})^{\frac{1}{2}} - e^{-\pi_l}]^{-1}$$
$$\simeq e^{-\pi\varepsilon_l} \gg 1.$$

The dominant amplitude therefore shifts from the external to the quasi-bound region as illustrated in Fig. 4.4. The effects of this resonant behaviour on the form of the total cross-section is discussed in Section 4.3.

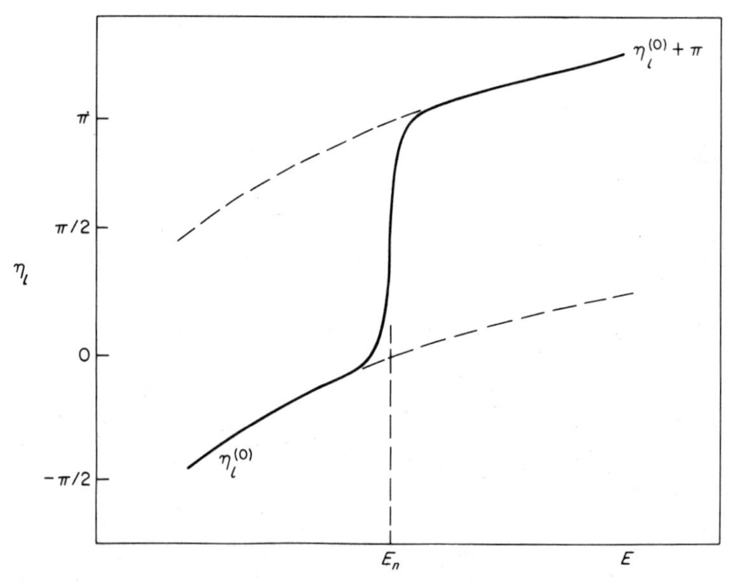

FIG. 4.3. A resonant change in the phase shift; $\eta_1^{(0)}$ is the non-resonant part of η_1.

As a rough measure of the resonance width, the energy may be taken such that

$$(n + \tfrac{1}{4})\pi < \eta_l^{(r)} < (n + \tfrac{3}{4})\pi,$$

so that at the limits

$$\tfrac{1}{2} e^{-2\pi|\varepsilon_l|} \tan \delta_l = \pm 1. \tag{4.53}$$

Hence on expanding δ_l in the form

$$\delta_l \simeq (n + \tfrac{1}{2})\pi + \left(\frac{\partial \delta_l}{\partial E}\right)(E - E_n), \tag{4.54}$$

and noting that (using (4.52))

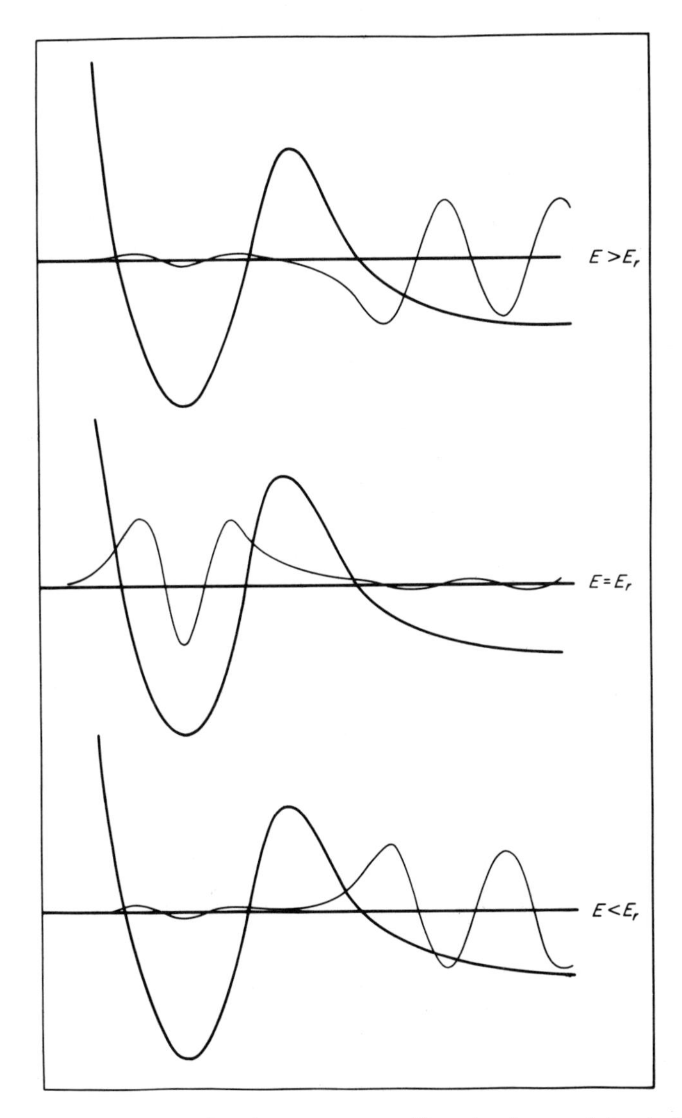

FIG. 4.4. Changes in the wavefunction at resonance. Note the sharp variation in the relative external and internal amplitudes and the phase change of π as E passes through E_r, due to the creation of an additional node within the barrier region.

$$\left(\frac{\partial \delta_l}{\partial E}\right) = \left(\frac{\partial \delta_l}{\partial n}\right)\left(\frac{\partial n}{\partial E}\right) = \pi/\hbar\omega \tag{4.55}$$

where $\hbar\omega$ is the local quasi-bound level spacing, it follows that

$$E \simeq E_n \pm \left(\frac{\hbar\omega}{2\pi}\right) e^{-2\pi|\varepsilon_l|}. \tag{4.56}$$

For another view of this width and its relation to the lifetime of the state, we may anticipate a general argument more fully developed in section 4.3. The technique is to apply an outgoing boundary condition ($C'' = 0$ in (4.36c)), corresponding to purely outgoing flux from the quasi-bound state; this leads to a complex energy eigenvalue

$$E = E_n - \frac{i}{2}\Gamma_n, \tag{4.57}$$

and hence to an exponential decay in the time evolution of the system;

$$\exp(-iEt/\hbar) = \exp(-iE_n t/\hbar)\exp(-\Gamma_n t/2\hbar). \tag{4.58}$$

The quasi-bound state therefore has real energy E_n and a lifetime determined by

$$\tau_n = \hbar/\Gamma_n. \tag{4.59}$$

Thus on applying the outgoing boundary condition in the present case, we require according to (4.43) that

$$C'^*(1 + e^{-2\pi|\varepsilon_l|})^{\frac{1}{2}} e^{-i\delta_l} + e^{-\pi|\varepsilon_l|} e^{i\delta_l} = 0, \tag{4.60}$$

and hence that

$$\begin{aligned}\delta_l &= -\frac{i}{2}\ln\left[-(1 + e^{-2\pi|\varepsilon_l|})^{\frac{1}{2}}\right] \\ &= (n + \tfrac{1}{2})\pi - \frac{i}{4}\ln(1 + e^{-2\pi|\varepsilon_l|}).\end{aligned} \tag{4.61}$$

This means, in energy terms, using (4.54) and (4.55), that

$$E = E_n - \frac{i}{2}\left(\frac{\hbar\omega}{2\pi}\right)\ln(1 + e^{-2\pi|\varepsilon_l|}), \tag{4.62}$$

so that by comparison with (4.57)

$$\Gamma_n = \left(\frac{\hbar\omega}{2\pi}\right)\ln(1 + e^{-2\pi|\varepsilon_l|}) \tag{4.63}$$

$$\simeq \left(\frac{\hbar\omega}{2\pi}\right) e^{-2\pi|\varepsilon_l|} \tag{4.63}$$

for $2\pi\varepsilon_l \ll -1$. The width Γ_n determined from the lifetime therefore coincides with the energy range spanned by (4.56).

4.3 Quasi-bound States

The existence of a quasi-bound state implies some distortion of the continuum due to interaction with an otherwise bound state. This leads to resonance features in the scattering, which are said to be of shape or Feshbach type according to whether the wavefunctions for the bound and continuum states belong to the same or to different internal states of the system (see Fig. 4.5).

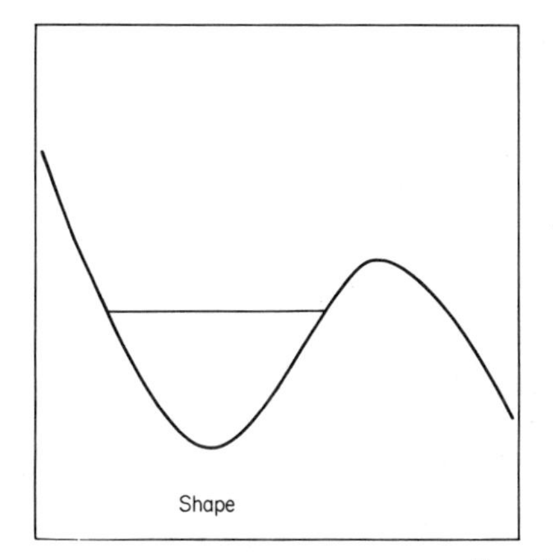

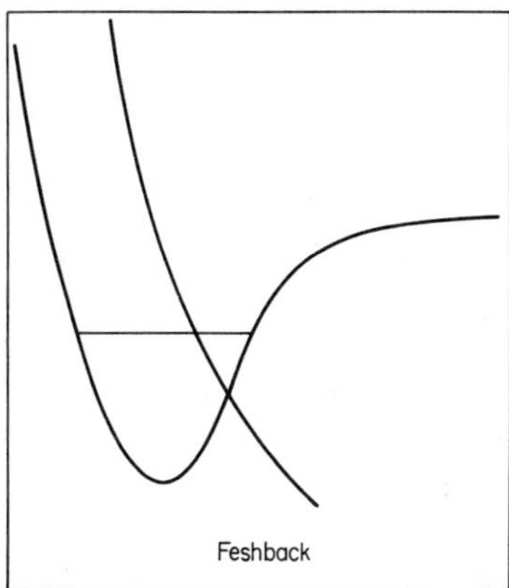

FIG. 4.5.

Particular examples are discussed in detail in sections 4.2, 7.4, and 8.4, but their general characteristics are conveniently discussed at this point, by means of an extension of the theory in the 'shape' case to allow for complex values of the wavenumber k (and energy E).

At real energies, according to appendix A, the bounded wavefunctions must be real apart from a multiplicative factor, with the physical result that its form in the asymptotic region

$$\begin{aligned}\psi_l(r) &\sim C\sin(kr - l\pi/2 + \eta_l)\\ &\sim e^{-ikr} - (-1)^l S_l(k)\, e^{ikr},\end{aligned} \tag{4.64}$$

where†

† S_l in (4.65), which represents the effect of the scattering potential on the relative amplitudes of the incoming and outgoing parts of the wave, is termed the scattering matrix; it has in general the same number of columns and rows as there are incoming and outgoing channels in the problem (see Section 6.2).

$$S_l(k) = e^{2i\eta_l}, \tag{4.65}$$

contains equal incoming and outgoing parts. This is in accordance with the conditions of an elastic scattering experiment. The introduction of a complex wavenumber and a complex phase shift allows the possibility of a nett incoming or nett outgoing solution. Thus if

$$k = k_1 + ik_2 \tag{4.66}$$

and

$$\eta_l = \mu_l + i\nu_l \tag{4.67}$$

the outgoing current density becomes,

$$j_r = -\frac{i\hbar}{2m}\left(\psi_l^* \frac{d\psi_l}{dr} - \psi_l \frac{d\psi_l^*}{dr}\right),$$

$$= \frac{\hbar}{2m}[-k_1 \sinh(2k_2 r + 2\nu_l) + k_2 \sin(2k_1 r + 2\mu_l)]. \tag{4.68}$$

This means, since the first term is dominant as $r \to \infty$, that the sign of j_r is given by the sign of the product $k_1 k_2$, with the implication that the energy

$$E = \frac{\hbar^2}{2m}(k_1 + ik_2)^2 = \frac{\hbar^2}{2m}(k_1^2 - k_2^2 + 2ik_1k_2), \tag{5.69}$$

contains a positive or a negative imaginary part according to whether $\psi_l(r)$ has a nett incoming or outgoing character. Imagine now a system prepared in a quasi-bound state and allowed to leak away, a situation closely approximated by optical predissociation. The corresponding solution, known as a Siegert (1939) state must, by (4.69), belong to a complex eigenvalue

$$E = E_{nl} - i\Gamma_{nl}/2, \tag{4.70}$$

lying in the lower half of the complex energy plane. Furthermore since the state is by hypothesis purely outgoing in character the amplitude S_l (or S matrix) in (4.64) must become infinite. It follows by expanding about this eigenvalue and noting that $|S_l| = 1$ for real energies, that S_l may be written

$$S_l = \left\{\frac{E - E_{nl} - i\Gamma_{nl}/2}{E - E_{nl} + i\Gamma_{nl}/2}\right\} \exp(2i\eta_l^{(0)}). \tag{4.71}$$

The phase shift in (4.65) therefore takes the Breit-Wigner form

$$\eta_l = \eta_l^{(0)} + \arctan[\Gamma_{nl}/2(E_{nl} - E)]. \tag{4.72}$$

Provided that $\eta_l^{(0)}$ varies smoothly with E, a condition which will be satisfied if the resonances are well separated compared with their widths Γ_{nl}, this means that the phase shift increases by π in passing through the resonance point, $E = E_{nl}$, with a rate of increase determined by Γ_{nl}.

As a direct consequence of (4.72), the contribution of the lth partial wave to the scattering amplitude of (3.32) and to the total cross-section of (3.34) shows pronounced resonance features;

$$f_l(\theta) = \frac{(2l+1)}{2ik} e^{2i\eta_l} P_l(\cos\theta) = \frac{(2l+1)}{2ik} e^{2i\eta_l^{(0)}} P_l(\cos\theta)\left[1 - \frac{i\Gamma_{nl}}{E - E_{nl} + i\Gamma_{nl}/2}\right] \tag{4.73}$$

$$\sigma_l = \frac{4\pi}{k^2}(2l+1)\sin^2\eta_l$$

$$= \frac{4\pi}{k^2}(2l+1)\left\{\sin^2\eta_l^{(0)} + \frac{\Gamma_{nl}^2\cos 2\eta_l^{(0)} + 2\Gamma_{nl}(E - E_{nl})\sin 2\eta_l^{(0)}}{4(E - E_{nl})^2 + \Gamma_{nl}^2}\right\}, \tag{4.74}$$

the resonance width being determined by the imaginary part of E in (4.70), while the shape of σ_l depends also on the local value of $\eta_l^{(0)}$ as shown in Fig. 4.6.

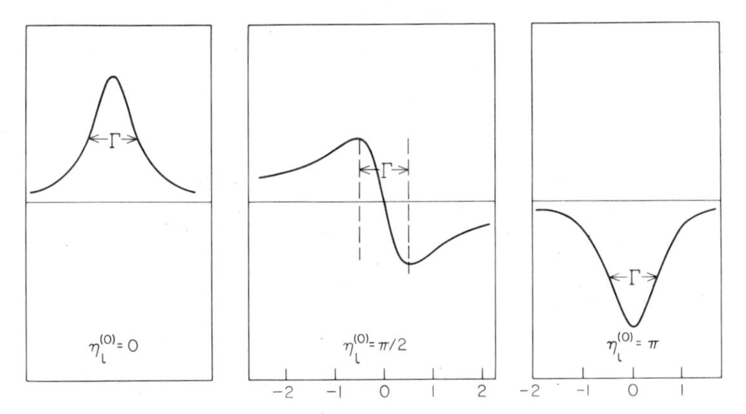

FIG. 4.6. Resonance line shapes.

As might be expected there is a direct connection between the resonance width Γ_{nl} and the lifetime of the quasi-bound state, which is seen most directly from the form of the time evolution factor for the wavefunction at the energy given by (4.70);

$$\exp(-iEt/\hbar) = \exp(-iE_{nl}t/\hbar)\exp(-\Gamma_{nl}t/2\hbar). \tag{4.75}$$

The probability that the system remains intact therefore decreases exponentially, with a time constant

$$\tau = \hbar/\Gamma_{nl}. \tag{4.76}$$

Equally, supposing for the sake of definiteness that in (4.66), $k_1 = \text{Re}\, k > 0$, and $k_2 = \text{Im}\, k < 0$, the wavefunction $\psi_l(r)$ in (4.64) is seen to increase exponentially with r

$$\exp(ikr) = \exp(ik_1 r)\exp(|k_2|r), \tag{4.77}$$

the reason being that the amplitude of $\psi_l(r)$ at the point r corresponds to emission at an earlier instant of time, $t = -r/v$, where $v = k_1\hbar/m$. This is confirmed by noting that according to (4.69) and (4.70), $\Gamma_{nl} = 2\hbar^2 k_1|k_2|/2m$; hence the amplitude of the time evolution factor in (4.75) at this time is

$$\exp(-\Gamma_{nl}t/2h) = \exp(|k_2|r) \tag{4.78}$$

in exact agreement with (4.77).

Turning finally to the scattering experiment it may be shown on quite general grounds (Wigner, 1955) that the time delay due to a collision process is given by

$$\tau = 2\hbar(\partial\eta_l/\partial E). \tag{4.79}$$

Hence in the present case, it follows on substituting from (4.72) at the energy $E = E_{nl}$ that

$$\tau = \left(\frac{\partial\eta_l^{(0)}}{\partial E}\right) + \frac{4\hbar}{\Gamma_{nl}}, \tag{4.80}$$

in which the term $4\hbar/\Gamma_{nl}$ may be attributed to capture into the quasi-bound state. This emphasizes yet again the relation between Γ_{nl} and the lifetime of the state.

4.4 Zero Energy Phase Shift: Levinson's Theorem

A direct relation between the phase shift at zero energy and the number of bound states supported by the scattering potential $U(r)$ may also be derived from the analytical properties of the S matrix, for complex values of the wavenumber k.

The argument is given in full by Alfaro and Regge (1965). It rests on use of a similar expression for the asymptotic wave function to that employed in (4.64), namely

$$\psi_l(r) \sim e^{-ikr} - (-1)^l[f_l(k)/f_l(-k)]\, e^{ikr} \tag{4.81}$$

so that

$$f_l(k)/f_l(-k) = S_l(k) = \exp[2i\eta_l(k)] \tag{4.82}$$

with the Jost functions $f_l(\pm k)$ being defined by Eqn (A.10). These may be shown to be analytic functions of k, subject to the general relation

$$f_l(-k) = [f_l(k^*)]^*, \tag{4.83}$$

so that, according to (4.82), for real values of k

$$\arg f_l(\pm k) = \pm\eta_l(k). \tag{4.84}$$

Furthermore, for potentials such that the integrals

$$\int_0^\infty rU(r)\,\mathrm{d}r \quad \text{and} \quad \int_0^\infty r^2U(r)\,\mathrm{d}r$$

are finite,

$$\lim_{|k|\to\infty} f_l(k) = 1; \tag{4.85}$$

hence $\eta_l(k)$ may be defined such that

$$\lim_{k\to\infty} \eta_l(k) = 0 \tag{4.86}$$

We are now in a position to examine the relation between the phase shift and the bound states. This follows from the familiar exponential decrease in the asymptotic bound state wave function

$$\psi_l(r) \sim \mathrm{e}^{-\gamma r}, \tag{4.87}$$

which leads by comparison with (4.81) to the conclusion that the bound states, with energies $E = -\gamma^2\hbar^2/2m$, correspond to zeros of $f_l(k)$ on the negative imaginary axis, at points $k = -i\gamma$. It is this knowledge, combined with equations (4.83)–(4.86), which leads by contour integration to the relation between the zero energy phase shift $\eta_l(0)$ and the number of the bound states.

Consider the integral

$$\int_c \frac{f_l{}'(k)}{f_l(k)}\,\mathrm{d}k = \ln\,[f_l(k)]_{C'} \tag{4.88}$$

taken around the closed contour in Fig. 4.7.

The integrand disappears on the large semi-circle, as its radius $\kappa \to \infty$, by virtue of (4.85), while the contribution from the real axis is obtained with the help of (4.84), using the identity (4.83) in the range $-\infty < k < 0$, and noting that $\eta_l\infty = 0$. The result is

$$\int_c \frac{f_l'(k)}{f_l(k)}\,\mathrm{d}k = 2i\eta_l(0) \tag{4.89}$$

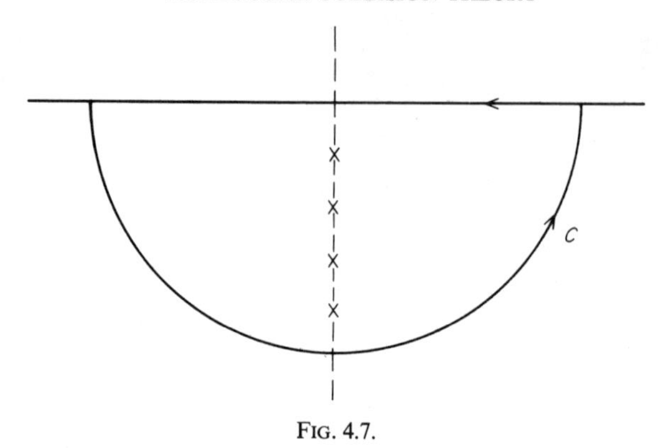

FIG. 4.7.

Since however the integrand has a pole at every zero of $f_l(k)$ it follows by the residue theorem that for n bound states

$$\int_C \frac{f_l'(k)}{f_l(k)} = 2\pi i n, \tag{4.90}$$

unless $f_l(0) = 0$, in which case

$$\int_C \frac{f_l'(k)}{f_l(k)} = 2\pi i(n + \tfrac{1}{2}). \tag{4.91}$$

The conclusion from (4.89)–(4.91) is therefore that

$$\eta_l(0) = n\pi, \tag{4.92}$$

unless there is a bound state at $k = 0$, in which case

$$\eta_l(0) = (n + \tfrac{1}{2})\pi. \tag{4.93}$$

These results are known as Levinson's theorem (Levinson (1949)).

A simple illustration is provided by the $l = 0$ states of the square well potential

$$\begin{aligned} U(r) &= -k_0^2 \qquad 0 < r < a \\ &= 0 \qquad r > a \end{aligned} \tag{4.94}$$

for which the function $f_0(k)$ is obtained by joining the solutions

$$\begin{aligned} f_0(k, r) &= e^{-ikr}, \qquad r > a \\ &= Ae^{ik'r} + Be^{-ikr} \qquad 0 < r < a \end{aligned} \tag{4.95}$$

where $k'^2 = k^2 + k_0^2$. Taking the limit as $r \to 0$,

$$f_0(k) = \lim_{r \to 0} f_0(k, r) = -[\cos k'a + i(k/k') \sin k'a] \exp(-ika), \tag{4.96}$$

and the phase shift becomes

$$\eta_0(k) = \arg[f_0(k)] = \arctan[(k/k')\tan k'a] - ka, \tag{4.97}$$

with the value $\eta_0(0) = s\pi$ where s is an integer. The bound states, corresponding with the roots of $f_0(k)$, are therefore given by

$$\cos k'a - (\gamma/k')\sin k'a = 0, \tag{4.98}$$

where $\gamma = -ik$, as may be confirmed by elementary methods. Fig. 4.8 (calculated for $k_0a = 7$) shows by means of contours of arg $f_0(k)$ how the zeros of $f_0(k)$ act as sources for the variation of $\eta_0(k)$ along the real axis.

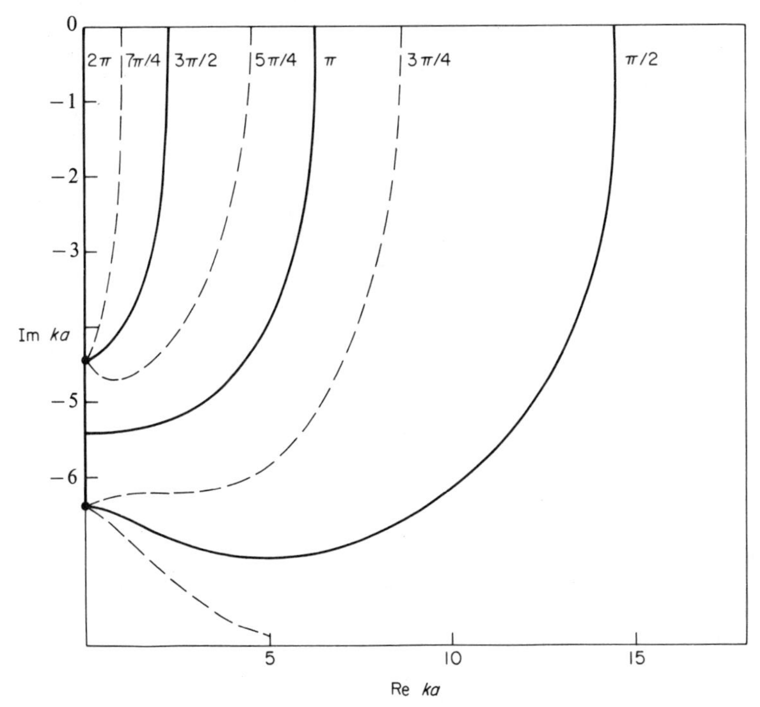

FIG. 4.8. Contours of arg $f_0(k)$ in the complex ka plane for $k_0a = 7$.

CHAPTER 5

Semi-Classical Elastic Scattering

Given the fundamental formulae (see (3.4), (3.30)–(3.32)),

$$\frac{d\sigma}{d\Omega} = I(\theta) = |f(\theta)|^2, \tag{5.1}$$

$$f(\theta) = \frac{1}{2ik} \sum_{l=0}^{\infty} (2l + 1)\, e^{2i\eta_l} P_l(\cos\theta), \qquad \theta \neq 0 \tag{5.2a}$$

$$= \frac{1}{k} \sum_{l=0}^{\infty} (2l + 1) \sin \eta_l\, e^{i\eta_l}, \qquad \theta = 0, \tag{5.2b}$$

$$\sigma = 2\pi \int_0^{\pi} I(\theta) \sin\theta\, d\theta = \frac{4\pi}{k^2} \sum_{l=0}^{\infty} (2l + 1) \sin^2 \eta_l \tag{5.3a}$$

$$= \frac{4\pi}{k} \operatorname{Im} f(0), \tag{5.3b}$$

and the validity of the JWKB phase shift (see (4.30) and (4.48)), it is a matter of simple computation to determine the total and differential cross-sections for any assumed potential $V(r)$. The only problem is the large number (100–1000) of partial waves which contribute to the scattering in a typical molecular system.

However, as first fully demonstrated by Ford and Wheeler (1959a and 1959b), and later consolidated by Bernstein (1966), the very magnitude of this sum allows approximations to (5.1)–(5.3) which offer insight into the physical nature of scattering on a molecular scale. These underline on one hand a close connection with the classical picture, as developed in terms of the classical trajectory in Chapter 2, except that the concept of additive scattering intensity (see (2.16)) gives way to quantum mechanical interference between the scattering amplitudes associated with different classical trajectories at the same scattering angle. This interference pattern leads to important quantum corrections to the rainbow and glory effects discussed in Section 2.2. On the other hand, as anticipated by the discussion in Section 2.4, there is a complete breakdown of the classical theory under orbiting conditions (see Sections 2.2 and 4.2) and at low scattering angles (see Sections 2.3 and 3.2). The latter leads in particular to a finite total cross-section for a potential of long range form C/r^n, $n > 2$.

The general theory is developed in Sections 5.1–5.5 and applied to determination of the scattering potential from experimental data in Section 5.6.

5.1 Simple Semi-classical Scattering

The term "simple semi-classical" will be used to describe the scattering at angles well separated (in a sense to be clarified below) from the classical rainbow and glory singularities. This allows the use of a variant of the saddle point method (or the method of steepest descents) in approximating the sum in (5.2). Known as the stationary phase approximation, it relies on replacing the sum by an integral, on the grounds that the number of contributing terms is large; and performing a quadratic expansion for the phase of the resulting integrand about its stationary value. This is based on the argument that any rapid variation of phase with l must lead to destructive interference.

The first step is to cast the Legendre functions in (5.2a) into an appropriate form, by the approximation (Abramowitz and Stegun, 1965)

$$P_l(\cos\theta) \simeq \left[\frac{2}{\pi l \sin\theta}\right]^{\frac{1}{2}} \cos\left[(l+\tfrac{1}{2})\theta - \pi/4\right], \tag{5.4}$$

valid for $l \sin\theta \gg 1$.† Hence (5.2) may be written

$$f(\theta) \simeq \frac{1}{k[2\pi \sin\theta]^{\frac{1}{2}}} \int_0^\infty l^{\frac{1}{2}}\left[e^{i\phi_+(l)} + e^{i\phi_-(l)}\right] dl, \tag{5.5}$$

where

$$\phi_\pm(l) = 2\eta_l - \pi/2 \pm (l\theta - \pi/4). \tag{5.6}$$

It follows that for a given angle θ one or other of the phases $\phi_\pm(l)$ will be stationary at a point l defined by

$$(\partial\phi_\pm/\partial l) = 2(\partial\eta_l/\partial l) \pm \theta = 0. \tag{5.7}$$

Before proceeding further it is convenient for the sake of a uniform description to replace the observable scattering angle θ which is defined to be positive, $0 < \theta < \pi$, by the classical deflection function Θ; $\Theta = \theta$ or $\Theta = -\theta$ for repulsive or attractive potentials respectively. Furthermore it

† This condition invalidates the resulting expression in the glory region.

is seen from Fig. 3.1 that the sign of $(\partial\eta_l/\partial l)$ also depends on the nature of the potential; it is positive for a repulsive trajectory, and negative in the attractive case. The conclusion is that the stationary phase point lies in the ϕ_+ or ϕ_- branches of (5.5) according to whether the potential has an attractive or repulsive effect, but that in either case

$$\Theta = 2\left(\frac{\partial\eta_l}{\partial l}\right). \tag{5.8}$$

It now remains, by substituting the semi-classical form of η_l from (4.30) to see that, with $l' = l + \frac{1}{2}$,

$$\frac{\partial\eta_l}{\partial l} = \int_{l/k}^{\infty} \frac{l'\,\mathrm{d}r}{[k^2 - l'^2/r^2]^{\frac{1}{2}}} - \int_{a_l}^{\infty} \frac{l'\,\mathrm{d}r}{r^2[k^2 - U(r) - l'^2/r^2]^{\frac{1}{2}}}, \tag{5.9}$$

This means after the substitutions

$$\left.\begin{aligned} L &= l'\hbar = mvb = k\hbar b \\ E &= k^2\hbar^2/2m, \qquad V(r) = U(r)\hbar^2/2m \end{aligned}\right\} \tag{5.10}$$

that

$$\Theta(E, L) = \pi - 2b\int_a^{\infty} \frac{\mathrm{d}r}{r^2[1 - V(r)/E - b^2/r^2]^{\frac{1}{2}}}, \tag{5.11}$$

in exact agreement with the classical result (2.5).†

In other words the point of stationary phase, l_θ in (5.5) for a given deflection, may be identified with the classical impact parameter, b, leading to the same scattering angle. Fig. 5.1 illustrates in more concrete terms how the quantum mechanical interference about the stationary phase point follows the classical trajectory.

This general behaviour rests however on a smooth local variation of η_l and on the validity (4.30) for the phase shift and hence cannot apply if there is a sharp discontinuity in the potential (Cowley *et al.*, 1969) or in the orbiting situation discussed in Section 5.4.

We might also note in passing that a relation of similar form to (5.8) may also be obtained, via the JWKB phase shift, between the classical collision time, (2.7), and the energy derivative of the phase shift,

$$2\hbar\left(\frac{\partial\eta}{\partial E}\right)_L = \frac{2}{v}\left(\frac{\partial\eta}{\partial k}\right)_L = \frac{2}{v}\left\{\int_a^{\infty} \frac{\mathrm{d}r}{[1 - V(r)/E - b^2/r^2]^{\frac{1}{2}}} - \int_b^{\infty} \frac{\mathrm{d}r}{[1 - b^2/r^2]^{\frac{1}{2}}}\right\} = \tau \tag{5.12}$$

† This result is foreshadowed by the connection between the radial action $\Delta(E, L)$ of (2.6) with $\Theta(E, L)$ on one hand (Eqn 2.8) and the semi-classical phase shift (Eqn 4.30) on the other.

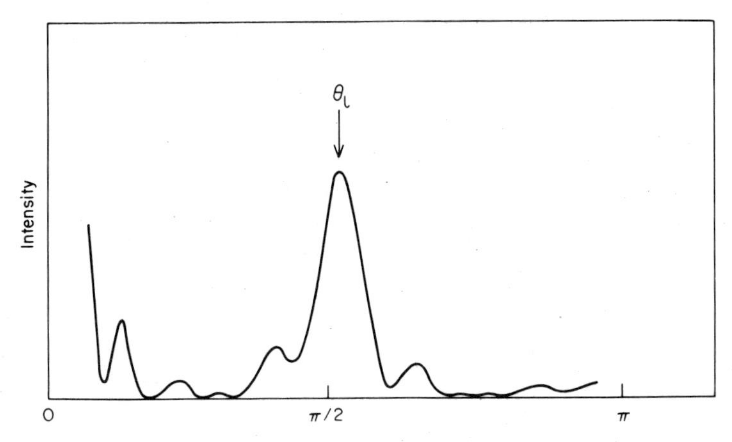

FIG. 5.1. Semi-classical interference about the classical trajectory. The figure gives the scattering intensity derived from the partial waves $l = 45$–65 for a Lennard-Jones potential $V(r) = 4\varepsilon[(r_0/r)^{12} - (r_0/r)^6]$ at energy $E = 2\varepsilon$ and $kr_0 = 80$. θ_l denotes the corresponding classical deflection angle for $\beta = (b/r_0) = (\bar{l}/kr_0) = 0{\cdot}69$.

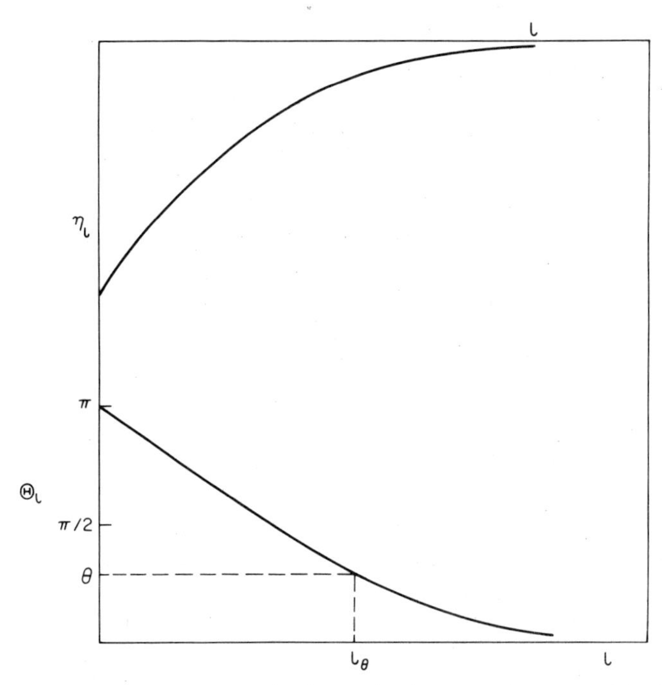

FIG. 5.2. Phase shifts and classical deflection angles for a repulsive potential, showing the relation $\Theta_l = 2(\partial\eta_l/\partial l)$.

Combined with (5.8), this means that (F. T. Smith, 1960, 1962)

$$d\eta = \left(\frac{\partial \eta}{\partial E}\right)_L dE + \left(\frac{\partial \eta}{\partial L}\right)_E dL = \frac{1}{2\hbar}\{\tau(E, L)\, dE + \Theta(E, L)\, dL\}. \tag{5.13}$$

The above connection between the stationary phase value of the angular momentum and its classical counterpart is the first important semi-classical result. For the second we turn to the relation between the intensity of constructive interference and the classical differential cross-section. Consider first a purely repulsive potential, leading to a negative phase shift, monotonically increasing to zero as $l \to \infty$ and hence in view of (5.8) to a monotonically increasing deflection function as shown in Fig. 5.2. There is then a single stationary phase point, $l = l_\theta$, lying (since the potential is repulsive) in the ϕ_- branch of (5.5), for each scattering angle. On expanding ϕ_- about this point

$$\phi_-(l) \simeq \phi_-(l_\theta) + \tfrac{1}{2}\phi''_-(l_\theta)(l - l_\theta)^2, \tag{5.14}$$

noting that

$$\phi''_-(l) = 2(\partial^2 \eta_l/\partial l^2)_{l=l_\theta} = -|\partial\Theta/\partial l|_{l=l_\theta}, \tag{5.15}$$

and employing the standard integrals

$$\int_{-\infty}^{\infty} e^{\pm iax^2}\, dx = [\pi/a]^{\frac{1}{2}} \exp^{\pm i\pi/4}, \qquad \text{for} \quad a > 0, \tag{5.16}$$

we find that $f(\theta)$ in (5.5) may be written†

$$f(\theta) = [I_{cl}(\theta)]^{\frac{1}{2}}\, e^{i\gamma}, \tag{5.17}$$

where

$$I_{cl}(\theta) = \frac{l_\theta}{\pi k^2 |\phi''_-(l_\theta)| \sin\theta} = \frac{b}{\sin\theta\, |d\theta/db|},$$

(after using (5.10) to replace 1 by b) and

$$\gamma = \phi_-(l_\theta) - \pi/4 = 2\eta_{l_\theta} - l_\theta\theta - \pi/2. \tag{5.18}$$

It follows, by comparison with (2.15), with the proviso $l_\theta \sin\theta \gg 1$ set by the validity of (5.4), that the classical and quantum mechanical cross-sections are identical for a purely repulsive potential.

The validity of this result requires that the width, $\Delta l \simeq |\partial\Theta/\partial l|^{-\frac{1}{2}}$, of the stationary phase region should be large compared with unity in order to justify the integral approximation employed in (5.5). Especial care may therefore be required in interpreting the elastic scattering in systems with a

† Note that the dominant contribution to the integral in (5.5) comes from a range of width $(\phi''_-)^{\frac{1}{2}}$, which is according to (5.15) just that set by the validity of the uncertainty principle, $\Delta L \Delta\theta = (\Delta L)^2(\Delta\theta/\Delta L) \simeq \hbar$.

low reduced mass, because the relatively small number of significant partial waves may lead non-semi-classical oscillations in the differential cross-section even in the presence of a purely repulsive potential.

True semi-classical oscillations become however directly apparent when the intermolecular potential has the typical short range repulsive, long range attractive form. There are then several stationary phase regions (corresponding to several classical trajectories) for a given scattering angle, as shown in Fig. 5.3, and it follows by direct generalization of (5.14)–(5.18) that, if the scattering angle θ in Fig. 5.3 is chosen so that the stationary phase regions

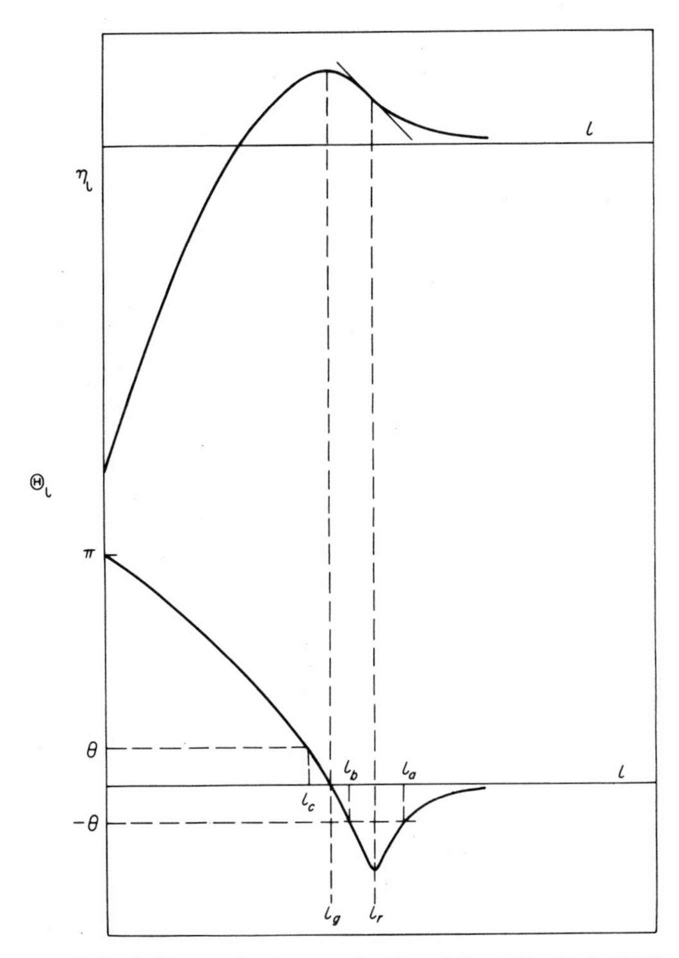

FIG. 5.3. The semi-classical connection between the phase shift and the classical deflection angle for a Lennard-Jones potential. The glory and rainbow angles correspond according to equation (5.8) to a maximum and to a point of inflexion in η_l respectively.

around l_a, l_b and l_c do not overlap, ($|l_a - l_b| \gg |\partial\Theta/\partial l|^{-\frac{1}{2}}$), the scattering amplitude reduces to a sum of terms

$$f(\theta) = \sum_{i=a,b,c} f_i(\theta), \tag{5.20}$$

where

$$f_i(\theta) = [I_{\mathrm{cl}}^{(i)}(\theta)]\, \mathrm{e}^{i\gamma_i}, \tag{5.21}$$

$I_{\mathrm{cl}}^{(i)}(\theta)$ being the classical contribution from the ith branch and†

$$\begin{aligned} \gamma_a &= 2\eta_{l_a} + l_a\theta - \pi/2, \\ \gamma_b &= 2\eta_{l_b} + l_b\theta - \pi, \\ \gamma_c &= 2\eta_{l_c} - l_c\theta - \pi/2. \end{aligned} \tag{5.22}$$

The difference between the classical and semi-classical theories is therefore that whereas in the former the full differential cross-section (2.16) appears simply a sum of terms from the contributing branches, it is the scattering amplitude in the latter which is obtained as a sum; this leads to an oscillatory variation in $I(\theta)$ about the classical value, as shown in Fig. 5.4.

Since this figure includes the special interference effects associated with overlap between the stationary phase region centred on l_a and l_b as θ approaches the rainbow angle θ_r, and those due to overlap between l_b and l_c as $\theta \to 0$ (glory effect), general analysis of this pattern to yield the scattering potential, $V(r)$, is deferred to Section 5.6. We note at present simply that in the small angle range, limited by the inequality $l_a \sin\theta \gg 1$ set by the validity of (5.4), $I(\theta)$ becomes increasingly dominated by the l_a branch because $(\partial\theta/\partial l) \to 0$ as $\theta \to 0$, with the result, as discussed after (2.36), that a log–log plot of $\theta \sin |f(\theta)|^2$ against $E\theta$ may be used to determine the parameters of a potential of long range form C/r^s. Recall that according to (5.1), $|f(\theta)|^2$ gives the differential cross-section.

5.2 Rainbow Scattering

The rainbow effect, arising quantum mechanically from a confluence of the stationary phase points l_a and l_b in Fig. 5.3, depends on the presence of a minimum (or maximum), at $l = l_r$, in the classical deflection function, and hence, via (5.8), to a point of inflexion in the phase shift. Near $l = l_r$ therefore

$$\Theta = \Theta_r + q(l - l_r)^2, \tag{5.23}$$

† These phase terms differ in form, first because l_a and l_b, which correspond to attractive trajectories, contribute to the ϕ_+ branch of (5.5) while l_c contributes to the ϕ_- branch, and secondly because $(\partial\Theta/\partial l) < 0$ at l_b and l_c, but $(\partial\Theta/\partial l) > 0$ at l_a.

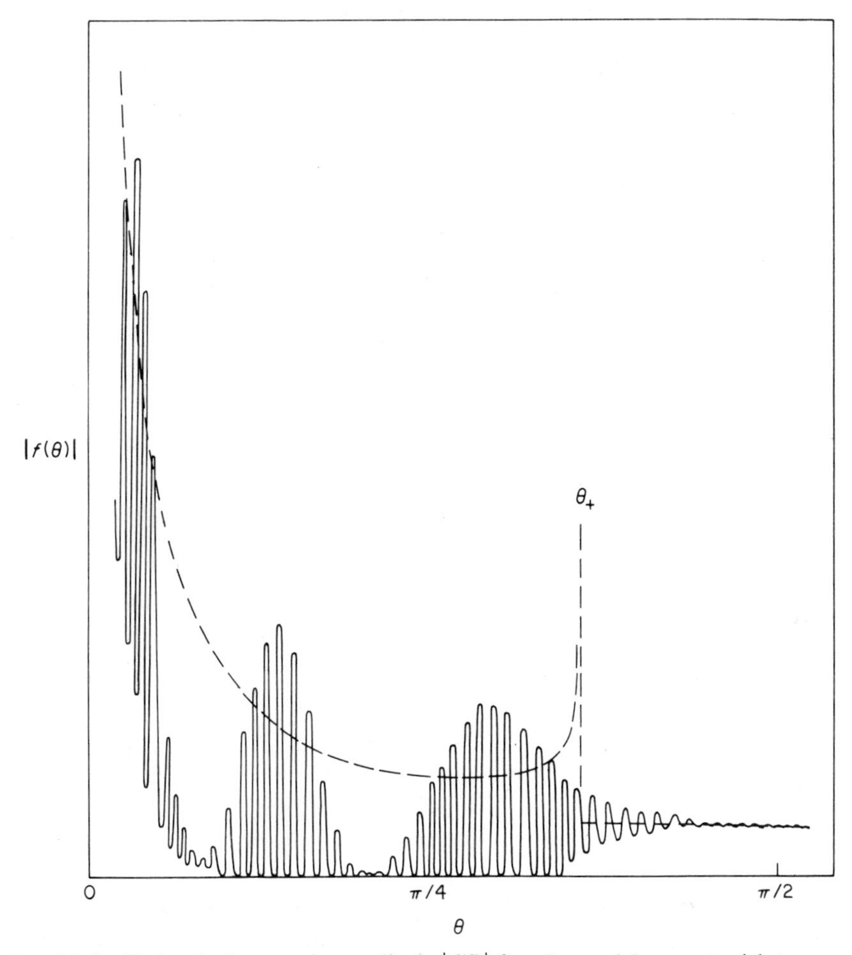

FIG. 5.4. Oscillations in the scattering amplitude, $|f(\theta)|$, for a Lennard-Jones potential at energy $E = 2\varepsilon$ and $kr_0 = 80$. The dashed line follows the classical function $[I_{cl}(\theta)]^{\frac{1}{2}}$ given by (2.15).

and

$$\eta_l = \eta_r + \tfrac{1}{2}\int_{l_r}^{l} \Theta \, dl = \eta_r + \tfrac{1}{2}\Theta_r(l - l_r) + \tfrac{1}{6}q(l - l_r)^3. \qquad (5.24)$$

Hence (using (5.6)), the starting point for the previous analysis, (namely Eqn (5.14)), must be replaced by

$$\phi_+(l) = \phi_+(l_r) + (\theta - \theta_r)(l - l_r) + \tfrac{1}{3}q(l - l_r)^3, \qquad (5.25)$$

where

$$\phi_+(l_r) = 2\eta_r + l_r\theta - \pi/4,$$

the ϕ_+ branch being chosen, and Θ_r being replaced by $-\theta_r$ because of the attractive effect of the potential. It follows on combining (5.5) and (5.25) and introducing a new integration variable $t = q^{\frac{1}{3}}(l - l_r)$ that the rainbow contribution to the scattering amplitude may be expressed in the form†

$$f_r(\theta) = \frac{1}{k}\left[\frac{2\pi l_r}{\sin\theta}\right]^{\frac{1}{2}} q^{-\frac{1}{3}}\, e^{i\phi_+(l_r)}\, \mathrm{Ai}[(\theta - \theta_r)/q^{\frac{1}{3}}], \tag{5.26}$$

where Ai(x) is the Airy integral integral (Abramowitz and Stegun, 1965).

$$\mathrm{Ai}(x) = \frac{1}{2\pi}\int_{-\infty}^{\infty} \exp\,(ixt + \tfrac{i}{3}t^3)\,\mathrm{d}t = \frac{1}{\pi}\int_{0}^{\infty} \cos\,(xt + \tfrac{1}{3}t^3)\,\mathrm{d}t,$$

with the form depicted in Fig. 5.5.

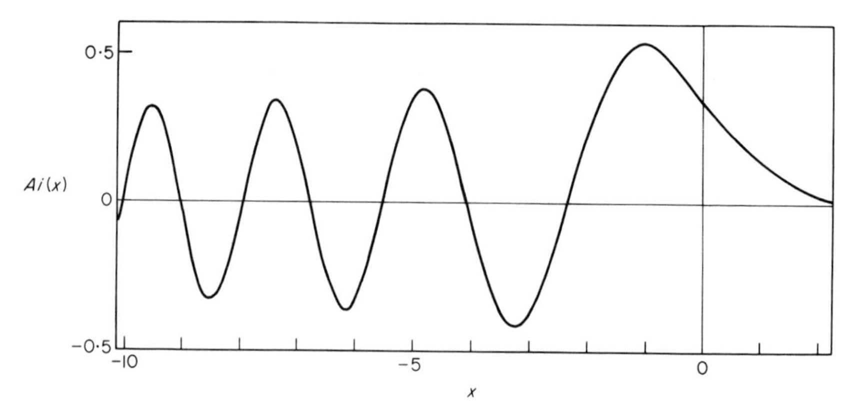

FIG. 5.5. The Airy function, Ai(x).

The quantum theory therefore smooths out the classical singularity in $\sigma(\theta)$ due to disappearance of the term $(\partial\theta/\partial l)$, replacing it by an exponential decrease for $\theta > \theta_r$ and an oscillatory variation into the low angle region. The principal maximum occurs at

$$x = (\theta - \theta_r)/q^{\frac{1}{3}} \simeq -1{\cdot}019,$$

with subsidiary maxima (supernumerary rainbows) in $[\mathrm{Ai}(x)]^2$ at $x = -3{\cdot}248$, $-4{\cdot}820$, $-6{\cdot}163\ldots$ The rainbow angle θ_r lies at a point of inflexion. This

† Berry (1966) has given a more accurate 'uniform' version of (5.26), (see also equations (9.57)–(9.68)),

$$|f_r(\theta)|^2 = \pi(I_a^{\frac{1}{2}} + I_b^{\frac{1}{2}})^2\zeta^{\frac{1}{2}}\mathrm{Ai}^2(-\zeta) + \pi(I_a^{\frac{1}{2}} - I_b^{\frac{1}{2}})^2\zeta^{-\frac{1}{2}}\mathrm{Ai}'^2(-\zeta),$$

where, with the designations a and b taken to refer to the stationary points of $\phi_+(1)$,

$$\tfrac{4}{3}\zeta^{\frac{3}{2}} = 2(\eta_a - \eta_b) + (l_a - l_b)\theta$$

and I_a and I_b are the classical forms which appear in (5.17).

comparatively low frequency oscillation, superimposed on a rising background due to increasing dominance of the l_a branch at small angles, is clearly apparent in Fig. 5.4.

The detailed form of this rainbow pattern may be used in practice to determine the well-depth, ε, and the range parameter, r_0, for an assumed two parameter form of scattering potential. In the first place the classical rainbow angle, given by the outermost point of inflexion in the pattern is characteristic, for a given form of potential, of the reduced collision energy (E/ε), and values of ε obtained in this way for various assumed forms of potential differ by only 10% (Bernstein and Muckerman, 1967).

Secondly the wavelength of the supernumerary oscillations may be used to determine the size parameter r_0. The first step is to determine the constant q by comparison between (5.26) and the observed oscillations. This is then compared with Q in the computed expansion

$$\Theta = \Theta_r + Q(\beta - \beta_r)^2, \tag{5.27}$$

where β is the reduced impact parameter

$$\beta = (b/r_0) = (l/kr_0). \tag{5.28}$$

Here Q, which depends for a given potential form only on the now known energy ratio (E/ε), is obtained from the classical Eqn (2.5). The final result derived from (5.23), (5.27), and (5.28) is that

$$r_0 = \frac{1}{k}\left(\frac{Q}{q}\right)^{\frac{1}{2}}. \tag{5.29}$$

Finally we demonstrate the direct connection between the rainbow pattern and the oscillations due to interference between $f_a(\theta)$ and $f_b(\theta)$ in (5.20). The argument is based on the standard asymptotic form of the Airy function (Abramowitz and Stegun, 1965).

$$\mathrm{Ai}(x) \sim \pi^{-\frac{1}{2}}(-x)^{-\frac{1}{4}} \sin\left[\tfrac{2}{3}(-x)^{\frac{3}{2}} + \pi/4\right], \tag{5.30}$$

for $x \ll -1$, and on two identities derived from (5.23); first

$$l - l_r = \pm[(\theta_r - \theta)/q]^{\frac{1}{2}} \tag{5.31}$$

so that on substituting for $(l - l_r)$ in (5.25)

$$\begin{aligned}\phi_+(l) &= \phi_+(l_r) \pm (\theta - \theta_r)[(\theta_r - \theta)/q]^{\frac{1}{2}} \pm \tfrac{1}{3}q[(\theta_r - \theta)/q]^{\frac{3}{2}}\\ &= \phi_+(l_r) \pm \tfrac{2}{3}(-x)^{\frac{3}{2}},\end{aligned} \tag{5.32}$$

and secondly

$$(\partial\theta/\partial l) = \mp 2[q(\theta_r - \theta)]^{\frac{1}{2}}, \tag{5.33}$$

from which

$$(-x)^{-\frac{1}{4}}q^{-\frac{1}{4}} = [\tfrac{1}{2}|\partial\theta/\partial l|]^{-\frac{1}{2}}, \tag{5.34}$$

the upper or lower signs in (5.31)–(5.33) being taken to refer to branches a and b respectively. It follows, on combining (5.26) with (5.30)–(5.34) that

$$f_r(\theta) = \frac{1}{k}\left[\frac{l_r}{\sin\theta|\partial\theta/\partial l|}\right]^{\frac{1}{2}}[e^{i\gamma_a} + e^{i\gamma_b}], \tag{5.35}$$

in agreement with the sum $(f_a(\theta) + f_b(\theta))$ in (5.20), to the extent that $l_r = l_a = l_b$.†

5.3 Small Angle and Glory Scattering

Two corrections to the general theory of Section 5.1 are required in the small angle region, the first arising from the breakdown of the asymptotic approximation (5.4) for the Legendre function, and the second from the coincidence of the stationary phase points l_b and l_c in Fig. 5.3 as $\theta \to 0$. The latter implies that the scattering amplitude contains an a branch contribution, $f_a(\theta)$, derived from the high angular momentum waves, and a second (glory) contribution, $f_g(\theta)$, due to the confluence of the b and c branches;

$$\begin{aligned} f(\theta) &= f_a(\theta) + f_g(\theta) \\ &= [I_a(\theta)]^{\frac{1}{2}} e^{i\gamma_a} + [I_g(\theta)]^{\frac{1}{2}} e^{i\gamma_g}. \end{aligned} \tag{5.36}$$

In practice $f_a(\theta) \gg f_g(\theta)$.

The derivation of the dominant term, $f_a(\theta)$, is similar to the treatment of the hard sphere case (Eqns (3.47)–(3.49)), except that the random phase approximation employed in (3.45) is no longer valid. $P_l(\cos\theta)$ is expanded about $\theta = 0$.

$$P_l(\cos\theta) = 1 - l^2\theta^2/4\ldots, \tag{5.37}$$

and the Jeffreys–Born approximation (4.18)

$$\eta_l = a_s l^{1-s}, \tag{5.38}$$

valid at large l for an asymptotic inverse s power potential, is employed for *all* phase shifts, because apart from the glory contribution derived below, the errors introduced in the low angular momentum terms of $f(\theta)$ are negligible in comparison with the contribution from the large number of (small phase shift) terms for which (5.38) is valid. The result using (5.2) is that

† The uniform approximation obtained by Berry (1966), and given in the footnote on p 68 is designed to go over exactly to the simple semi-classical form given by (5.20).

$$\operatorname{Im} f_a(\theta) \simeq \frac{2}{k}(J_1 - \tfrac{1}{4}J_2\theta^2) = \frac{2J_1}{k}\left[1 - \frac{\theta^2}{4}\left(\frac{J_2}{J_1}\right)\right], \tag{5.39}$$

$$\operatorname{Re} f_a(\theta) \simeq \frac{1}{k}(J_3 - \tfrac{1}{4}J_4\theta^2) = \frac{J_3}{k}\left[1 - \frac{\theta^2}{4}\left(\frac{J_4}{J_3}\right)\right], \tag{5.40}$$

where the integrals

$$\begin{aligned} J_1 &= \int_0^\infty l \sin^2 \eta_l \, dl, \qquad J_2 = \int_0^\infty l^3 \sin^2 \eta_l \, dl, \\ J_3 &= \int_0^\infty l \sin 2\eta_l \, dl, \qquad J_4 = \int_0^\infty l^3 \sin 2\eta_l \, dl, \end{aligned} \tag{5.41}$$

may be expressed, after the substitution, $x = \eta_l$, in terms of the standard forms (Dwight (1961)).

$$\int_0^\infty \frac{\sin^2 x \, dx}{x^{p+1}} = \frac{1}{p}\int_0^\infty \frac{\sin 2x \, dx}{x^p} = \frac{\pi 2^{p-2}}{p \sin(p\frac{\pi}{2})\Gamma(p)}. \tag{5.42}$$

It is convenient for comparison with the hard sphere result (3.49), to use the optical theorem (3.32) to relate J_1 to the total cross-section contribution from the a branch,

$$J_1 = \frac{k}{2} \operatorname{Im} f_a(0) = \frac{k^2 \sigma_a}{8\pi}, \tag{5.43}$$

and to express J_2, J_3 and J_4 in terms of J_1 in the form (Bernstein, 1963)

$$\begin{aligned} (J_2/J_1) &= J_1(2/\pi) \tan[\pi/(s-1)] \{\Gamma[2/(s-1)]\}^2/\Gamma[4/(s-1)], \\ (J_3/J_1) &= 2 \tan[\pi/(s-1)], \\ (J_4/J_3) &= J_1(2/\pi) \tan[2\pi/(s-1)] \{\Gamma[2/(s-1)]\}^2/\Gamma[4/(s-1)]. \end{aligned} \tag{5.44}$$

Hence on combining (5.39), (5.40), (5.43) and (5.44)

$$\begin{aligned} I_a(\theta) = |f_a(\theta)|^2 &= \left(\frac{k\sigma_a}{4\pi}\right)^2 q(s)\left[1 - c(s)\frac{k^2\sigma_a\theta^2}{8\pi} + \dots\right] \\ &\simeq \left(\frac{k\sigma_a}{4\pi}\right)^2 q(s) \exp\left[-c(s)\frac{k^2\sigma_a\theta^2}{8\pi}\right], \end{aligned} \tag{5.45}$$

where

$$q(s) = 1 + \tan^2[\pi/(s-1)], \tag{5.46}$$

$$c(s) = \frac{1}{2\pi} \tan\left[\frac{2\pi}{s-1}\right] \frac{\{\Gamma[2/(s-1)]\}^2}{\Gamma[4/(s-1)]}, \tag{5.47}$$

and

$$\gamma_a(\theta) = \arctan\left\{\frac{2J_1}{J_3}\left[\frac{1 - J_2\theta^2/4J_1 \ldots}{1 - J_4\theta^2/4J_3 \ldots}\right]\right\}. \tag{5.48}$$

It follows by comparison with (3.49) that the low angle scattering intensity, $I_a(\theta)$, differs from the hard sphere expression by virtue of the terms $q(s)$ and $c(s)$, both of which tend to unity as $s \to \infty$. A second point of later importance is that the phase term for forward scattering reduces, with the help of (5.44), to an expression

$$\gamma_a(\theta) = \arctan(2J_1/J_3) = \frac{\pi(s-3)}{2(s-1)} \tag{5.49}$$

which depends only on the potential exponent s.

The glory contribution, $f_g(\theta)$, to the scattering amplitude is more simply obtained. We note, according to (5.8) and Fig. 5.3, that the phase shift has a maximum at $l = l_g$,

$$\eta_l = \eta_g + \tfrac{1}{2}\eta_g''(l - l_g)^2 + \ldots, \tag{5.50}$$

where

$$\eta_g'' = \frac{1}{2}\left(\frac{\mathrm{d}\Theta}{\mathrm{d}l}\right)_{l=l_g} < 0. \tag{5.51}$$

Furthermore the Legendre functions are almost independent of l in the small angle region ($P_1(\cos\theta) \simeq 1$). Hence the stationary phase approximation is again applicable in the form

$$\begin{aligned} f_g(\theta) &\simeq -ik^{-1}\, l_g\, P_{l_g}(\cos\theta)\int_{-\infty}^{\infty} \exp[i\eta_g''(l - l_g)^2]\,\mathrm{d}(l - l_g), \\ &= k^{-1}[2\pi/|\mathrm{d}\theta/\mathrm{d}l|]^{\frac{1}{2}} l_g P_{l_g}(\cos\theta)\exp(2i\eta_g - 3i\pi/4). \end{aligned} \tag{5.52}$$

Note that since the maximum phase shift η_{l_g} may be shown (see Eqn (5.72)) to decrease with energy, the phase term

$$\gamma_g = 2\eta_g - 3\pi/4, \tag{5.53}$$

is energy dependent, in contrast with the term $\gamma_a(0)$ in (5.49).

That $f_g(\theta)$ in (5.52) indeed arises from a confluence between the b and c branches in Fig. 5.3 is readily confirmed by replacing $P_{l_g}(\cos\theta)$ by its asymptotic form (5.4) as θ increases;

$$f_g(\theta) \simeq \frac{1}{k}\left[\frac{l_g}{\sin\theta|\mathrm{d}\theta/\mathrm{d}l|}\right]^{\frac{1}{2}}[\exp(2i\eta_g + l_g\theta - \pi) + \exp(2i\eta_g - l_g\theta - \pi/2)], \tag{5.54}$$

in close agreement with the sum $f_b(\theta) + f_c(\theta)$ in (5.20).

This completes analysis of the two components of $f(\theta)$. Since in practice $|f_a(\theta)| \gg |f_g(\theta)|$, the interference between them leads to a small angle scattering intensity of the form

$$I(\theta) = |f(\theta)|^2 = I_a(\theta)\{1 + 2[I_g(\theta)/I_a(\theta)]^{\frac{1}{2}} \cos(\gamma_g - \gamma_a)\}, \tag{5.55}$$

where $I_a(\theta)$, $I_g(\theta)$, γ_a and γ_b are determined by (5.45), (5.48), (5.52) and (5.53).

The conclusion is that the forward scattering intensity, $I(0)$, is finite, in contrast with the classical result, (2.34), and that its value oscillates as a function of energy. Moreover since according to (5.48) and (5.53),

$$\gamma_g(0) - \gamma_a(0) = 2\eta_g - \frac{(2s-3)\pi}{4(s-1)}, \tag{5.56}$$

this oscillation frequency is governed entirely by the energy variation of the maximum phase shift η_g.

5.4 Orbiting

Attention in the quantum theory of orbiting centres on the role of the quasi-bound states in Fig. 5.6, rather than on the particular "trapped" classical trajectory for which the energy coincides with the maximum in the effective potential function $V_l(r)$. Indeed the important classical singularity in the deflection function (2.12), disappears, and the exponential contributions (2.19) to the classical cross-section are replaced by characteristic resonance terms, (4.73) and (4.74).

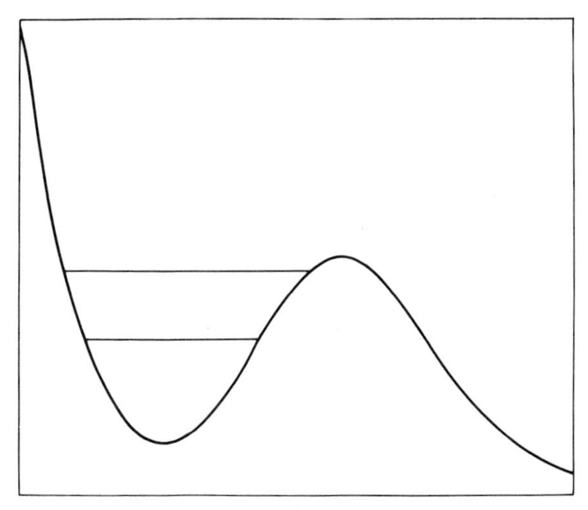

FIG. 5.6. Quasi-bound levels.

The theory follows directly from the form of the orbiting phase shift given by (4.48),

$$\eta_l = \lim_{r\to\infty}\left\{\int_{c_l}^{r} k_l(r)\,dr - kr + (l+\tfrac{1}{2})\pi/2 + \eta_l^{(r)} - \tfrac{1}{2}\phi_l\right\}, \tag{5.57}$$

where

$$\eta_l^{(r)} = \arctan\left\{\left[\frac{(e^{2\pi\varepsilon_l}+1)^{\frac{1}{2}}-1}{(e^{2\pi\varepsilon_l}+1)^{\frac{1}{2}}+1}\right]\tan\left[\int_{a_l}^{b_l} k_l(r)\,dr - \tfrac{1}{2}\phi_l\right]\right\}. \tag{5.58}$$

The first point is that η_l varies smoothly with l at the classical orbiting point, $\varepsilon_l = 0$, although on taking the derivative with respect to l, the integrals in (5.57) and (5.58) both yield singularities of the classical orbiting type (2.12)

$$\tfrac{1}{2}\beta \ln|(l - l^*)/l^*|, \tag{5.59}$$

where $\beta = (b^*/r_0^2)[2E/\kappa]^{\frac{1}{2}} = (l^*\hbar/mr_0^2)[m/\kappa]^{\frac{1}{2}}$, as l approaches the critical value l^*; these singularities are however exactly cancelled by the terms $-\frac{1}{2}(\partial\phi_l/\partial l)$, because, according to (C.50) and (4.42)

$$\phi_l = \arg\Gamma(\tfrac{1}{2} + i\varepsilon_l) - \varepsilon_l \ln|\varepsilon_l| + \varepsilon_l$$

where

$$\varepsilon_l = \frac{E - V_l^{\max}}{h\nu^*} = \frac{V_{l^*}^{\max} - V_l^{\max}}{h\nu^*} = \frac{(l^2 - l^{*2})\hbar[m/\kappa]^{\frac{1}{2}}}{2mr_0^2}, \tag{5.60}$$

the important contribution to $(\partial\phi_l/\partial l)$ being the final term $(\partial\varepsilon_l/\partial l)\ln|\varepsilon_l|$. The result is that the stationary phase method of Section 5.2 may be used to relate $(\partial\eta_l/\partial l)$ to the deflection function Θ by (5.8), but that Θ no longer diverges at l^*. The associated exponential terms, (2.19), in the cross-section therefore disappear.

The resonant changes in $\eta^{(r)}$ in (5.58), with widths given according to (4.63) by

$$\Gamma_{nl} = \frac{\hbar\omega}{2\pi}e^{-2\pi|\varepsilon_l|} \tag{5.61}$$

where $\hbar\omega$ is the energy spacing between the quasi-bound levels in Fig. 5.6, lead however by the arguments of Section 4.3 to a scattering amplitude of the form (4.73),

$$f(\theta) = f^{(0)}(\theta) + f^{(r)}(\theta) \tag{5.62}$$

where

$$f^{(0)}(\theta) = \frac{1}{2ik}\sum_{l=0}^{\infty}(2l+1)(e^{2i\eta_l^{(0)}} - 1)P_l(\cos\theta), \tag{5.63a}$$

$$f^{(r)}(\theta) = -\frac{1}{2k}\sum_{l=0}^{\infty}(2l+1)\left[\frac{\Gamma_{nl}}{E - E_{nl} + i\Gamma_{nl}/2}\right]e^{2i\eta_l^{(0)}}P_l(\cos\theta), \tag{5.63b}$$

and $\eta_l^{(0)}$ is the smoothly varying, non-resonant part of the phase shift. $f^{(0)}(\theta)$ is therefore amenable to the methods of Section 5.1–5.3 but analysis of $f^{(r)}(\theta)$ is more difficult; it depends on the number of quasi-bound levels lying within the resonance width at the energy in question.

Since $P_l(\cos\theta)$, which determines the angular dependence of the lth term in $f^{(r)}(\theta)$, oscillates according to (5.4) within the envelope $\pm(\sin\theta)^{\frac{1}{2}}$, a single isolated resonance will contribute only a roughly uniform background to the scattering amplitude. Interference between several such resonances at the same energy, arising in molecular terms from overlapping rotational levels of different vibrational states, might however lead to a small perturbation of $f^{(0)}(\theta)$ in a given direction. The effects of a broad resonance, spanning several neighbouring l values is more interesting. Suppose that the resonant energies are approximated in the molecular form

$$E_{nl} = E_n + l(l+1)\hbar^2/2I \tag{5.64}$$

and that the level width is independent of l over a small range. The stationary phase approximation to $f^{(r)}(\theta)$, then leads to a deflection angle, (5.8), of the form

$$\begin{aligned}\Theta &= \frac{\partial}{\partial l}\{2\eta_l^{(0)} + \arctan[\Gamma_n/2(E_{nl} - E)]\}\\ &= \Theta^{(0)} - \frac{(2l+1)\hbar^2}{4I}\left[\frac{\Gamma_{nl}}{(E - E_{nl})^2 + \Gamma_n^2/4}\right],\end{aligned} \tag{5.65}$$

which passes through a minimum value, which may be represented in the form

$$\Theta_{\min} \simeq \Theta^{(0)} - \frac{(2l+1)\hbar^2}{\Gamma_n I} = \Theta^{(0)} - 2\dot{\theta}\tau_n, \tag{5.66}$$

after identifying $(l\hbar/I)$ with the angular velocity $\dot{\theta}$, and $\hbar/\Gamma_n$ with the resonance lifetime τ_n. The result would be a rainbow feature centred on $\theta_{\min}$ with a strength governed by the number of l values contributing to the resonance. In view of the low tunnelling efficiency of all but the lightest atoms, it is likely that sharp resonances will predominate in practice, and that the observable effects of orbiting on the differential cross-section will be small.

Orbiting contributions to the total cross-section are discussed in Section 5.5 below.

5.5 The Total Cross-Section

At energies above the classical orbiting limit (see Fig. 2.4) the phase shift is a smoothly varying function of l, and the total cross-section is most conveniently obtained by the optical theorem, (3.32);

$$\sigma = \frac{4\pi}{k} \operatorname{Im} f(0), \tag{5.67}$$

the forward scattering amplitude $f(0)$ being given in Section 5.3, as a combination of *a* branch and glory scattering terms.

$$f(0) = f_a(0) + f_g(0) \tag{5.68}$$

It follows from (5.39) and (5.52) (after evaluating the integral J_1 with the help of (5.42)), that the total cross-section for a potential with asymptotic form $-C/r^s$ may be written†

$$\begin{aligned}\sigma &= \frac{4\pi}{k}[\operatorname{Im} f_a(0) + \operatorname{Im} f_g(0)] \\ &= p(s)[C/\hbar v]^{2/(s-1)} + \frac{4\pi}{k}[I_g(0)]^{\frac{1}{2}} \sin(2\eta_g - 3\pi/4),\end{aligned} \tag{5.69}$$

where

$$p(s) = \pi^2[2f(s)]^{2/(s-1)}/\sin[\pi/(s-1)]\Gamma[2/(s-1)],$$

$f(s)$ being given by (4.19), and

$$I_g(0) = 2\pi l_g/k^2|d\theta/dl|. \tag{5.70}$$

The overall result is a cross-section function $\sigma(E)$ of the form shown in Fig. 5.7, the analysis of which yields important information about the scattering potential. In the first place the variation of the glory phase shift η_g is directly related to the parameters of an assumed form of potential. According to the footnote to p 36, the phase shift for a two parameter potential may be expressed in reduced form $\eta_l = (mvr_0/\hbar)\eta^*(K, \beta)$, and at energies $K \gg 1$ $(K = E/\varepsilon)$ where the Born approximation becomes applicable, $\eta^*(K, \beta)$ varies inversely with K. In the Lennard-Jones case for example, it is readily shown by including both repulsive and attractive contributions to the Born phase shift, (4.18), that η_l reaches its maximum (glory) value,

$$\eta_g = \left(\frac{mvr_0}{K\hbar}\right)\left(\frac{120}{121}\right)\frac{f^2(6)}{f(12)}\left[\frac{11f(12)}{5f(6)}\right]^{\frac{1}{6}} = 0{\cdot}947\,(\varepsilon r_0/\hbar v), \tag{5.72}$$

at the point $\beta = [11f(12)/5f(6)]^{\frac{1}{6}}$, when $K \gg 1$. This means that the positions of the glory maxima in Fig. 5.7, which occur, according to (5.69), at the points

† The first (and dominant) term in (5.69) may be compared with the equivalent impact parameter, $b^* = l^*\hbar$, at which the classical action, measured as $\eta_{l*}\hbar$, becomes of order $\hbar$; using (4.18)

$$\pi b^{*2} = (f(s)\,C/\hbar v)^{2/(s-1)}. \tag{5.71}$$

This reinforces the argument of Section 2.4 that a collision can be defined as such, and hence can contribute to the total cross-section only if the classical action is sufficiently large to span the limits imposed by the uncertainty principle.

$$\eta_g = (N - \tfrac{3}{8})\pi, \qquad N = 1, 2, 3 \ldots. \tag{5.73}$$

may be used to determine the product εr_0. Secondly a log–log plot of the energy averaged background cross-section in the intermediate energy region will provide the parameters $C^{(s)}$ and s for a potential of asymptotic form $-C^{(s)}/r^s$; a similar analysis may in principle be applied in the very high energy region to determine the repulsive part of the potential in the form $C^{(t)}/r^t$.

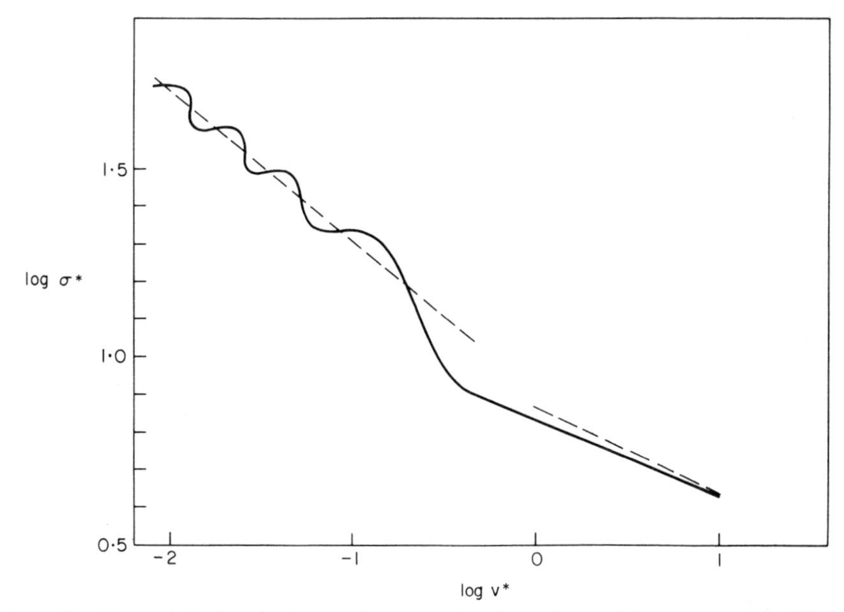

FIG. 5.7. Glory oscillations in the total cross-section for a Lennard-Jones potential, $V(r) = 4\varepsilon[(r_0/r)^{12} - (r_0/r)^6]$, with $B = 2m\varepsilon r_0^2/\hbar^2 = 500$. σ^* denotes the total cross-section, scaled by the square of the range parameter, r_0, and $v^* = (\hbar v/4\varepsilon r_0)$. The dashed lines follow the low and high energy classical forms given by the first term in (5.69) with $s = 6$ and $s = 12$ respectively.

Bernstein (1962, 1963) has emphasized an interesting connection between the maxima in Fig. 5.7 and the capacity of the scattering potential to support bound states. This arises from the behaviour of the glory phase shift $\eta_g(k)$ which increases monotonically as $k \to 0$, while the l value at the glory point tends to zero in this limit. Consequently there is an upper limit on $\eta_g(k)$, set by the s-wave zero energy phase shift $\eta_0(0)$, which by Levinson's theorem (4.92) takes the value $n\pi$ where n is the number of bound states with $l = 0$. This means, since the maxima occur when $\eta_g = (N - \frac{3}{8})\pi$ that the number of possible maxima of this type, in what has been termed the impact spectrum (Bernstein, 1963), is equal to the number of rotationless bound states supported by $V(r)$.

Experimental observation of such glory maxima in the low energy region is however complicated by the possibility of orbiting resonances of the type discussed by Section 5.4. As a result, according to (4.74)

$$\sigma = \sigma^{(0)} + \sigma^{(\text{res})} \tag{5.75}$$

where the non-resonant term $\sigma^{(0)}$ derived from the smoothly varying part of the phase shift $\eta_l^{(0)}$ is given by (5.69), and the resonant term takes the form

$$\sigma^{(\text{res})} = \frac{4\pi}{k^2} \sum_{l=0}^{\infty} (2l+1) \left\{ \frac{\Gamma_{nl}^2 \cos 2\eta_l^{(0)} + 2\Gamma_{nl}(E_{nl} - E) \sin 2\eta_l^{(0)}}{4(E_{nl} - E)^2 + \Gamma_{nl}^2} \right\}, \tag{5.76}$$

the linewidths being given by (4.63);

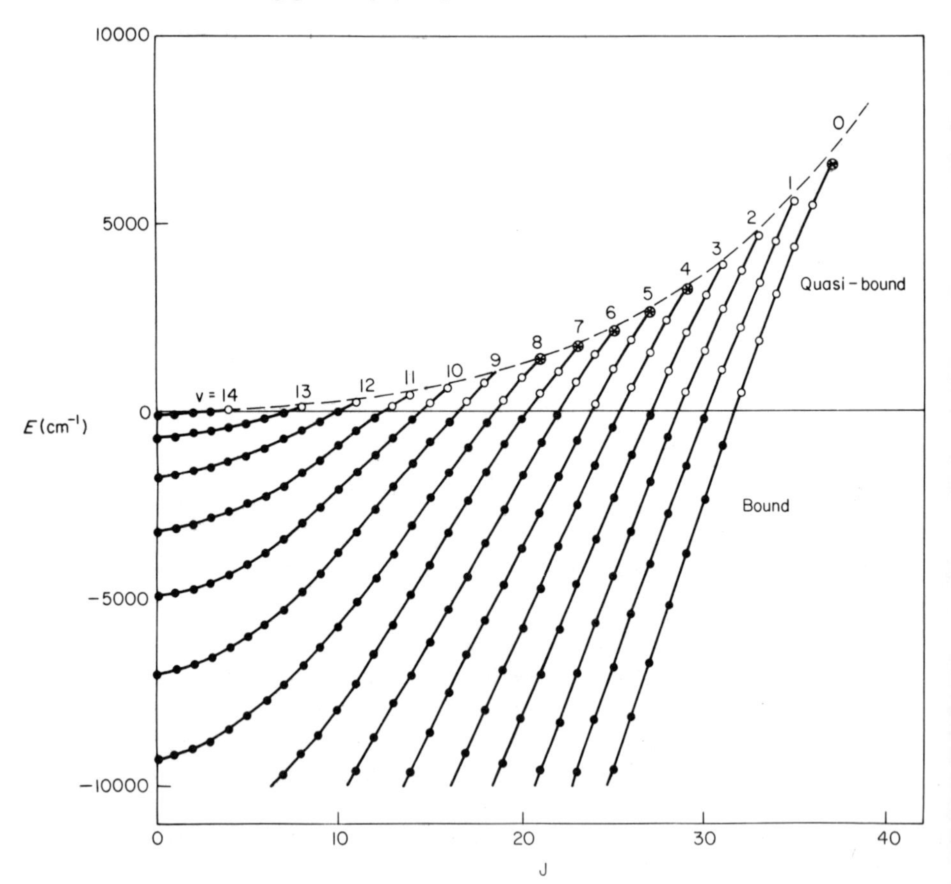

FIG. 5.8. Bound (solid circles) and quasi-bound (open circles) levels for the ground, $\times(^1\Sigma_g^+)$, state of H_2 (Le Roy (1971)). The dashed line follows the locus of the centrifugal barrier maximum, at which classical orbiting occurs (compare Fig. 2.4). Quasi-bound levels with widths in excess of 25 cm^{-1} are marked ⊛.

$$\Gamma_{nl} = \frac{\hbar\omega}{2\pi}\exp\left[\frac{-2\pi(V_l^{\max} - E)}{\hbar\omega^*}\right], \tag{5.77}$$

where $\hbar\omega$ is the local energy spacing between the quasi-bound states and $i\omega^*$ is the imaginary frequency derived from the negative curvature at $V_l^{\max}$. Some insight into possible observable effects of $\sigma^{(\mathrm{res})}$ may be obtained with the help of Fig. 5.8.

In view of the energy band width in any experiment only the broadest resonances will be readily detectable. The most favourable circumstances therefore arise with systems of low reduced mass. Furthermore in view of the sharp drop in tunnelling efficiency as the energy falls below the barrier (see (5.77)) any experimentally significant effects will be confined to the points (marked by asterisks in Fig. 5.8) at which the quasi-bound levels are pushed into the continuum. These correspond to the points of limiting predissociation in spectroscopic language. Hence Bernstein (1966b) was able to use the known spectrum of the HgH molecule to predict the positions of orbiting resonances too diffuse for detection by spectroscopic methods, but subsequently confirmed in a scattering experiment by Schutte *et al.* (1972). The pattern of orbiting resonances in the total cross-section for the scattering of ground state hydrogen atoms has also been computed by Gersh and Bernstein (1969).

5.6 Determination of the Scattering Potential

The important features of the scattering potential, $V(r)$, are most directly determined by comparison between the experimental scattering cross-sections and theoretical predictions for a potential of assumed functional form. This is the method commonly applied in practice. More ambitiously one can attempt to proceed without prior assumptions, working either with the scattering data directly, or from knowledge of the phase shift $\eta_l(k)$ as a function of k or l. The latter problem is reviewed at length by Mott and Massey (1965). Two relatively simple techniques are discussed below, the first based on the Born approximation (4.14) and the second on the classical (or semi-classical) approximation.

We review first the mathematical and physical background to the various comparison methods which have been applied (see Amdur, 1966, 1968; Bernstein and Muckerman, 1967; Pauly and Toennies, 1968). Some types of experiment give information about the characteristic length r_0,† and well depth ε of an assumed two parameter form of potential.

$$V(r) = \varepsilon W(r/r_0), \tag{5.78}$$

† r_0 is preferred to the conventional symbol σ in this context in order to avoid confusion with the total cross-section.

an important example of which is the Lennard-Jones expression

$$V(r) = 4\varepsilon\left[\left(\frac{r_0}{r}\right)^{12} - \left(\frac{r_0}{r}\right)^{6}\right]. \tag{5.79}$$

Other experiments are dominated by the long range attraction or short range repulsion terms

$$V(r) \simeq -C^{(s)}/r^s \text{ for } (r/r_0) \gg 1 \tag{5.80}$$

$$\simeq C^{(t)}/r^t \text{ for } (r/r_0) < 1 \tag{5.81}$$

say. Since the results of a collision experiment depend on the strength of the interaction potential in relation to the available energy, it is natural to think in terms of the reduced collision energy, $K = (E/\varepsilon)$.

The first point, developed in Section 5.1, is that the low resolution differential cross-section at all but very small scattering angles may be analysed by classical methods. In particular the low angle scattering at ordinary energies, $K \simeq 1$, which is dominated by the long range form of the potential, determines with the help of Eqn (2.34) the long range parameters s and $C^{(s)}$ in (5.80). The same theory could in principle be applied at very high energies to determine the repulsive constants t and $C^{(t)}$ in (5.81). Experimentally however t and $C^{(t)}$ have been obtained (Amdur, 1966, 1968) by comparing the cross-section deduced from very high energy attenuation experiments with the classical differential cross-section integrated over a range $\theta_d < \theta < \pi$ where θ_d, which is determined by the detection system, is known to exceed the non-classical near-forward scattering region.

At lower energies, the width of the non-classical region increases (see for example the discussion after (3.49)), and it becomes feasible to measure the true total cross-section σ given by (5.69). Quantum corrections are responsible for the finite value of σ and for the glory oscillations illustrated in Fig. 5.7. According to (5.71) the first term in (5.69) is essentially a measure of the (large) impact parameter at which it becomes impossible by virtue of the uncertainty principle to define a collision. The smoothed energy variation of σ may therefore be analysed for the long range form of the potential. The oscillatory pattern superimposed on this background may be traced via the optical theorem (3.32) to interference between high angular momentum and glory contributions to the forward scattering amplitude (see Section 5.4). The qualitative connection is that σ itself is a measure of the depletion of the initial plane wave state by interference with the forward scattered term; hence any interference within this term is reflected in the total cross-section. The observed oscillation frequency therefore depends on the energy variation of the glory phase shift, which depends in turn, according to (5.72) on the magnitude of the available angular momentum mvr_0, and a term which decreases as (ε/E). Equations (5.72) and (5.73) may therefore be applied to determine the product εr_0. Total cross-section experiments of this type have been reviewed by Bernstein and Muckerman (1967).

The rainbow pattern in the differential cross-section illustrated in Figs 5.4 and 5.5 also contains important information. In the first place the position of the classical rainbow angle θ_r (given by the outermost point of inflexion in Fig. 5.5) which is determined by a balance between the attractive and repulsive effects of the potential, is seen from Table 5.1 to depend for $K > 1$

TABLE 5.1. Rainbow angles. Values are given in radians for the potentials—(a) exp-6: $V(r) = [1 - (6/\alpha)]^{-1} \{(6/\alpha) \exp(-\alpha[1 - (r/r_0)]) - (r_0/r)^6\}$ (b) Lennard-Jones: $V(r) = 4\varepsilon[(r_0/r)^{12} - (r_0/r)^6]$.

K	exp-6 $\alpha = 12$	exp-6 $\alpha = 13$	exp-6 $\alpha = 14$	exp-6 $\alpha = 15$	Lennard-Jones
1	3·0	3·3	3·7	4·4	3·1
1·2	2·2	2·3	2·5	2·7	2·3
1·6	1·5	1·5	1·6	1·6	1·5
2·0	1·1	1·2	1·2	1·2	1·1
3·0	0·70	0·72	0·74	0·75	0·72
4·0	0·52	0·53	0·54	0·56	0·53
5·0	0·41	0·42	0·42	0·44	0·42
10·0	0·20	0·21	0·21	0·21	0·21

almost entirely on the reduced collision energy $K = (E/\varepsilon)$ for a potential of given long range form. Values of ε determined in this way however differ by up to 10% according to the assumed form of the potential. Secondly, according to (5.27)–(5.29) the oscillation frequency (or with less precision the width of the principal rainbow peak) is directly related to the range parameter r_0. The reason is that the interference between the two relevant branches of the scattering amplitude is governed by the variation of the phase shift around the rainbow point, which is in turn related by (5.8) to the variation with l of the deflection angle Θ. For an assumed form of potential and a given value of $K = (E/\varepsilon)$, however, the variation of Θ with reduced impact parameter $\beta = (l/kr_0)$ is readily calculated, using (2.5). Comparison between the experimental and theoretical patterns therefore determines r_0. (See Bernstein and Muckerman, 1967, Pauly and Toennies, 1968).

We consider now two possible approaches to direct inversion of the scattering data. The first is based on the Born approximation, (4.14)

$$f(k, \theta) = -\int_0^\infty \frac{\sin Kr}{Kr} U(r)\, r^2 \,\mathrm{d}r, \tag{5.82}$$

where

$$U(r) = 2mV(r)/\hbar^2,$$

and

$$K = 2k \sin \theta/2.$$

This implies that

$$rU(r) = \frac{-2}{\pi} \int_0^\infty Kf(k, \theta) \sin Kr \, \mathrm{d}K; \tag{5.83}$$

hence within the validity of the Born approximation, knowledge of the scattering amplitude as a function of K is sufficient to determine $U(r)$. The determination of $f(k, \theta)$ from the differential cross-section is simplified by the knowledge given by (5.82) that $f(k, \theta)$ is real, hence,

$$f(k, \theta) = \pm [I(k, \theta)]^{\frac{1}{2}}, \tag{5.84}$$

but the oscillatory nature of the differential cross-section due to a potential of realistic non-monotonic form (see Fig. 5.4) must present serious problems in practice.

The second method due to Firsov (1953) relies on manipulation of the classical deflection function, given by (2.5)

$$\Theta(b) = \pi - \int_a^\infty \frac{2b \, \mathrm{d}r}{r^2[1 - V(r)/E - b^2/r^2]^{\frac{1}{2}}}, \tag{5.85}$$

to determine a function

$$X(r) = r^2(1 - V(r)/E) \tag{5.86}$$

from which the potential function $V(r)$ is later extracted. The theory is closely related to the Rydberg (1931) and Klein (1932) analysis of spectroscopic data. The first steps are to substitute X as integration variable in (5.85), and to note that in the case $V(r) = 0$

$$\int_{b^2}^\infty \frac{b \, \mathrm{d}X}{X[X - b^2]^{\frac{1}{2}}} = \pi. \tag{5.87}$$

Hence, after eliminating π from (5.85) and (5.87), it may be verified that

$$\Theta(b) = \int_{b^2}^\infty \frac{b(\mathrm{d}Y/\mathrm{d}X) \, \mathrm{d}X}{[X - b^2]^{\frac{1}{2}}} \tag{5.88}$$

where

$$Y(X) = \ln(X/r^2) = \ln(1 - V(r)/E). \tag{5.89}$$

With this definition, $Y \to 0$ as $X \to \infty$. The value of Y at a typical point, $X_0 = b_0^2$ say, may in favourable circumstances, now be determined by multiplying both sides of (5.89) by $(b^2 - X_0)^{-\frac{1}{2}}$ and integrating with respect to b from $Mb_0 = X_0^{\frac{1}{2}}$ to infinity,

$$\int_{b_0}^{\infty} \frac{\Theta(b)\,\mathrm{d}b}{[b^2 - X_0]^{\frac{1}{2}}} = \frac{1}{2}\int_{X_0}^{\infty}\left\{\int_{b^2}^{\infty} \frac{(\mathrm{d}Y/\mathrm{d}X)\,\mathrm{d}X}{[(X - b^2)(b^2 - X_0)]^{\frac{1}{2}}}\right\}\mathrm{d}b^2 \tag{5.90}$$

the right-hand side of which may be reduced by reversing the order of integration to the form

$$\int_{b_0}^{\infty} \frac{\Theta(b)\,\mathrm{d}b}{[b^2 - X_0]} = \frac{1}{2}\int_{X_0}^{\infty}\left(\frac{\mathrm{d}Y}{\mathrm{d}X}\right)\left\{\int_{X_0}^{X} \frac{\mathrm{d}b^2}{[(X - b^2)(b^2 - X_0)]^{\frac{1}{2}}}\right\}\mathrm{d}X = \frac{\pi}{2}\,Y(X_0). \tag{5.91}$$

Finally, therefore in view of (5.89)

$$r(X_0) = X_0^{\frac{1}{2}} \exp\left[\frac{1}{\pi}\int_{X_0^{\frac{1}{2}}}^{\infty} \frac{\Theta(b)\,\mathrm{d}b}{[b^2 - X_0]^{\frac{1}{2}}}\right]; \tag{5.92}$$

inversion of the resulting function $r(X)$ yields on comparison with (5.86) the potential $V(r)$.

It is clear from the derivation that a necessary condition on the validity of (5.92) is that $(\mathrm{d}Y/\mathrm{d}X)$ in (5.90) should be finite over the integration domain (Fig. 5.9), because reversal of the order of integration depends on this. The

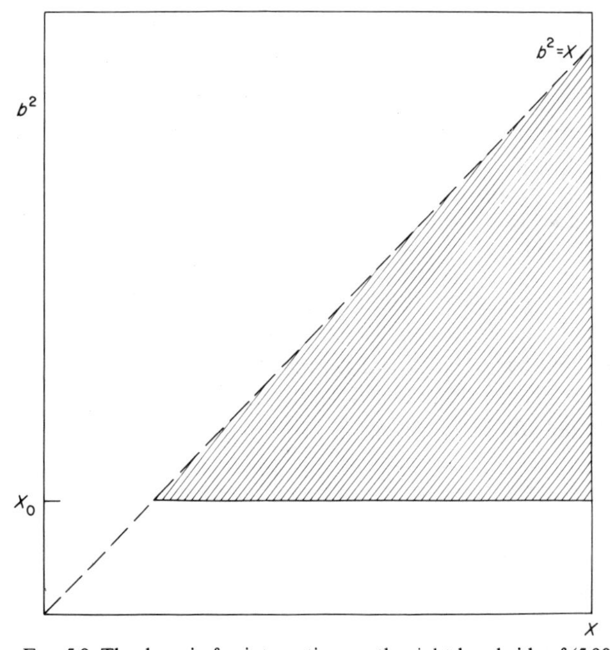

FIG. 5.9. The domain for integration on the right-hand side of (5.90).

implication, in view of (5.89) is that $X(r)$ must be a monotonic function, or equivalently (see Fig. 5.10) that $r(X)$ should be single valued for all values of X. Since $r(X)$ is given by (5.86) this requires that the equation

$$1 - (V(r)/E) - (X/r^2) = 0 \tag{5.93}$$

should have a single root for all X; in other words, replacing X by b^2 and comparing with Eqn (2.4), the energy E must lie above the orbiting limit.

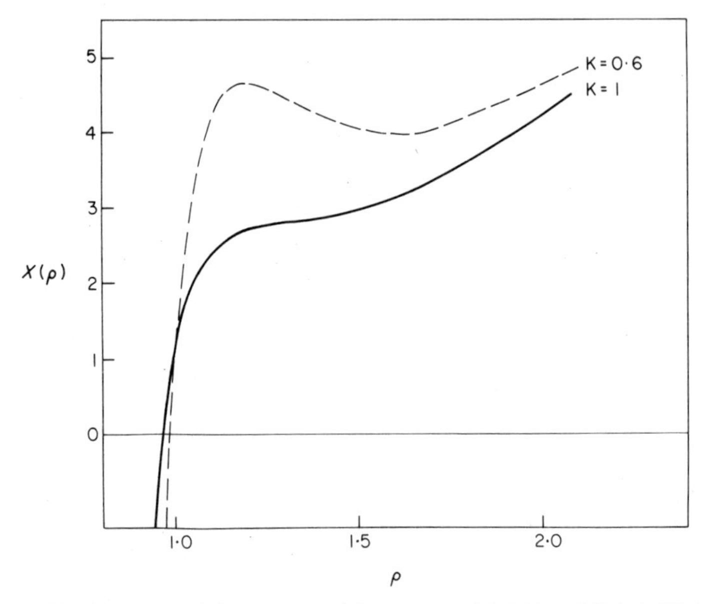

FIG. 5.10. The function $X(r)$ for a Lennard-Jones potential, $V(r) = 4\varepsilon[(r/r_0)^{-12}(r/r_0)^{-6}]$ at $K = E/\varepsilon = 0{\cdot}6$ and $1{\cdot}0$. $\rho = r/r_0$.

A second more practical difficulty arises from ignorance of the function $\Theta(b)$. If the potential is known to be purely repulsive in character, $\Theta(b)$ is identical with the observed scattering angle $\theta(b)$ which is uniquely determined by the differential cross-section. The procedure is first to determine $b(\theta)$, by rearrangement of (2.15), to give

$$b(\theta) = \left[2\int_{\theta}^{\pi} \mathrm{I}(\theta)\sin\theta\,\mathrm{d}\theta\right]^{\frac{1}{2}}. \tag{5.94}$$

$\theta(b) = \Theta(b)$ is then obtained by simple inversion since $b(\theta)$ is monotonic under these conditions.

More typically however the potential contains an attractive well, and $b(\theta)$ is the multivalued at angles less than the rainbow value. Under semi-classical conditions $\Theta(b)$ must then be derived with the help of (5.8) from the JWKB phase shift given by (4.30). Alternatively the potential itself may be obtained directly from the phase shift by a variant of the Firsov (1953) technique due to Vollmer (1969), the solution being given in the form

$$
\begin{aligned}
V(r) &= E(1 - \exp[-Q(t)]) \\
r(t) &= t \exp[Q(t)/2E] \\
Q(t) &= \frac{4E}{\pi k} \frac{1}{t} \frac{\mathrm{d}}{\mathrm{d}t} \int_t^\infty \frac{b\eta(b)\,\mathrm{d}b}{(b^2 - t^2)^{\frac{1}{2}}}
\end{aligned}
\qquad (5.95)
$$

where $b = (l + \frac{1}{2})/k$. It is readily verified that t here is simply the square root of X in (5.86) and that $Q(t)$ is directly proportional to $Y(X)$ in (5.89).

Systematic methods for adjustments to the phase shift function $\eta(b)$ in order to reproduce the differential cross-section (5.1) have been described by Buck (1971) and Rich *et al.* (1971). Both depend on first reproducing the rainbow pattern, upon which higher frequency oscillations due to the repulsive branch of the potential are later superimposed. The method of Rich *et al.* (1971) differs from that of Buck (1971) in replacing the infinite sum (5.2) by a finite sum, containing between five and twenty terms, over closely grouped S matrix poles (see Sections 4.3–4.4) in the first quadrant of the l (or b) plane. This allows a relatively simple fit to the rainbow pattern.

CHAPTER 6

General Theory of Inelastic Collisions

In the case of scattering between particles with structure it is convenient to distinguish between the relative position vector, $\mathbf{r}$, of the collision partners and the remaining 3N-3 internal degrees of freedom, $\boldsymbol{\rho}$. The task is then to extend the theory to take into account a possible change in the internal state $\phi_i(\boldsymbol{\rho})$ of the system, where $\phi_i(\boldsymbol{\rho})$ typically represents a product of rotational-, vibrational-electronic functions for the two colliding atoms or molecules, evaluated at infinite separation. Each such state is said to define a different channel of the system.

This chapter outlines the general structure of the theory. The first step is to set up the equations of relative motion on the assumption of a known set of internal states $\phi_i(\boldsymbol{\rho})$ and a known interaction potential $V(\mathbf{r}, \boldsymbol{\rho})$. The form of the boundary conditions and their relation to the collision cross-sections is then discussed. This leads to the introduction of a scattering matrix $\mathbf{S}$ to relate the outgoing probability amplitudes in the different channels to their possible incoming values. As such $\mathbf{S}$ is the direct generalization of the phase shift. Finally expressions for the scattering amplitudes $f_{ij}(\theta)$ are obtained in terms of the matrix elements of $\mathbf{S}$, first in the presence of a spherically symmetric field and then in the general case.

6.1 Equations of Relative Motion

With the Hamiltonian conveniently written

$$H = H_{\text{int}}(\boldsymbol{\rho}) - \frac{\hbar^2}{2m}\nabla_r^2 + V(\mathbf{r}, \boldsymbol{\rho}), \tag{6.1}$$

where m is the reduced mass of the collision partners and the interaction potential $V(\mathbf{r}, \boldsymbol{\rho})$ is defined to vanish at infinity,

$$\lim_{r\to\infty} V(\mathbf{r}, \boldsymbol{\rho}) = 0, \tag{6.2}$$

the time independent Schrödinger equation

$$H\Phi(\mathbf{r}, \boldsymbol{\rho}) = E\Phi(\mathbf{r}, \boldsymbol{\rho}) \tag{6.3}$$

may be cast into two limiting forms for the relative motion, according to whether the potential $V(\mathbf{r}, \boldsymbol{\rho})$, or a variation in the form of the internal state, appears as a perturbation to the system. Other possible coupling schemes have also been suggested (Levine, 1968; Levine *et al.*, 1969).

In the potential coupling scheme the full wavefunction is expanded in terms of the unperturbed (diabatic) orthonormal eigenstates $\phi_j(\boldsymbol{\rho})$ of $H_{\text{int}}(\boldsymbol{\rho})$,

$$H_{\text{int}}(\boldsymbol{\rho})\, \phi_j(\boldsymbol{\rho}) = E_j \phi_j(\boldsymbol{\rho}). \tag{6.4}$$

In other words

$$\Phi(\mathbf{r}, \boldsymbol{\rho}) = \sum_j \Psi_j(\mathbf{r})\, \phi_j(\boldsymbol{\rho}), \tag{6.5}$$

so that after substitution in Eqn (6.3), multiplication by $\phi_i^*(\boldsymbol{\rho})$ and integration over the internal variables,

$$[\nabla_r^2 + k_i^2]\, \Psi_i(\mathbf{r}) = \sum_j U_{ij}(\mathbf{r})\, \Psi_j(\mathbf{r}) \tag{6.6}$$

where

$$k_i^2 = 2m\,(E - E_i)/\hbar^2$$

$$U_{ij}(\mathbf{r}) = (2m/\hbar^2) \int \phi_i^*(\boldsymbol{\rho})\, V(\mathbf{r}, \boldsymbol{\rho})\, \phi_j(\boldsymbol{\rho})\, \mathrm{d}\boldsymbol{\rho} \tag{6.7}$$

Possible analytical approaches to the solution of (6.6) are described in Chapters 7 and 8. Systems to which these do not apply may be treated by the numerical methods of Gordon (1969), Chan *et al.* (1968) or Secrest and Johnson (1966, 1968). The latter include results for the important model cases of scattering by a harmonic oscillator and by a rigid diatomic rotor.

From a physical viewpoint the diagonal terms $U_{ii}(\mathbf{r})$ in (6.6) contribute directly to the elastic scattering; by distorting the $\Psi_i(\mathbf{r})$ from their plane wave forms they also affect the inelastic cross-section (see for example Section 7.2). The off-diagonal terms $U_{ij}(\mathbf{r})$, which couple different channels together, are responsible for any inelastic processes which occur and for polarization contributions to the elastic scattering.

Equations (6.6) represent what is termed the diabatic formulation of the problem. In problems, such as those concerned with electronic excitation and others for which relaxation of the internal motion may be assumed to be fast compared with the relative velocity, an adiabatic formulation is preferable. This means that (6.5) is replaced by an expansion in terms of the adiabatic internal states $\chi_j(\mathbf{r}; \boldsymbol{\rho})$, which are the orthonormal eigenfunctions of the full internal part of the hamiltonian at a given point r;

$$[H_{\text{int}}(\boldsymbol{\rho}) + V(\mathbf{r}, \boldsymbol{\rho})]\, \chi_j(\mathbf{r}; \boldsymbol{\rho}) = W_j(\mathbf{r})\chi_j(\mathbf{r}; \boldsymbol{\rho}). \tag{6.8}$$

$\chi_j(\mathbf{r}; \boldsymbol{\rho})$ and $W_j(\mathbf{r})$ therefore depend parametrically on r. Since $V(\mathbf{r}, \boldsymbol{\rho})$ vanishes at infinity, they go over at long range to their counterparts $\phi_j(\boldsymbol{\rho})$ and E_j defined by (6.4).

$$\begin{aligned} \chi_j(\mathbf{r}; \boldsymbol{\rho}) &\overset{r\to\infty}{\sim} \phi_j(\boldsymbol{\rho}) \\ W_j(\mathbf{r}) &\overset{r\to\infty}{\sim} E_j \end{aligned} \tag{6.9}$$

The expansion

$$\Phi(\mathbf{r}; \boldsymbol{\rho}) = \sum_j \tilde{\Psi}_j(\mathbf{r})\, \chi_j(\mathbf{r}; \boldsymbol{\rho}) \tag{6.10}$$

now leads on substitution in (6.3) to the following equations for $\tilde{\Psi}_j(\mathbf{r})$

$$[\nabla_r^2 + k_i^2(\mathbf{r})]\, \tilde{\Psi}_i(\mathbf{r}) = \sum_j [\mathbf{X}_{ij}(r) . \nabla_r + Y_{ij}(r)]\, \Psi_j(\mathbf{r}), \tag{6.11}$$

where

$$k_i^2(\mathbf{r}) = 2m[E - W_i(\mathbf{r})]/\hbar^2$$

$$\mathbf{X}_{ij}(\mathbf{r}) = -2\int \chi_i^*(\mathbf{r}; \boldsymbol{\rho})\, \nabla_r \chi_j(\mathbf{r}; \boldsymbol{\rho})\, \mathrm{d}\boldsymbol{\rho} = -2\langle \chi_i | \nabla_r | \chi_j \rangle$$

$$Y_{ij}(\mathbf{r}) = -\int \chi_i^*(\mathbf{r}; \boldsymbol{\rho})\, \nabla_r^2 \chi_j(\mathbf{r}; \boldsymbol{\rho})\, \mathrm{d}\boldsymbol{\rho} = -\langle \chi_i | \nabla_r^2 | \chi_j \rangle.$$

This is the adiabatic, or kinetic, coupling scheme.

Non-adiabatic effects may arise from both the radial and angular terms in $X_{ij}(r)$ and $Y_{ij}(r)$, but apart from certain high angular velocity effects at short range in relatively high energy processes (see for example Thorson (1961, 1963)), the radial contributions are normally the most important for molecular systems. As a measure of the magnitude of these radial terms, we require for a perturbation treatment that the energy difference $k_i^2(r) - k_j^2(r)$ should dominate the interaction term $\langle \chi_i | \mathrm{d}/\mathrm{d}r | \chi_j \rangle\, \mathrm{d}/\mathrm{d}r$.† Since under semi-classical conditions the derivative (d/dr) extracts the reduced momentum ($mv/\hbar$), this means that for an essentially adiabatic description

$$2[k_i^2(r) - k_j^2(r)]\, \delta R(mv/\hbar)^{-1} = [W_i(r) - W_j(r)]\, \delta R/\hbar v \gg 1, \tag{6.12}$$

where the inverse of the characteristic length $(\delta R)^{-1}$ denotes the integral $\langle \chi_i | \mathrm{d}\chi_j/\mathrm{d}r \rangle$. The ratio $(W_i - W_j)\, \delta R/\hbar v$ is called the Massey parameter. Factors favouring an adiabatic description are therefore a large adiabatic energy difference $(W_i - W_j)$ and a low collision velocity v.

† The second derivative term $\langle \chi_i | \mathrm{d}^2/\mathrm{d}r^2 | \chi_j \rangle$ is relatively unimportant in practice.

As an alternative to (6.11) the coupling terms may also be written as matrix elements of ∇V or $\nabla^2 V$. Thus on taking the gradient of (6.8) multiplying through by $\chi_i^*(\mathbf{r}, \boldsymbol{\rho})$ and integrating over $\boldsymbol{\rho}$, for example, we find that

$$[W_i(\mathbf{r}) - W_j(\mathbf{r})] \langle \chi_i | \nabla_r | \chi_j \rangle = \nabla W_j \delta_{ij} - \langle \chi_i | \nabla_r V | \chi_j \rangle \tag{6.13}$$

Hence

$$\langle \chi_i | \nabla_r | \chi_j \rangle = -\frac{\langle \chi_i | \nabla_r V | \chi_j \rangle}{W_i(\mathbf{r}) - W_j(\mathbf{r})} \quad \text{for} \quad i \neq j; \tag{6.14}$$

the diagonal term $\langle \chi_i | \nabla_r | \chi_i \rangle$ vanishes by virtue of the normalization condition, $\langle \chi_i | \chi_i \rangle = 1$, because

$$\langle \chi_i | \nabla_r | \chi_i \rangle = \tfrac{1}{2} \nabla_r \langle \chi_i | \chi_i \rangle = 0$$

Similarly, on taking the divergence of (6.13), it turns out after some manipulation that, for $i \neq j$

$$\langle \chi_i | \nabla_r^2 | \chi_j \rangle = 2 \sum_{k \neq i,j} \frac{\langle \chi_i | \nabla_r V | \chi_k \rangle . \langle \chi_k | \nabla_r V | \chi_j \rangle}{(W_i - W_j)(W_k - W_j)}$$

$$+ 2(\nabla_r W_i - \nabla_r W_j) . \frac{\langle \chi_i | \nabla_r V | \chi_j \rangle}{(W_i - W_j)^2} - \frac{\langle \chi_i | \nabla_r^2 V | \chi_j \rangle}{W_i - W_j} \tag{6.15a}$$

and

$$\langle \chi_i | \nabla_r^2 | \chi_i \rangle = - \langle \nabla_r \chi_i | . \nabla_r \chi_i \rangle = - \sum_{k \neq i} \frac{\langle \chi_i | \nabla_r V | \chi_k \rangle . \langle \chi_k | \nabla_r V | \chi_i \rangle}{(W_i - W_k)^2}. \tag{6.15b}$$

Levine and Johnson (1972) have also given a numerical algorithm for the computation of these elements.

Yet another form for the equations of motion may be obtained by transforming in the adiabatic formulation to a rotating axis system. This is particularly convenient when considering electronic motion because the adiabatic states $\chi_i(\mathbf{r}, \boldsymbol{\rho})$ are normally obtained in a system based set of axes (x', y', z'). In this form the non-adiabatic coupling terms appear identical with those responsible for vibrational-rotational-electronic effects in molecular spectroscopy (see Van Vleck, 1951). For a diatomic problem for example, with z' along the diatomic axis

$$H(\mathbf{r}, \boldsymbol{\rho}) = H_0(\boldsymbol{\rho}') + V(\mathbf{r}', \boldsymbol{\rho}') + H'(\mathbf{r}', \boldsymbol{\rho}'),$$

$$H'(\mathbf{r}', \boldsymbol{\rho}') = \frac{\hbar^2}{2mr^2}\left[-\frac{\partial}{\partial r'}\left(r'^2 \frac{\partial}{\partial r'} \right) + (\hat{J}_{x'} - \hat{P}_{x'})^2 + (\hat{J}_{y'} - \hat{P}_{y'})^2 \right], \tag{6.16}$$

where $\hat{\mathbf{J}}$ denotes the total angular momentum and $\hat{\mathbf{P}}$ the internal (orbital plus spin) angular momentum operators for the system. (There can be no term $(J_{z'} - P_{z'})^2$ in (6.16) because $J_{z'}$ is identical with $P_{z'}$ in the diatomic case).

It must be remembered (Van Vleck, 1951) that the components of $\hat{\mathbf{J}}$ with respect to rotating axes satisfy the commutation relations

$$[\hat{J}_{x'}, \hat{J}_{y'}] = -i\hbar \hat{J}_{z'} \tag{6.17}$$

The use of (6.16) has the advantage over (6.11) that the origin of inelastic effects is clearly and concisely displayed; radial contributions arise in adiabatic description from $V(\mathbf{r}', \boldsymbol{\rho}')$ or in the adiabatic formulation from the first term of $H'(\mathbf{r}, \boldsymbol{\rho}')$ in (6.16), while angular effects come from the products $(J_{x'}P_{x'} + J_{y'}P_{y'})$. On the other hand, as discussed in Section 6.4, the use of a rotating coordinate system complicates the boundary conditions to some extent.

It is sufficient for the present to note that the boundary conditions on the three forms for the equations of motion, (6.6), (6.11) and those derived from (6.16), may all be expressed in the same form, because different formulations of the problem cannot affect its physical nature. In a space-fixed axis system, the conditions of a scattering experiment require, by extension of the arguments of Section 3.1, a solution with the asymptotic form of a plane wave together with an outgoing term in the entrance channel, and outgoing waves in all other channels:

$$\Psi_i(r) \overset{r\to\infty}{\sim} e^{ik_i z} + f_{ii}(\theta, \phi)\, e^{ik_i r}/r$$

$$\Psi_j(r) \overset{r\to\infty}{\sim} f_{ij}(\theta, \phi)\, e^{ik_j r}/r \tag{6.18}$$

$f_{ij}(\theta, \phi)$ here is interpreted as the probability amplitude for scattering from channel i to channel j.

In converting these scattering amplitudes to cross-sections we must remember that any energy transfer between internal and relative motion must lead to a change in the limiting velocity of the system. This means that the inelastic differential cross-section, obtained by comparison between the number of particles, $v_j r^{-2}|f(\theta, \phi)|^2\, dS$, scattered per unit time through the area $dS = r^2 \sin\theta\, d\theta\, d\phi$, and the incident flux v_i, contains a factor (v_j/v_i) in addition to the obvious term $|f_{ij}(\theta, \phi)|^2$, because by definition

$$(d\sigma_{ij}/d\Omega)\, d\Omega = I_{ij}(\theta, \phi) \sin\theta\, d\theta\, d\phi = (v_j/v_i)|f_{ij}(\theta, \phi)|^2 \sin\theta\, d\theta\, d\phi \tag{6.19}$$

from which

$$d\sigma_{ij}/d\Omega = I_{ij}(\theta, \phi) = (k_j/k_i)|f_{ij}(\theta, \phi)|^2, \tag{6.20}$$

because $v_i = (k_i\hbar/m)$. The total inelastic cross-section therefore becomes

$$\sigma_{ij} = (k_j/k_i)\int_0^{2\pi}\int_0^{\pi} |f_{ij}(\theta, \phi)|^2 \sin\theta\, d\theta\, d\phi. \tag{6.21}$$

The next step is to obtain expressions for the $f_{ij}(\theta, \phi)$ analogous to the elastic scattering amplitude of (3.28). It is convenient for this purpose to define a scattering matrix **S** (or a related transition matrix **T**) which takes the place of the phase shift, upon which the theory of purely elastic scattering is based. Specific expressions for the $f_{ij}(\theta, \phi)$ in the cases of scattering in a spherical field and scattering by a diatomic rotor are obtained in sections 6.3 and 6.4 respectively.

6.2 Scattering Matrices

As an introduction to the scattering matrix **S** it is convenient to consider the simple one dimensional problem of potential coupling between two channels $\psi_0(r)$ and $\psi_1(r)$ in exact resonance (in the sense that $k_0^2 - U_{00}(r) = k_1^2 - U_{11}(r) = k^2 - U(r)$ say), for which an exact analytical solution is readily available. This situation is typified by a direct energy exchange between identical molecules. The implied equations

$$\left[\frac{d^2}{dr^2} + k_0^2 - U_{00}(r)\right]\psi_0(r) = U_{01}(r)\,\psi_1(r) \tag{6.22}$$

$$\left[\frac{d^2}{dr^2} + k_1^2 - U_{11}(r)\right]\psi_1(r) = U_{10}(r)\,\psi_0(r)$$

are readily decoupled, if the internal states are chosen so that $U_{01}(r)$ and $U_{10}(r)$ are real and hence equal, by introduction of the functions

$$\psi_\pm(r) = \frac{1}{\sqrt{2}}\left[\psi_0(r) \pm \psi_1(r)\right], \tag{6.23}$$

to give

$$\left[\frac{d^2}{dr^2} + k^2 - U_\pm(r)\right]\psi_\pm(r) = 0, \tag{6.24}$$

where $U_\pm(r) = U(r) \pm U_{01}(r)$. These are identical with the elastic scattering equations (3.9). Hence the asymptotic form of the general bounded solutions of (6.24) may be described in terms of two phase shifts $\eta_\pm$ and two arbitrary constants $C_\pm$,†

$$\psi_\pm(r) \overset{r\to\infty}{\sim} C_\pm \sin(kr - l\pi/2 + \eta_\pm). \tag{6.25}$$

Now in the scattering problem we wish to determine the amplitudes of the outgoing parts of the solution which corresponds to unit incident amplitude

† The terms $l\pi/2$ are conventionally retained to remove any purely centrifugal contribution from $U_{00}(r)$ and $U_{11}(r)$, to the S matrix.

in a particular entrance channel. The coefficients $C_\pm$ in (6.25) are therefore chosen to eliminate incoming terms in the other channel. Specifically the choice

$$C_\pm = \pm(\sqrt{2})i\,\mathrm{e}^{i\eta_\pm}, \tag{6.26}$$

leads after inversion of (6.23) to a solution $\psi^{(0)}(r)$, incoming in channel zero, the components of which have asymptotic forms

$$\begin{aligned} \psi_0^{(0)}(r) &\sim \mathrm{e}^{-ikr+il\pi/2} - S_{00}\,\mathrm{e}^{ikr-il\pi/2} \\ \psi_1^{(0)}(r) &\sim \qquad\qquad\quad - S_{01}\,\mathrm{e}^{ikr-il\pi/2} \end{aligned} \tag{6.27}$$

where

$$\begin{aligned} S_{00} &= \mathrm{e}^{i(\eta_+ + \eta_-)}\cos(\eta_+ - \eta_-) \\ S_{01} &= i\,\mathrm{e}^{i(\eta_+ + \eta_-)}\sin(\eta_+ - \eta_-) \end{aligned} \tag{6.28}$$

Similarly the choice

$$C_\pm = \mp(\sqrt{2})\,i\,\mathrm{e}^{i\eta_\pm} \tag{6.29}$$

gives

$$\begin{aligned} \psi_0^{(1)}(r) &\sim \qquad\qquad\quad -S_{10}\,\mathrm{e}^{ikr-il\pi/2} \\ \psi_1^{(1)}(r) &\sim \mathrm{e}^{-ikr+il\pi/2} - S_{11}\,\mathrm{e}^{ikr-il\pi/2} \end{aligned} \tag{6.30}$$

where S_{10} and S_{11} make up the second row of the S matrix.

$$\mathbf{S} = \begin{pmatrix} \mathrm{e}^{i(\eta_+ + \eta_-)}\cos(\eta_+ - \eta_-), & i\,\mathrm{e}^{i(\eta_+ + \eta_-)}\sin(\eta_+ - \eta_-) \\ i\,\mathrm{e}^{i(\eta_+ + \eta_-)}\sin(\eta_+ - \eta_-), & \mathrm{e}^{i(\eta_+ + \eta_-)}\cos(\eta_+ - \eta_-) \end{pmatrix} \tag{6.31}$$

Note that **S** here is unitary and symmetric;

$$\begin{aligned} \mathbf{S}\mathbf{S}^\dagger &= \mathbf{I} \\ \tilde{\mathbf{S}} &= \mathbf{S} \end{aligned} \tag{6.32}$$

where $\tilde{\mathbf{S}}$ and $\mathbf{S}^\dagger$ are the transpose and the hermitian conjugate of **S** respectively. It is shown below that these are general properties of the **S** matrix, with important physical implications. At a more particular level, it is clear since the phase shift difference, $\eta_+ - \eta_-$, is a variable function of k, given within the validity of the Born approximation, (4.12), by

$$\eta_+ - \eta_- \simeq \frac{-2}{k}\int_0^\infty U_{01}(r)\,[krj_l(kr)]^2\,\mathrm{d}r, \tag{6.33}$$

that the inelastic transition probability $|S_{01}|^2$ is an oscillatory function of energy as shown in Fig. 6.1.

We now turn to the general properties of the S matrix in cases of interest to us, and suppose that Eqn (6.6) has been reduced by an expansion in appropriate angular functions, such as those given in Sections 6.3 or 6.4, to the set of N coupled radial equations

$$\left[\frac{\mathrm{d}^2}{\mathrm{d}r^2} + k_i^2\right]\psi_i(r) = \sum_j U_{ij}(r)\,\psi_j(r). \tag{6.34}$$

The S matrix is defined, by generalization of (6.27) and (6.30), in terms of the N bounded independent solutions of (6.33), $\boldsymbol{\psi}^{(i)}(r)$ say, with components chosen to have the asymptotic forms

$$\begin{aligned} \psi_i^{(i)}(r) &\sim f_{i-}(r) - S_{ii} f_{i+}(r) \\ \psi_j^{(i)}(r) &\sim \quad\quad\quad - S_{ij} f_{i+}(r) \end{aligned} \tag{6.35}$$

where

$$f_{i\pm}(r) = k_i^{-\frac{1}{2}} \exp\left(\pm ik_i r \mp il_i\pi/2\right).$$

This solution is seen by evaluation of the outward flux density given by (3.35), to represent an outward flux of $|S_{ii}|^2 - 1$ in the entrance channel, and $|S_{ij}|^2$ in other channels j.

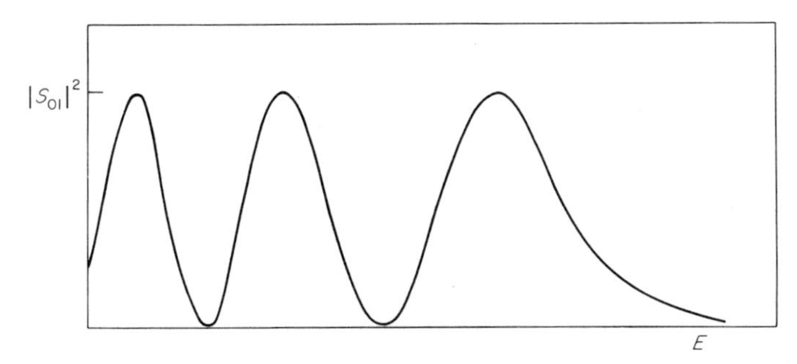

FIG 6.1. Oscillations in the transition probability, $P_{01} = |S_{01}|^2$, for the resonant two level problem.

In order to establish the general symmetry and unitarity of $\mathbf{S}$ it is convenient to write (6.33) in matrix form

$$\boldsymbol{\psi}'' = (\mathbf{U} - \mathbf{k}^2)\boldsymbol{\psi} \tag{6.36}$$

where $\mathbf{k}^2$ is a diagonal matrix with elements k_i^2, and to express the general solution $\boldsymbol{\psi}$ in terms of the independent solutions of (6.34) by means of a matrix $\mathbf{M}$, the columns of which are these independent solutions $\boldsymbol{\psi}^{(i)}$;

$$\boldsymbol{\psi} = \tilde{\mathbf{a}}\mathbf{M} \tag{6.37}$$

where $\tilde{\mathbf{a}}$ is an arbitrary constant vector. The choice of $\tilde{\mathbf{a}}$ as the ith unit vector therefore generates $\boldsymbol{\psi}^{(i)}$. Equation (6.36) then becomes

$$\mathbf{M}'' = \mathbf{M}(\mathbf{U} - \mathbf{k}^2). \tag{6.38}$$

subject to the boundary conditions

$$\begin{aligned} &\mathbf{M}(0) = 0 \\ &\mathbf{M}(r) \overset{r\to\infty}{\sim} \mathbf{f}_-(r) - \mathbf{S}\mathbf{f}_+(r) \end{aligned} \tag{6.39}$$

where $\mathbf{f}_\pm(r)$ are diagonal matrices.

The unitarity of $\mathbf{S}$ now follows from the hermitian character of $\mathbf{U} - \mathbf{k}^2$, since according to (6.38)

$$\mathbf{M}\mathbf{M}''^\dagger - \mathbf{M}''\mathbf{M}^\dagger = \mathbf{M}(\mathbf{U} - \mathbf{k}^2)^\dagger\mathbf{M}^\dagger - \mathbf{M}(\mathbf{U} - \mathbf{k}^2)\mathbf{M}^\dagger = 0. \tag{6.40}$$

It follows on integrating by parts and using (6.39) that

$$\begin{aligned} \int_0^\infty (\mathbf{M}\mathbf{M}''^\dagger - \mathbf{M}''\mathbf{M}^\dagger)\,\mathrm{d}r &= \mathbf{M}(\infty)\mathbf{M}'^\dagger(\infty) - \mathbf{M}'(\infty)\mathbf{M}^\dagger(\infty), \\ &= (\mathbf{f}_- - \mathbf{S}\mathbf{f}_+)(\mathbf{f}'^*_- - \mathbf{f}'^*_+\mathbf{S}^\dagger) - (\mathbf{f}'_- - \mathbf{S}\mathbf{f}'_+)(\mathbf{f}^*_- - \mathbf{f}^*_+\mathbf{S}^\dagger) = 0 \end{aligned} \tag{6.41}$$

from which, after using (6.35)

$$\mathbf{S}\mathbf{S}^\dagger = \mathbf{I} \tag{6.42}$$

This means in particular that

$$\sum_j S_{ij}S^\dagger_{ji} = \sum_j |S_{ij}|^2 = 1;$$

in other words, in physical terms, the total fluxes associated with the incoming and outgoing parts of (6.34) are equal. The unitarity of $\mathbf{S}$ therefore ensures that the number of particles in the system is conserved.

The symmetry of $\mathbf{S}$, which follows by similar arguments from the identity

$$\mathbf{M}\tilde{\mathbf{M}}'' - \mathbf{M}''\tilde{\mathbf{M}} = 0 \tag{6.43}$$

when $\mathbf{U}$ is real and hence symmetric, is associated with the behaviour of the system under time reversal. The reality of $\mathbf{U}$ implies that the internal states $\phi_i(\boldsymbol{\rho})$ are real and hence invariant to time reversal, while the symmetry of $\mathbf{S}$ ensures that the flux scattered from the ith into the jth channel is equal to that scattered in the reverse direction. We shall see however in Sections 6.3 and 6.4 that this symmetry does not apply to the scattering cross-sections.

In obtaining expressions for the scattering amplitudes $f_{ij}(\theta, \phi)$ in (6.18) it is often convenient to replace $\mathbf{S}$ by the closely related transition matrix $\mathbf{T}$, such that

$$\mathbf{S} = \mathbf{I} - i\mathbf{T} \tag{6.44}$$

This substitution provides a separation of (6.35) into an outgoing part on one hand, and a plane wave component rather than a purely incoming part on the other;

$$\begin{aligned} \psi_i^{(i)}(r) &\sim k_i^{-\frac{1}{2}}\,[\sin(k_i r - l\pi/2) - \tfrac{1}{2}T_{ii}\exp(ik_i r - il\pi/2)] \\ \psi_j^{(i)}(r) &\sim \qquad\qquad\qquad\qquad -\tfrac{1}{2}k_j^{-\frac{1}{2}}T_{ij}\exp(ik_j r - il\pi/2) \end{aligned} \tag{6.45}$$

Another matrix, termed the reactance matrix **R** is also sometimes preferred to **S**. It is defined in terms of a set of independent standing wave solutions $\phi^{(i)}(r)$, such that

$$\begin{aligned} \phi_i^{(i)}(r) &\sim k_i^{-\frac{1}{2}}[\sin(k_i r - l\pi/2) + R_{ii}\cos(k_i r - l\pi/2)] \\ &\sim i/2[(1 - iR_{ii})f_{i-}(r) - (1 + iR_{ii})f_{i+}(r)]. \\ \phi_j^{(i)}(r) &\sim k_j^{-\frac{1}{2}}R_{ij}\cos(k_j r - l\pi/2) \\ &\sim \frac{i}{2}[-iR_{ij}f_{j-}(r) - iR_{ij}f_{j+}(r)], \end{aligned} \tag{6.46}$$

rather than the travelling wave solutions $\psi^{(i)}(r)$ of (6.35). The necessary linear relation

$$\phi^{(i)} = \sum_k C_{ik}\,\psi^{(k)} \tag{6.47}$$

establishes the connection between **R** and **S**. On combining (6.35) and (6.46) this identity is seen to require that

$$C_{ik} = \frac{1}{2i}(\delta_{ik} - iR_{ik})$$

and hence

$$\sum_k (\delta_{ik} - iR_{ik})S_{kj} = \delta_{ij} + iR_{ij}. \tag{6.48}$$

In other words

$$(\mathbf{I} - i\mathbf{R})\mathbf{S} = (\mathbf{I} + i\mathbf{R}) \tag{6.49}$$

or

$$\mathbf{S} = (\mathbf{I} - i\mathbf{R})^{-1}(\mathbf{I} + i\mathbf{R})$$

It follows from the first of these equations that

$$(\mathbf{I} - i\mathbf{R})\mathbf{S}\mathbf{S}^\dagger(\mathbf{I} + i\mathbf{R}^\dagger) = (\mathbf{I} + i\mathbf{R})(\mathbf{I} - i\mathbf{R}^\dagger) \tag{6.50}$$

and hence, since $\mathbf{S}\mathbf{S}^\dagger = \mathbf{I}$, that the matrix **R** is hermitian, $\mathbf{R} = \mathbf{R}^\dagger$. Thus if the potential matrix **U** in (6.33) is real, the elements R_{ij} are real and **R** is symmetric.

The advantages of working with **R** rather than **S** are first that the reality of **R** frequently simplifies the computation; secondly any approximation to **R** which retains the symmetry of the matrix automatically ensures that **S** is unitary and hence that the number of particles in the system is conserved.

Another representation of **S** with the latter properties is the exponential form introduced by Heisenberg (1943)

$$\mathbf{S} = \exp i\mathbf{\Delta}, \tag{6.51}$$

where $\mathbf{\Delta}$ is a hermitian matrix, and (6.51) carries the implication

$$\mathbf{S} = \mathbf{I} + i\mathbf{\Delta} + \tfrac{1}{2}(i\mathbf{\Delta})^2 + \ldots \tag{6.52}$$

It is readily verified (since the hermitian character of $\mathbf{\Delta}$, $\mathbf{\Delta} = \mathbf{\Delta}^\dagger$, removes any commutation difficulties in evaluating the product $\mathbf{SS}^\dagger$) that **S** is unitarity, $\mathbf{SS}^\dagger = \mathbf{I}$. Moreover any hermitian approximation to $\mathbf{\Delta}$ ensures that **S** retains this property. Another convenient property arising from this hermitian character is that $\mathbf{\Delta}$ may be reduced to diagonal form by a unitary matrix **X**,

$$\mathbf{\Delta} = \mathbf{X}\boldsymbol{\delta}\mathbf{X}^\dagger, \tag{6.53}$$

and that the elements δ_k of $\boldsymbol{\delta}$ are real. This means on substitution in (6.52) that

$$\begin{aligned} \mathbf{S} &= \mathbf{I} + \mathbf{X}\, i\boldsymbol{\delta}\mathbf{X}^\dagger + \tfrac{1}{2}\mathbf{X}(i\delta)^2\mathbf{X}^\dagger + \ldots \\ &= \mathbf{X}\exp(i\boldsymbol{\delta})\,\mathbf{X}^\dagger, \end{aligned} \tag{6.54}$$

where the matrix $\exp(i\boldsymbol{\delta})$ is also diagonal with elements $\exp(i\delta_k)$. Hence **S** may be characterized by N real eigen-phase-shifts, $\frac{1}{2}\delta_k$, and a transformation matrix **X**. The implication is that the system has N eigen-channels, defined by the columns of **X**, along which it will acquire a simple phase change exp $i\delta_k$. For an example of this behaviour we may turn to Eqn (6.31) where

$$\delta_1 = 2\eta_+, \delta_2 = 2\eta_-,$$

and

$$\mathbf{X} = \frac{1}{\sqrt{2}}\begin{pmatrix} 1 & 1 \\ 1 & -1 \end{pmatrix}.$$

6.3 Scattering in a Central Field

The relation between a particular scattering amplitude $f_{ij}(\theta, \phi)$ and the S matrix follows by generalization of the partial wave analysis given in Section 3.2.

Account must however be taken of the two possible sources of angular momentum in a molecular scattering problem; one, the rotational part, with quantum number j is associated with the internal motion, and the other, the orbital component labelled by l, arises from the relative motion of the collision partners. These are in general coupled together, because only the magnitude and one component (about a fixed axis) of the total angular momentum are strictly conserved. There can however be no such coupling if the potential is spherically symmetric, $V(\mathbf{r}, \boldsymbol{\rho}) = V(r, \boldsymbol{\rho})$; hence the orbital and rotational angular momenta are both constants of the motion.

A simple partial wave decomposition

$$\Psi_j(\mathbf{r}) = \frac{1}{r} \sum_{l=0}^{\infty} A_l \psi_{jl}(r) P_l(\cos\theta), \tag{6.55}$$

of the translational functions in Eqn (6.6) is therefore permissible. This leads to equations for the radial functions $\psi_{jl}(r)$,

$$\left[\frac{d^2}{dr^2} + k_i^2 - \frac{l(l+1)}{r^2}\right]\psi_{il}(r) = \sum_j U_{ij}(r)\psi_{jl}(r), \tag{6.56}$$

of the same form at (6.22). (Note that the coefficient A_l in (6.55) is independent of the channel label j).

Each orbital angular momentum quantum number l therefore gives rise to a different set of coupled equation and a different S matrix, $\mathbf{S}^{(l)}$. Since the boundary condition (6.18) requires a plane wave component in channel i together with outgoing terms, we select according to (6.45) that solution of (6.56) which contains the appropriate plane wave component,

$$\begin{aligned} \psi_{il}^{(i)}(r) &\sim \sin(k_i r - l\pi/2) - \tfrac{1}{2} T_{ii}^{(l)} e^{ik_i r - il\pi/2} \\ \psi_{jl}^{(l)}(r) &\sim \qquad\qquad -\tfrac{1}{2}(k_i/k_j)^{\frac{1}{2}} T_{ij}^{(l)} e^{ik_j r - il\pi/2}, \end{aligned} \tag{6.57}$$

and choose the coefficient in (6.55), now labelled $A_l^{(i)}$, such that the incoming part of $\Psi_i(\mathbf{r})$ behaves as the incoming part of a plane wave. This means by comparison with (3.23) and (3.26) that

$$A_l^{(i)} = (2l+1)\, i^l/k_i. \tag{6.58}$$

Eqns (6.55) then take the required asymptotic form,

$$\Psi_j(\mathbf{r}) = \delta_{ij}\, e^{ik_j z} + f_{ij}(\theta)\, e^{ik_j r}/r, \tag{6.59}$$

where

$$\begin{aligned} f_{ij}(\theta) &= -\frac{1}{2[k_i k_j]^{\frac{1}{2}}} \sum_{l=0}^{\infty} (2l+1)\, T_{ij}^{(l)} P_i(\cos\theta) \\ &= \frac{1}{2i[k_i k_j]^{\frac{1}{2}}} \sum_{l=0}^{\infty} (2l+1)(S_{ij}^{(l)} - \delta_{ij}) P_l(\cos\theta), \end{aligned} \tag{6.60}$$

the second line being derived from the identity (6.44). Equation (6.60) is the required expression for the scattering amplitude in terms of the transition, **T**, and scattering, **S**, matrices of Section 6.2. It is apparent that the scattering in a central field is cylindrical symmetric.

The differential and total elastic scattering cross-sections are therefore given, on combining (6.20), (6.21) and (6.60), and applying the orthogonality condition (3.24), by

$$\frac{\mathrm{d}\sigma_{ij}}{\mathrm{d}\Omega} = I_{ij}(\theta) = (k_j/k_i)|f_{ij}(\theta)|^2$$

$$= \frac{1}{4k_i^2}\left|\sum_{l=0}^{\infty}(2l+1)\,T_{ij}^{(l)}\,P_l(\cos\theta)\right|^2, \tag{6.61}$$

and

$$\sigma_{ij} = (k_j/k_i)\int_0^{2\pi}\int_0^{\pi}|f_{ij}(\theta)|^2\sin\theta\,\mathrm{d}\theta\,\mathrm{d}\phi$$

$$= \frac{\pi}{k_i^2}\sum_{l=0}^{\infty}(2l+1)|T_{ij}^{(l)}|^2. \tag{6.62}$$

Note, in view of the symmetry of the S and T matrices established by (6.43), that the scattering amplitude is symmetric in the labels i and j, $f_{ij}(\theta) = f_{ji}(\theta)$, but that

$$k_i^2 I_{ij}(\theta) = k_j^2 I_{ji}(\theta)$$

$$k_i^2\sigma_{ij} = k_j^2\sigma_{ji}. \tag{6.63}$$

These are termed the equations of detailed balance. The factors k_i^2 and k_j^2 which destroy the symmetry of the cross-section may be attributed (see for example, Landau and Lifshitz, 1965 Section 141) to the degeneracy of the entrance channel, the number of quantum states in any element of phase space with momenta ($p_i = k_i\hbar$) between p_i and $p_i + \mathrm{d}p_i$ being proportional to $p_i^2\,\mathrm{d}p_i$.

It is convenient here in order to emphasize the relation between (6.60) and the scattering amplitude, (3.28) in the purely elastic case, and in order to establish certain bounds on the total cross-sections, to express the diagonal element of the scattering matrix in terms of a complex phase shift.

$$S_{ii}^{(l)} = \exp\left[2i(\lambda_l + i\mu_l)\right], \tag{6.64}$$

with $\mu_l > 0$ because $|S_{ii}^{(l)}|^2 < 1$. The imaginary part of the central bracket then accounts for the necessary depletion from the initial channel. This casts the elastic scattering amplitude into a form analogous to (3.28)

$$f_{ii}(\theta) = \frac{1}{2ik_i}\sum_{l=0}^{\infty}(2l+1)\{\exp\left[2i(\lambda_l + i\mu_l)\right] - 1\}\,P_l(\cos\theta), \tag{6.65}$$

and the total elastic cross-section becomes

$$\sigma_{el} = \sigma_{ii} = \frac{\pi}{k_i^2} \sum_{l=0}^{\infty} (2l+1)[|S_{ii}^{(l)}|^2 + 1 - 2 \operatorname{Re} S_{ii}^{(l)}]$$

$$= \frac{2\pi}{k_i^2} \sum_{l=0}^{\infty} (2l+1) e^{-2\mu_l} [\cosh 2\mu_l - \cos 2\lambda_l]. \tag{6.66}$$

A complete total inelastic cross-section may also be defined by summing over all final states, which may also be expressed in view of the conservation theorem $\sum_j |S_{ij}^{(l)}|^2 = 1$, in terms of the diagonal elements $S_{ii}^{(l)}$,

$$\sigma_{in} = \sum_{j \neq i} \sigma_{ij} = \frac{\pi}{k_i^2} \sum_{l=0}^{\infty} (2l+1) \sum_{j \neq i} |S_{ij}^{(l)}|^2$$

$$= \frac{\pi}{k_i^2} \sum_{l=0}^{\infty} (2l+1)(1 - |S_{ii}^{(l)}|^2) \tag{6.67}$$

$$= \frac{\pi}{k_i^2} \sum_{l=0}^{\infty} (2l+1)(1 - e^{-4\mu_l}).$$

Thus the full total cross-section becomes

$$\sigma_{\text{tot}} = \sigma_{\text{el}} + \sigma_{\text{in}} = \frac{2\pi}{k_i^2} \sum_{l=0}^{\infty} (2l+1)(1 - \operatorname{Re} S_{ii}^{(l)})$$

$$= \frac{2\pi}{k_i^2} \sum_{l=0}^{\infty} (2l+1)(1 - e^{-2\mu_l} \cos 2\lambda_l). \tag{6.68}$$

If we now recall the discussion following the optical theorem, (3.32), for purely elastic scattering we must expect, on grounds of particle conservation, a connection between the total scattering of all kinds, and that part of the forward elastic scattering amplitude, $f_{ii}(0)$, which accounts for the necessary (by interference depletion) of the incident plane wave. We find, indeed, by comparison between (6.65) and (6.68) that

$$\sigma_{\text{tot}} = \frac{4\pi}{k_i^2} \operatorname{Im} f_{ii}(0) \tag{6.69}$$

which differs from, (3.32), only by the appearance of σ_{tot} in place of σ.

It is also apparent from (6.66)–(6.68) that the presence of inelastic scattering requires that $\mu_l \neq 0$ for some values of l, and this implies that $\sigma_{\text{el}} \neq 0$ since $\cosh 2\mu_l > \cos 2\lambda_l$. In other words there must be a simultaneous contribution to the elastically scattered wave.

Finally each term in (6.66)–(6.68) is bounded above

$$\begin{aligned} \sigma_{\text{el}}^l &\leqslant 4\pi(2l+1)/k_i^2, \\ \sigma_{\text{in}}^l &\leqslant \pi(2l+1)/k_i^2, \\ \sigma_{\text{tot}}^l &\leqslant 4\pi(2l+1)/k_i^2. \end{aligned} \tag{6.70}$$

The equalities in the first and last of these equations arise only for $\mu_l = 0$, in which case $\sigma^l_{\text{in}} = 0$, while the equality in the second equation requires that $\mu_l \to \infty$, in which case $\sigma^l_{\text{el}} = \pi(2l + 1)/k_i^2$ also.

The first conclusion is that the maximum value of σ^l_{tot} and σ^l_{el} may be seen to be four times the equivalent classical term, because by (3.11), $4\pi(2l + 1)/k^2 = 4 \times 2\pi b(\mathrm{d}b/\mathrm{d}l)$. A similar factor of four has already been encountered in Eqn (3.44). Secondly, not only can there be no inelastic without elastic scattering, but the maximum inelastic effects are obtained when the inelastic and elastic cross-sections are equal.

6.4 Scattering by an Anisotropic Potential

The analysis in the previous section is complicated in the general case by angular momentum coupling problems. The total angular momentum and its component about the incident direction are of course strictly conserved, but angular momentum may be exchanged between internal and translational (orbital) motion. Two descriptions of the problem have been proposed. The first, which employs a fixed quantization axis in the direction of incident motion is conceptually simpler but the second, due originally to Jacob and Wick (1959), which allows the quantization axis to rotate with the interparticle vector, leads to more tractable results. We consider the application of both methods to the problem of scattering between an atom and a rigid diatomic rotor, as illustrated in Fig. 6.2.

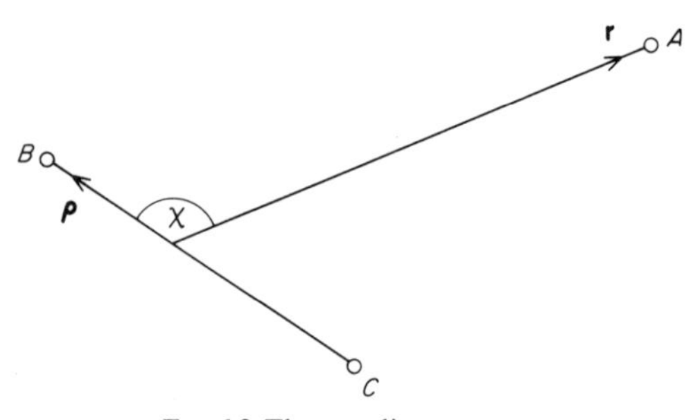

FIG. 6.2. The coordinate system.

In the fixed axis picture (Arthurs and Dalgarno (1960)), the Hamiltonian is taken in the straightforward form

$$H = -\frac{\hbar^2}{2m}\nabla_r^2 + \frac{\hat{\mathbf{j}}^2}{2I} + V(r, \chi), \tag{6.71}$$

where $\hat{\mathbf{j}}$ denotes the rotational angular momentum vector. The total wavefunction is expanded in eigenstates of the total angular momentum

$$\Phi_J^M(\mathbf{r},\hat{\boldsymbol{\rho}}) = \frac{1}{r}\sum_{j'l'} \psi_{j'l'}^J(r)\mathscr{Y}_{J_{j'}l'}^M(\hat{\mathbf{r}},\hat{\boldsymbol{\rho}}), \tag{6.72}$$

where $\hat{\mathbf{r}}$, $\hat{\boldsymbol{\rho}}$ denote the angular coordinates of $\mathbf{r}$ and $\boldsymbol{\rho}$, and $\mathscr{Y}_{J_{j'}l'}^M(\hat{\mathbf{r}},\hat{\boldsymbol{\rho}})$ is the proper combination of orbital, $|l', m_{l'})$ and rotational $|j', m_{j'})$ functions;

$$\mathscr{Y}_{J_{j'}l'}^M(\hat{\mathbf{r}},\hat{\boldsymbol{\rho}}) = \sum_{m_{l'}m_{j'}} (j'l'm_{j'}m_{l'}|j'l'JM)\, Y_{j'm_{j'}}(\hat{\boldsymbol{\rho}})\, Y_{l'm_{l'}}(\hat{\mathbf{r}}). \tag{6.73}$$

The necessary matrix of the potential which is found to be diagonal in J and independent of M may be derived by means of the expansion

$$V(r,\ \ ) = \sum_n V_n(r)\, P_n(\cos\chi), \tag{6.74}$$

so that

$$\int [\mathscr{Y}_{J_jl}^M(\hat{\mathbf{r}},\hat{\boldsymbol{\rho}})]^*\, V(r,\chi)\mathscr{Y}_{J_{j'}l'}^M(\hat{\mathbf{r}},\hat{\boldsymbol{\rho}})\,\mathrm{d}^2\hat{\mathbf{r}}\,\mathrm{d}^2\hat{\boldsymbol{\rho}}$$

$$= \frac{\hbar^2}{2m}\langle jl;J|U|j'l';J\rangle$$

$$= \sum_n V_n(r)\,\langle jl;J|P_n(\cos\theta)|j'l';J\rangle. \tag{6.75}$$

where, according for example to Eqn (5.14) of Brink and Satchler (1968),†

$$\langle jl;J|P_n(\cos\theta)|j'l';J\rangle$$

$$= (-1)^{J+n}[(2j+1)(2j'+1)(2l+1)(2l'+1)]^{\frac{1}{2}}\begin{pmatrix} j & j' & n \\ 0 & 0 & 0\end{pmatrix}\begin{pmatrix} l & l' & n \\ 0 & 0 & 0\end{pmatrix} W(jj'll';nJ). \tag{6.76}$$

It follows in the light of (6.71) and (6.75) that the radial components in (6.72) must satisfy equations of the same general form as (6.34);

$$\left[\frac{\mathrm{d}^2}{\mathrm{d}r^2} + k_{j'}^2 - \frac{l'(l'+1)}{r^2}\right]\psi_{j'l'}^J(r) = \sum_{j''l''}\langle j'l';J|U|j''l'';J\rangle\,\psi_{j''l''}^J(r), \tag{6.77}$$

where

$$k_{j'}^2 = \frac{2m}{\hbar^2}\left(E - \frac{j'(j'+1)\hbar^2}{2I}\right). \tag{6.78}$$

Note that the form of radial wave equation, like the potential matrix $\mathbf{U}$ is independent of M; this justifies omission of an additional superscript M in

† Values of the Wigner 3j symbols $\begin{pmatrix} a & b & c \\ \alpha & \beta & \gamma\end{pmatrix}$ and of the Racah coefficients $W(a\,b\,c\,d;\,e\,f)$ are given by Rotenberg *et al.* (1959) and Simon, Vander Sluis, and Biedenharn (1954).

(6.72). We now define the Jth S matrix, $\mathbf{S}^J$ with components $(jl|S^J|j'l')$ by picking out that solution of (6.77) for which at total angular momentum J, only the entrance channel (j, l) contains an incoming term. In other words, with additional superscripts (jl) to distinguish this solution,

$$\psi^{Jjl}_{j'l'}(r) \overset{r\to\infty}{\sim} \delta_{jj'}\delta_{ll'}\,\mathrm{e}^{-ik_jr} - (k_j/k_{j'})^{\frac{1}{2}}\,i^{-l-l'}\,(jl|S^J|j'l')\,\mathrm{e}^{ik_{j'}r}. \tag{6.79}$$

It now remains to combine the corresponding partial waves defined by (6.72) (now labelled $\Phi^M_{Jjl}(r, \hat{\mathbf{r}}, \hat{\boldsymbol{\rho}})$) in such a way that the resultant

$$\Psi_{jm_j}(\mathbf{r}, \hat{\boldsymbol{\rho}}) = \sum_{J,l} A^{m_j}_{Jl}\,\Phi^{m_j}_{Jjl}(\mathbf{r}, \hat{\boldsymbol{\rho}}) \tag{6.80}$$

contains as its incoming part the incoming part of a plane wave, multiplied by $Y_{jm_j}(\hat{\boldsymbol{\rho}})$. Note that M in (6.72) may be identified with m_j because there can be no orbital angular momentum about the incident direction. Now according to the plane wave expansion (3.23),

$$\mathrm{e}^{ik_jz}\,Y_{jm_j}(\hat{\boldsymbol{\rho}}) \overset{r\to\infty}{\sim} \frac{\pi^{\frac{1}{2}}}{ik_jr}\sum_{l=0}^{\infty}(2l+1)^{\frac{1}{2}}\left[\mathrm{e}^{ikr} - (-1)^l\,\mathrm{e}^{-ikr}\right]Y_{l0}(\hat{\mathbf{r}})\,Y_{jm_j}(\hat{\boldsymbol{\rho}}) \tag{6.81}$$

because

$$P_l(\cos\theta) = \left[4\pi/(2l+1)\right]^{\frac{1}{2}}Y_{l0}(r).$$

Hence $A^{m_j}_{Jl}$ in (6.80) must be chosen in such a way as to project out from (6.72), (6.73) and (6.79) an incoming term containing only $Y_{l0}(\hat{\mathbf{r}})\,Y_{jm_j}(\hat{\boldsymbol{\rho}})$. The proper choice is

$$A^{m_j}_{Jl} = \frac{i\pi^{\frac{1}{2}}}{k_j}(2l+1)^{\frac{1}{2}}(-1)^l\,(jlm_j0|jlJm_j). \tag{6.82}$$

This leads to the final result

$$\Psi_{jm_j}(\mathbf{r}, \hat{\boldsymbol{\rho}}) \overset{r\to\infty}{\sim} \mathrm{e}^{ik_jz}\,Y_{jm_j}(\hat{\boldsymbol{\rho}}) + \sum_{j'm''} f(jm_j, j'm_{j'}|\theta, \phi)\,(\mathrm{e}^{ik_{j'}r}/r)\,Y_{j'm_{j'}}(\hat{\boldsymbol{\rho}}), \tag{6.83}$$

where

$$f(jm_j, j'm_{j'}|\theta, \phi)$$
$$= \sum_J\sum_{ll'}\sum_{m_{l'}}\left[\frac{(2l+1)\pi}{k_jk_{j'}}\right]^{\frac{1}{2}} i^{l-l'}(jlm_j0|jlJm_j)\,(jl|T^J|j'l')\,(j'l'Jm_j|j'l'm_{j'}m_{l'})$$
$$\times\, Y_{l'm_{l'}}(\theta, \phi) \tag{6.84}$$

$$(jl|T^J|j'l') = i[(jl|S^J|j'l') - \delta_{ll'}\delta_{jj'}]. \tag{6.85}$$

$f(jm_j, j'm_{j'}|\theta, \phi)$ therefore gives the probability amplitude for scattering from channel (jm_j) to $(j'm_{j'})$, from which the inelastic differential and total cross-sections may be calculated in the usual way (Arthurs and Dalgarno

(1960)). The resulting expressions are however quite cumbersome, particularly in the case of the differential cross-section.

Simpler results (given by (6.95)–(6.98) below) may be obtained by transforming to a rotating coordinate system such that the quantization axis coincides with the direction of the interparticle vector **r** (Jacob and Wick, 1959; Lawley and Ross, 1965). We then speak of scattering from a state of helicity M about the initial direction to a state of helicity M' about the final direction, rather than from a state (jm_j) to $(j'm_{j'})$ where m_j and $m_{j'}$ are the space fixed components of j and j' respectively. The main mathematical effect of this transformation is actually to remove terms with $m_{l'} \neq 0$ in the analogue of (6.84) because there can be no orbital angular momentum about **r** and the disappearance of these terms allows a substantially more compact formulation.

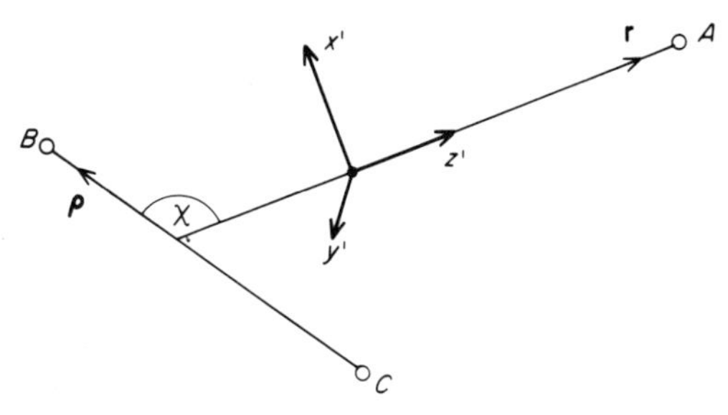

FIG. 6.3. The rotating axis system (x', y', z').

A convenient new set of system based coordinates (x', y', z') is illustrated in Fig. 6.3, with z' along r and the system in the $x'z'$ plane. Hence the transformation from (xyz) to $(x'y'z')$ may be regarded, in the conventions of Brink and Satchler (1968), as an Euler angle rotation $R = (\phi, \theta, \psi)$, where (θ, ϕ) denote the polar angles of **r** in the (xyz) system and ψ is the final rotation about z' required to bring $\boldsymbol{\rho}$ into the $x'z'$ plane. It follows that the polar coordinates of **r** and $\boldsymbol{\rho}$ in the new system may be written $\hat{\mathbf{r}}' = (0, 0)$ and $\hat{\boldsymbol{\rho}}' = (\chi, 0)$, and that the necessary transformations of $Y_{j'm_{j'}}(\hat{\boldsymbol{\rho}})$ and $Y_{l'm'_l}(\hat{\mathbf{r}})$ in (6.83) and (6.84) may be expressed in terms of the rotation matrices (Brink and Satchler, 1968),

$$D^{\lambda}_{\mu\mu'}(R) = D^{\lambda}_{\mu\mu'}(\phi, \theta, \psi) = e^{-i\mu\phi}\, d^{\lambda}_{\mu\mu'}(\theta)\, e^{-i\mu'\psi} = (-1)^{\mu-\mu'}[D^{\lambda}_{-\mu,-\mu'}(R)], \quad (6.86)$$

in the form

$$Y_{j'm_{j'}}(\hat{\boldsymbol{\rho}}) = \sum_{M'} [D^{j'}_{m_{j'}M'}(R)]^*\, Y_{j'M'}(\hat{\boldsymbol{\rho}}') = \sum_{M'} [D^{j'}_{m_{j'}M'}(R)]^*\, Y_{j'M'}(\chi, 0), \quad (6.87)$$

$$Y_{l'm_{l'}}(\hat{\mathbf{r}}) = \sum_{m_{l''}} [D^{l'}_{m_{l'}m_{l''}}(R)]^*\, Y_{l'm_{l''}}(\hat{\mathbf{r}}') = [(2l'+1)/4\pi]^{\frac{1}{2}}\, [D^{l'}_{m_{l'},0}(R)]^* \quad (6.88)$$

Here we have used the identity

$$Y_{l'm_{l'}}(0, 0) = [(2l + 1)/4\pi]^{\frac{1}{2}}\, \delta_{m_{l'},0} \tag{6.89}$$

which expresses the absence of orbital angular momentum about the direction of motion. One may also note the use of M' in (6.87) to denote the helicity of the state in question; the full internal wavefunction is represented by the product $Y_{j'M'}(\chi, 0) \exp iM'\psi$ in (6.87).

Two final steps are required to facilitate the transformation of (6.83) and (6.84). The first is to replace the Clebsch–Gordon coefficients by the more symmetrical $3j$ symbols

$$\begin{pmatrix} a & b & c \\ \alpha & \beta & \gamma \end{pmatrix} = (-1)^{a-b-\gamma}(2c + 1)^{-\frac{1}{2}}\, (ab\alpha\beta|abc-\gamma) \tag{6.90}$$

which vanish for $\alpha + \beta + \gamma \neq 0$ and acquire a factor $(-1)^{a+b+c}$ under any binary interchange of columns and under the sign reversal $(\alpha, \beta, \gamma) \Rightarrow (-\alpha, -\beta, -\gamma)$. The second is to recognize the identity (Brink and Satchler (1968))

$$\sum_{\alpha\beta} \begin{pmatrix} a & b & c \\ \alpha & \beta & \gamma \end{pmatrix} D^{a}_{\alpha\alpha'}(R)\, D^{b}_{\beta\beta'}(R) = \begin{pmatrix} a & b & c \\ \alpha' & \beta' & \gamma' \end{pmatrix} [D^{c}_{\gamma\gamma'}(R)]^* \tag{6.91}$$

Hence with M introduced in place of m_j, the relevant parts of (6.83) and (6.84) may be written

$$\sum_{m_{j'}} f(jm_j; j'm_j'|\theta, \phi)\, Y_{j'm_{j'}}(\hat{\mathbf{p}})$$

$$= \sum_J \sum_{ll'} \sum_{m_{l'}m_{j'}} \left[\frac{(2l + 1)\pi}{k_j k_{j'}}\right]^{\frac{1}{2}} i^{l-l'}\, (jlM0|jlJM)\, (jl|T^J|j'l')$$

$$\times\, (j'l'JM|j'l'm_{j'}m_{l'})\, Y_{l'm_{l'}}(\theta, \phi)\, Y_{j'm_{j'}}(\hat{\mathbf{p}})$$

$$= \sum_{M'} f(jM; j'M'|\theta, \phi)\, Y_{j'M'}(\hat{\mathbf{p}}')\, e^{iM'\psi} \tag{6.92}$$

where

$$f(jM; j'M'|\theta, \phi) = e^{-iM'\psi} \sum_J (-1)^{-j-j'} \frac{(2J + 1)}{2(k_j k_{j'})^{\frac{1}{2}}}\, g^J(jM; j'M'|\theta, \phi), \tag{6.93}$$

$$g^J(jM; j'M'|\theta, \phi)$$

$$= \sum_{ll'} \sum_{m_j m_{j'}} [(2l + 1)(2l' + 1)]^{\frac{1}{2}}\, i^{l'-l}(jl|T^J|j'l') \begin{pmatrix} j & l & J \\ M & 0 & -M \end{pmatrix} \begin{pmatrix} j' & l' & J \\ m_{j'} & m_{l'} & -M \end{pmatrix}$$

$$\times\, [D^{j}_{m_{j'}M'}(R)\, D^{l}_{m_{l'}0}(R)]^*$$

$$= \sum_{ll'} [(2l+1)(2l'+1)]^{\frac{1}{2}} i^{l'-l} (jl|T^J|j'l') \begin{pmatrix} j & l & J \\ M & 0 & -M \end{pmatrix} \begin{pmatrix} j' & l' & J \\ M' & 0 & -M' \end{pmatrix} \times D^J_{-M-M'}(R)$$

$$= \sum_{ll'} [(2l+1)(2l'+1)]^{\frac{1}{2}} i^{l'-l} (-1)^{l+l'+j+j'} (jl|T^J|j'l') \begin{pmatrix} j & J & l \\ M & -M & 0 \end{pmatrix} \times \begin{pmatrix} j' & J & l' \\ M' & -M' & 0 \end{pmatrix} (-1)^{M-M'} [D^J_{MM'}(R)]^*$$

$$= \sum_{ll'} (-1)^{M-M'} i^{l-l'} (jJM-M|jJl0)(jl|T^J|j'l')(j'Jl'0|j'JM'-M')[D^J_{MM'}(R)] \tag{6.94}$$

The advantage of this expression over (6.84) is that the angular factor $D^J_{MM'}(R)$ is common to every term in the sum. Hence the transformed scattering amplitude, $f(jM;j'M'|\theta,\phi)$ defined by (6.93), may be written in a form closely analogous to the elastic scattering expression in (3.28):

$$f(jM;j'M'|\theta,\phi) = \frac{(-1)^{j-j'+M-M'}}{2(k_j k_{j'})^{\frac{1}{2}}} \sum_J (2J+1)(jM|T^J|j'M')[D^J_{MM'}(\phi,\theta,0)]^* \tag{6.95}$$

where the elements $(jM|T^J|j'M')$ of the T matrix in the new (jM) representation are obtained from those in the old (jl) by means of the equation

$$(jM|T^J|j'M') = \sum_{ll'} i^{l-l'} (jJM-M|jJl0)(jl|T^J|j'l')(j'Jl'0|j'JM'-M'). \tag{6.96}$$

This leads in view of the common ϕ dependence of the $D^J_{MM'}(\phi,\theta,0)$ given by (6.86) to a relatively simple formula for the differential cross-section for scattering from one helicity state to another:

$$\frac{d\sigma(jM;j'M')}{d\Omega} = I(jM;j'M'|\theta) = (k_{j'}/k_j)|f(jM;j'M'|\theta,\phi)|^2$$

$$= \frac{1}{4k_j^2} \left| \sum_{J=0}^{\infty} (2J+1)(jM|T^J|j'M') \, d^J_{MM'}(\theta) \right|^2. \tag{6.97}$$

The corresponding total cross-section is also readily deduced;

$$\sigma(jM;j'M') = \int_0^{2\pi} \int_0^{\pi} I(jM;j'M'|\theta) \sin\theta \, d\theta \, d\phi$$

$$= \frac{\pi}{k_j^2} \sum_{J=0}^{\infty} (2J+1)|(jM|T^J|j'M')|^2, \tag{6.98}$$

because (Brink and Satchler, 1968),

$$\int_0^{\pi} d^J_{MM'}(\theta)\, d^{J'}_{MM'}(\theta) \sin\theta\, \mathrm{d}\theta = [2/(2J+1)]\delta_{JJ'}. \tag{6.99}$$

It may also be verified from the unitarity of the S matrix, and the standardization $d^J_{MM'}(0) = \delta_{MM'}$ that the full total cross-section for scattering from the channel (jM) to all other channels satisfies the optical theorem

$$\sigma^{jM}_{\text{tot}} = \sum_{j'M'} \sigma(jM; j'M') = \frac{4\pi}{k_j} \operatorname{Im}[f(jM; fM|0,0)], \tag{6.100}$$

and that the mean cross-section for scattering from channel j to channel j' regardless of helicity, satisfies the principle of detailed balance

$$\begin{aligned} \bar{\sigma}(j;j') &= \frac{1}{(2j+1)} \sum_{MM'} \sigma(jM; j'M') \\ &= \frac{\pi}{(2j+1)k_j^2} \sum_{J=0}^{\infty} \sum_{MM'} (2J+1)|(jM|T^J|j'M')|^2 \\ &= \frac{\pi}{(2j+1)k_j^2} \sum_{J=0}^{\infty} \sum_{ll'} (2J+1)|(jl|T^J|j'l')|^2 = \frac{(2j'+1)}{(2j+1)} \left(\frac{k_{j'}^2}{k_j^2}\right) \bar{\sigma}(j';j), \end{aligned} \tag{6.101}$$

the factors $(2j+1)$ being included to account for rotational degeneracy of the entrance channel.

The conclusion is that the rotating axis scattering amplitude, $f(jM; j'M'|\theta, \phi)$ given by (6.95), has certain formal advantages over its fixed axis counterpart, $f(jm_j, j'm'_j|\theta, \phi)$ of (6.84): namely that the relative simplicity of its structure admits ready manipulation to yield the scattering cross-sections. However, as the theory stands at present, the necessary T matrix elements must be derived by solution of the equations (6.77) (which apply to the fixed axis formulation) to obtain the elements $(jl|T^J|j'l')$ defined by (6.79) and (6.85): these are then transformed by (6.96) to the rotating axis (jM) representation. This is a well-defined and perfectly respectable procedure, provided the internal states in question are readily quantized in the fixed axis frame. On the other hand cases may arise, particularly in the field of electronic excitation, where the internal states are most conveniently defined in a rotating coordinate system. The problem is then more readily tackled by solution of the equations of motion in a rotating frame (see for example (6.16)) to obtain the "transformed" elements $(jM|T^J|j'M')$ directly.

Although, as mentioned above, this approach is most commonly applicable to electronic inelasticity, we shall endeavour to maintain contact with the previous discussion by outlining its application to the excitation of a rigid rotor.

The required form for the hamiltonian follows from the relation

$$\mathbf{v}_i = \dot{\mathbf{r}}_i + \boldsymbol{\omega} \wedge \mathbf{r}_i \tag{6.102}$$

where $\mathbf{r}_i$ denotes distance from the centre of mass, $\mathbf{v}_i$ and $\dot{\mathbf{r}}_i$ are velocities with respect to space fixed and rotating axes respectively and $\boldsymbol{\omega}$ is the angular velocity of the (x', y', z') frame. Thus the classical kinetic energy for the present system may be written

$$\begin{aligned} T &= \sum_i \tfrac{1}{2}m_i(\mathbf{v}_i \,.\, \mathbf{v}_i) \\ &= \tfrac{1}{2}m(\dot{\mathbf{r}} \wedge \dot{\mathbf{r}}) + \tfrac{1}{2}m(\boldsymbol{\omega} \wedge \mathbf{r}) \,.\, (\boldsymbol{\omega} \wedge \mathbf{r}) + \tfrac{1}{2}\mu(\boldsymbol{\omega} \wedge \boldsymbol{\rho}) \,.\, (\boldsymbol{\omega} \wedge \boldsymbol{\rho}) \\ &\quad + \tfrac{1}{2}\mu(\dot{\boldsymbol{\chi}} \wedge \boldsymbol{\rho}) . (\dot{\boldsymbol{\chi}} \wedge \boldsymbol{\rho}) + \mu(\boldsymbol{\omega} \wedge \boldsymbol{\rho}) . (\dot{\boldsymbol{\chi}} \wedge \boldsymbol{\rho}) \\ &= \tfrac{1}{2}m\, p_r^2 + \frac{1}{2mr^2}\left[J_{x'}^2 + J_{y'}^2 + J_{z'}^2 \cot^2 \chi + (J_{z'}J_{x'} + J_{x'}J_{z'}) \cot \chi - J_y p_\chi\right] \\ &\qquad + \frac{1}{2\mu\rho^2}\left[J_{z'}^2 \operatorname{cosec}^2 \chi + p_\chi^2\right]. \end{aligned} \tag{6.103}$$

where

$$\begin{aligned} p_r &= \frac{\partial}{\partial \dot{r}}(T - V) = \frac{\partial T}{\partial \dot{r}} \\ p_\chi &= \frac{\partial T}{\partial \dot{\chi}} \\ J_\alpha &= \frac{\partial T}{\partial \omega_\alpha} \qquad (\alpha = x', y', z') \end{aligned} \tag{6.104}$$

$\mathbf{J}$ therefore represents the total angular momentum; p_r and p_χ are the momenta conjugate to r and χ respectively. It now follows from the rules for constructing the quantum mechanical hamiltonian (see Margenau and Murphy, 1956 p. 299), and from the reversed commutation relation (Van Vleck, 1951),

$$[J_{z'}, J_{x'}] = -i\hbar J_{y'} \tag{6.105}$$

between the angular momentum operators in the rotating frame, that the hamiltonian takes the same general form as in (6.16),

$$\begin{aligned} H &= H_a(r,\chi) + \frac{1}{2mr^2}\left[-\hbar^2 \frac{\partial}{\partial r}\left(r^2 \frac{\partial}{\partial r}\right) + (\hat{J}_{x'} - \hat{j}_{x'})^2 + (\hat{J}_{y'} - \hat{j}_{y'})^2\right] \\ &= H_a(r,\chi) + \frac{1}{2mr^2}\left[-\hbar^2 \frac{\partial}{\partial r}\left(r^2 \frac{\partial}{\partial r}\right) + \hat{J}^2 + \hat{j}^2 - 2J_{z'}^2 - \hat{J}_+ \hat{j}_- - \hat{J}_- \hat{j}_+\right] \end{aligned} \tag{6.106}$$

where $H_a(r, \chi)$ denotes the adiabatic hamiltonian appropriate to a stationary frame

$$H_a(r, \chi) = \frac{\hat{j}^2}{2\mu\rho^2} + V(r, \chi)$$

and

$$\hat{j}^2 = -\frac{\hbar^2}{\sin\chi}\frac{\partial}{\partial\chi}\left(\sin\chi\frac{\partial}{\partial\chi}\right) + \frac{\hat{J}_{z'}^2}{\sin^2\chi} \tag{6.107}$$

$$\hat{J}_\pm = (\hat{j}_{x'} \pm i\hat{j}_{y'}) = \pm\hbar\frac{\partial}{\partial\chi} - \hat{J}_z' \cot\chi. \tag{6.108}$$

Note that, by the choice of coordinate system, $J_{z'}$ is inextricably associated with the motion of the molecule BC; this accounts for its appearance in both $\hat{j}^2$ and $\hat{j}_\pm$ in Eqn (6.108).

It is clear from the above form, that both $\hat{J}^2$ and the space fixed component $\hat{J}_z$ commute with H in (6.106). Thus the eigenfunctions of H are conveniently expanded in the appropriate angular momentum eigenstates, which are, according to Brink and Satchler (1968), the rotation matrices $[D^J_{MM'}(R)]^*$;

$$\begin{aligned}
\hat{J}^2[D^J_{MM'}(R)]^* &= J(J+1)\hbar^2[D^J_{MM'}(r)]^* \\
\hat{J}_z[D^J_{MM'}(R)]^* &= M\hbar[D^J_{MM'}(R)]^* \\
\hat{J}_{z'}[D^J_{MM'}(R)]^* &= M'\hbar[D^J_{MM'}(R)]^*.
\end{aligned} \tag{6.109}$$

This means that the required eigenfunctions of H may be written

$$\Phi^J_M(r, R, \chi) = \frac{1}{r}\sum_{M'} [D^J_{MM'}(R)]^* \, \xi^J_{M'}(r, \chi), \tag{6.110}$$

with the $\xi^J_{M'}(r, \chi)$ subject to

$$\left[H_a(r,\chi) - \frac{\hbar^2}{2m}\frac{\partial^2}{\partial r^2} + \frac{(J(J+1) - 2M'^2)\hbar^2}{2mr^2} + \frac{\hat{j}^2}{2mr^2} - E\right]\xi^J_{M'}(r,\chi)$$
$$= \frac{\hbar^2}{2Mr^2}[\lambda_+(J, M')\hat{j}_-\xi^J_{M'+1}(r,\chi) + \lambda_-(J, M')\hat{j}_+\xi^J_{M'-1}(r,\chi)] \tag{6.111}$$

in which the terms

$$\lambda_\pm(J, M') = [J(J+1) - M'(M' \pm 1)]^{\frac{1}{2}} \tag{6.112}$$

arise from the shift operations which follow from (6.105), namely

$$(J_{x'} \pm iJ_{y'})\,[D^J_{MM'}(R)]^* = \lambda_\mp(J, M')[D^J_{MM'\mp 1}(R)]^*. \tag{6.113}$$

Equations (6.111) were first derived in a slightly different phase convention by Curtiss and Adler (1952), who used group-theoretical arguments.

It remains to note that $V(r, \chi) \rightarrow 0$ and that all coupling terms in (6.111) disappear as $r \rightarrow \infty$. Hence by the form of $H_a(r, \chi)$ in (6.107), (6.111) has asymptotic solutions of the form

$$\xi^J_M(r, \chi) \overset{r\rightarrow\infty}{\sim} e^{\pm ik_j r}\, Y_{jM}(\chi, 0). \tag{6.114}$$

Thus the $\xi^J_M(r, \chi)$ may be expanded in any convenient set of normalized functions $\phi_{jM}(\chi)$,

$$\xi^J_M(r, \chi) = \frac{1}{r} \sum_j \psi^J_{jM}(r)\, \phi_{jM}(\chi) \tag{6.115}$$

such that†

$$\phi_{jM}(\chi) \overset{r\rightarrow\infty}{\sim} Y_{jM}(\chi, 0). \tag{6.116}$$

The independent solutions of the resulting equations of motion

$$\left[-\frac{\hbar^2}{2m}\frac{d^2}{dr^2} + \langle jM|V|jM\rangle - E\right]\psi^J_{jM}(r) = -\sum_{j'M'} \langle jM|V|j'M'\rangle\, \psi^J_{j'M'}(r) \tag{6.117}$$

where

$$\langle jM|V|j'M'\rangle = \left\langle \phi_{jM} \left| H_a + \frac{[J(J+1) - 2M^2]\hbar^2}{2mr^2} + \frac{\hat{\jmath}^2}{2mr^2} \right| \phi_{j'M'} \right\rangle \delta_{MM'}$$
$$- \frac{\hbar^2}{2m}\lambda_+(J, M)\, \langle \phi_{jM}|\hat{\jmath}_-|\phi_{j'M'}\rangle\, \delta_{M,M'-1} - \frac{\hbar^2}{2m}\lambda_-(J, M)\, \langle \phi_{jM}|\hat{\jmath}_+|\phi_{j'M'}\rangle$$
$$\times\, \delta_{M,M'+1}, \tag{6.118}$$

are then combined to determine a solution (with components labelled $\psi^{JjM}_{j'M'}(r)$) appropriate to the entrance channel (JjM):

$$\psi^{JjM}_{j'M'}(r) \overset{r\rightarrow\infty}{\sim} e^{-ik_j r} - \sum_{j'M'} (-1)^{J-j'+M-M'} (k_j/k_{j'})^{-\frac{1}{2}} (jM|S^J|j'M')\, e^{ik_{j'}r}. \tag{6.119}$$

It may be verified by following the transformation between the fixed axis total angular momentum state $\Phi^M_{Jjl}(\mathbf{r}, \boldsymbol{\rho})$ defined by (6.72) and (6.79), and the corresponding rotating axis form (now labelled $\phi^J_{jM}(r, R, \chi)$) given by (6.110) and (6.119) that the elements $(jM|S^J|j'M')$ in (6.119) are identical with those defined by the S matrix analogue of (6.96).

Finally we emphasize the significance of this discussion for electronic excitation problems in diatomic systems. Here we should replace $H_a(r, \chi)$

† Another convenient choice might be the eigenfunctions of the adiabatic hamiltonian $H_a(r, \hat{\boldsymbol{\rho}})$; the coupling terms in (6.117) would then contain contributions from the radial derivatives.

by the adiabatic electronic hamiltonian $H_a(r, \boldsymbol{\rho})$ and j by the total (orbital plus spin) electronic angular momentum. The helicity M is merely the familiar angular momentum component Λ about the diatomic axis. At low (molecular, collision energies the most convenient choice of expansion functions would be the adiabatic electronic eigenfunctions, $\phi_{n\Lambda}(\boldsymbol{\rho}; r)$ say, leading to two types of coupling terms in the analogue of (6.117); the first, due to the radial variation of $\phi_{n\Lambda}(\boldsymbol{\rho}; r)$ are responsible for homogeneous interactions in spectroscopic parlance (see Herzberg, 1950), while the second, due to the Coriolis terms $\hat{j}_{\pm}$ in (6.114), give rise to heterogeneous effects.

Extensions of the theory, both for the present model and for more complicated systems have been given in a long series of papers by Curtiss *et al.* references to which are given by Curtiss (1970).

CHAPTER 7

Quantum Inelastic Transition Probabilities

The framework established in the previous chapter provides the relation between the S or T matrices derived from the radial equations (6.56) (6.77) or (6.117) and the scattering amplitudes and collision cross-sections given by (6.60)–(6.62), (6.84), (6.95) or (6.97)–(6.101). The next problem is to solve the necessary equations for the elements of **S** or **T**. Four approximate analytical methods, appropriate to quantum mechanical molecular scattering conditions, are described below. Other methods based on solution of semi-classical approximations to the exact equations of motion are discussed in Chapter 8.

The Born approximation outlined in Section 7.1 applies when both the relevant elastic and inelastic partial cross-sections are small; for most realistic systems this will become true at high energies and large impact parameters. As a better approximation in most cases, the distorted wave method allows for strong elastic scattering by the diagonal, distortion potentials $U_{ii}(r)$ in (6.56) or (6.80), but it is again restricted by the condition that the inelastic scattering should be weak. As a further improvement the exponential approximation of Section 7.3, due to Levine (1971), offers a renormalization of the distorted wave results, provided that the polarization contribution to the elastic scattering, due to the off-diagonal terms $U_{ij}(r)$ is small and that the de Broglie wavelength is short. This means that even quite strong inelastic scattering may be handled. The final Section 7.4 is devoted to the polarization and Feshbach resonance processes associated with excitation of internal motion below the energy threshold for inelastic scattering.

While the main emphasis is on the physical and mathematical nature of the theory, relevant practical applications are included in the discussion. The classic Jackson and Mott (1932) treatment of vibrational excitation, which is central to the Schwartz *et al.* (1951) theory of vibrational relaxation is developed in detail. The relevance of the theory to the interpretation of oscillations in the differential inelastic cross-sections, as given by Olson and Smith (1971), to the evaluation of non-adiabatic effects in high energy elastic scattering, Thorson (1963), and to the theory of molecular predissociation, Child (1970), Ramsay and Child (1972) is also indicated.

Since many systems will not be amenable to analytical treatment, some indication of the relation between the theory and relevant numerical methods may be valuable. Gordon (1969) uses what is, apart from the boundary conditions, essentially the distorted wave method, with the equations transformed to remove the coupling terms at the midpoint of each integration step. With linear approximations applied elsewhere, the necessary integrals are obtained analytically and the step length may be taken to be large compared with the de Broglie wavelength. This is an important advantage in the molecular case. Chan, Light and Lin (1968) employ an exponential method due to Magnus (1954) similar in principle to the exponential approximation of Section 7.3. Indeed the basic equations (7.97) may be obtained by rearrangement of those of Chan *et al.* (1968). Finally Secrest and Johnson (1966) base their integration on a form of the amplitude density method used below to derive equation (7.96).

7.1 The Born Approximation

As an introduction to the Born approximation we may consider the case of scattering in a spherical field, for which Eqn (6.56) determines the motion;

$$\left[\frac{\mathrm{d}^2}{\mathrm{d}r^2} + k_i^2 - \frac{l(l+1)}{r^2}\right]\psi_{il}(r) = \sum_j U_{ij}(r)\psi_{jl}(r) \tag{7.1}$$

It is natural to seek a solution in terms of the outgoing Green's function derived in Appendix B,

$$\begin{aligned}\mathscr{G}_{il}^{(+)}(r, r') &= -k_i^{-1}[\psi_{il}^{(0)}(r)\,\psi_{il}^{(1)}(r') + i\psi_{il}^{(0)}(r)\,\psi_{il}^{(0)}(r')],\ r < r' \\ &= -k_i^{-1}[\psi_{il}^{(1)}(r)\,\psi_{il}^{(0)}(r') + i\psi_{il}^{(0)}(r)\,\psi_{il}^{(0)}(r'),\ r > r',\end{aligned} \tag{7.2}$$

where $\psi_{il}^{(0)}(r)$ and $\psi_{il}^{(1)}(r)$ satisfy (7.1) when $U_{ij}(r) = 0$; $\psi_{il}^{(0)}(r)$ is the regular solution, normalized in the form

$$\psi_{il}^{(0)}(r) \overset{r\to\infty}{\sim} \sin(k_i r - l\pi/2),$$

while

$$\psi_{il}^{(1)}(r) \overset{r\to\infty}{\sim} \cos(k_i r - l\pi/2). \tag{7.3}$$

Hence in view of the absence of potential terms on the left hand side of (7.1), $\psi_{il}^{(0)}(r)$ and $\psi_{il}^{(1)}(r)$ are given in terms of spherical Bessel functions (Abramowitz and Stegun, 1965),

$$\psi_{il}^{(0)}(r) = k_i r\, j_l(k_i r)$$

$$\psi_{il}^{(1)}(r) = -k_i r\, y_l(k_i r) \tag{7.4}$$

The role of this Green's function is to transform (7.1) into the corresponding integral equation,

$$\psi_{il}(r) = \delta_{i0}\,\psi_{0l}^{(0)}(r) + \sum_j \int_0^\infty \mathscr{G}_i^{(+)}(r, r')\, U_{ij}(r')\, \psi_{jl}(r')\, \mathrm{d}r'. \tag{7.5}$$

The Born approximation rests on the assumption that the motion is dominated by a plane wave component in the incident channel

$$\psi_{0l}(r) \simeq \psi_{0l}^{(0)}(r) \gg \psi_{jl}(r), \qquad (j \neq 0), \tag{7.6}$$

so that on retaining only the dominant term in the sum in (7.5)

$$\psi_{0l}(r) = \psi_{0l}^{(0)}(r) + \int_0^\infty \mathscr{G}_{0l}^{(+)}(r, r')\, U_{00}(r')\, \psi_{0l}^{(0)}(r')\, \mathrm{d}r' \overset{r\to\infty}{\sim} \sin(k_0 r - l\pi/2)$$
$$- k_0^{-1} \exp(ik_0 r - il\pi/2) \int_0^\infty \psi_{0l}^{(0)}(r)\, U_{00}(r)\, \psi_{0l}^{(0)}(r)\, \mathrm{d}r, \tag{7.7}$$

$$\psi_{il}(r) = \int_0^\infty \mathscr{G}_{il}^{(+)}(r, r')\, U_{i0}(r')\, \psi_{0l}^{(0)}(r')\, \mathrm{d}r'$$
$$\overset{r\to\infty}{\sim} -k_i^{-1} \exp(ik_i r - il\pi/2) \int_0^\infty \psi_{il}^{(0)}(r)\, U_{i0}(r)\, \psi_{0l}^{(0)}(r)\, \mathrm{d}r, \tag{7.8}$$

where the asymptotic forms in (7.7) and (7.8) have been obtained with the help of (7.2) and (7.3). By the definition of the $T^{(l)}$ matrix in (6.57), this means that

$$T_{0i}^{(l)} = 2(k_0 k_i)^{-\frac{1}{2}} \int_0^\infty \psi_{il}^{(0)}(r)\, U_{i0}(r)\, \psi_{0l}^{(0)}(r)\, \mathrm{d}r$$
$$= 2(k_0 k_i)^{-\frac{1}{2}} \int_0^\infty j_l(k_i r)\, j_l(k_0 r)\, U_{i0}(r) r^2\, \mathrm{d}r, \tag{7.9}$$

where the second line follows from (7.4). The corresponding scattering amplitudes $f_{0i}(\theta)$ and total cross-sections σ_{0i} are then given by (6.60) and (6.62) respectively

$$f_{0i}(\theta) = -\tfrac{1}{2}(k_0 k_i)^{-\frac{1}{2}} \sum_{l=0}^{\infty} (2l+1) T_{0i}^{(l)}\, P_l(\cos\theta)$$
$$\sigma_{0i} = \pi k_0^{-2} \sum_{l=0}^{\infty} (2l+1) \left| T_{0i}^{(l)} \right|^2. \tag{7.10}$$

With the above form for $T_{0i}^{(l)}$, $f_{0i}(\theta)$ may also be cast, by use of the identity (Abramowitz and Stegun, 1965)

$$\sum_{l=0}^{\infty} (2l + 1) j_l(k_0 r) j_l(k_i r) P_l(\cos\theta) = \sin Kr/Kr;$$

$$K^2 = k_0^2 + k_i^2 - 2k_0 k_i \cos\theta, \tag{7.11}$$

into the form

$$f_{0i}(\theta) = -\int_0^{\infty} \frac{\sin Kr}{Kr} U_{i0}(r)\, r^2\, \mathrm{d}r. \tag{7.12}$$

It is seen that Eqns (7.7) and (7.12) for the *elastic* scattering ($i = 0$) are identical with these given by (4.10) and (4.14) with $U_{00}(r)$ in place of $U(r)$. The *inelastic* amplitude $\psi_{il}(r)$ in (7.8) is equivalent to that obtained by approximating the system (7.1) in the form

$$\left[\frac{\mathrm{d}^2}{\mathrm{d}r^2} + k_0^2 - \frac{l(l+1)}{r^2}\right]\psi_{0l}(r) = 0$$

$$\left[\frac{\mathrm{d}^2}{\mathrm{d}r^2} + k_i^2 - \frac{l(l+1)}{r^2}\right]\psi_{il}(r) = U_{i0}(r)\psi_{0l}(r). \tag{7.13}$$

In assessing the validity of (7.8) and (7.9) for a given partial wave we must examine the assumptions that $|\psi_{il}(r)| \ll |\psi_{0l}(r)|$, and that $\psi_{0l}(r)$ is the appropriate plane wave component. The first requires that $T_{0i}^{(l)}$ in (7.9) is small and the second that the Born approximation (4.12) to the phase shift η_{0l}, due to the mean distortion potential $U_{00}(r)$ in the initial channel, is valid. We have seen in Section 4.1 that the validity of (4.12) typically improves as k_0 and l increase. The same is true for $T_{0i}^{(l)}$ in (7.9).

Consider as an illustration the exponential form of interaction potential

$$U_{i0}(r) = U\,\mathrm{e}^{-(r/R)}, \tag{7.14}$$

for which, since $j_0(z) = (\sin z/z)$, the s-wave transition amplitude becomes

$$\begin{aligned} T_{0i}^{(0)} &= 2U(k_0 k_i)^{-\frac{1}{2}} \int_0^{\infty} \sin k_i r \sin k_0 r\, \mathrm{e}^{-r/R}\, \mathrm{d}r \\ &= U(k_0 k_i)^{-\frac{1}{2}} \int_0^{\infty} [\cos(k_0 - k_i)r + \cos(k_0 + k_i)r]\, \mathrm{e}^{-r/R}\, \mathrm{d}r \\ &\simeq \frac{UR}{(k_0 k_i)^{\frac{1}{2}}[1 + (k_0 - k_i)^2 R^2]} \end{aligned} \tag{7.15}$$

where the contribution from the second term in the second line of (7.15) has been neglected on account of the relatively rapid oscillations of the integrand. $T_{0i}^{(0)}$ therefore depends on the Fourier component of $U_{i0}(r)$ at a frequency, $k_0 - k_i$, which decreases with increasing energy because,

$$k_0 - k_i = (k_0^2 - k_i^2)/(k_i + k_0) = m\Delta E_{i0}/\bar{k}h^2, \tag{7.16}$$

where $\bar{k} = \frac{1}{2}(k_i + k_0)$. After an initial divergence at threshold, which implies a breakdown of the approximation, $T_{0i}^{(0)}$ in (7.15) may be seen to pass through a region dominated by the energy difference ΔE_{i0} and the range parameter R, to reach a high energy limit where $T_{0i}^{(0)}$ decreases inversely with the mean relative velocity, $v = (k_0 k_i)^{\frac{1}{2}}/m\hbar$.

For a crude extension of these results to higher 1 values, we may simply change the origin of the oscillatory terms and the lower integration limit in (7.15) to the average classical turning point

$$\bar{a}_l = l(1/2k_0 + 1/2k_i) \tag{7.17}$$

because the spherical Bessel functions decrease rapidly for $r < a_l$, and oscillate roughly sinusoidally for $r > a_1$. In order of magnitude therefore

$$T_{0i}^{(l)} \simeq \frac{U\,e^{-(\bar{a}_l/R)}R}{(k_0 k_i)^{\frac{1}{2}}[1 + (k_0 - k_i)^2 R^2]}, \tag{7.18}$$

and $T_{0i}^{(l)}$ is seen to decrease with l.

The conclusion is that the Born approximation increases in validity as the energy and angular momentum increase.

Similar arguments may be employed to determine the scattering amplitude $f_{0i}(\theta, \phi)$, and hence σ_{0i} when the potential is anisotropic. It is convenient, rather than using the coupled representation in Section 6.4 to start from the three dimensional Eqn (6.6). On the assumption that the motion is dominated by a plane wave in the incident channel, this may be written

$$[\nabla_r^2 + k_i^2]\,\Psi_i(\mathbf{r}) = F(\mathbf{r}) \tag{7.19}$$

where

$$F(\mathbf{r}) \simeq U_{i0}(r)\exp(i\mathbf{k}_0 \,.\, \mathbf{r}).$$

Equation (7.19) reduces, on expanding $\Psi_i(\mathbf{r})$ and $F(\mathbf{r})$ in spherical harmonics

$$\Psi_i(\mathbf{r}) = \frac{1}{r}\sum_{l=0}^{\infty}\sum_{m=-l}^{l} \psi_{ilm}(r)\,Y_{lm}(\theta, \phi) \tag{7.20a}$$

$$F(\mathbf{r}) = \sum_{l=0}^{\infty}\sum_{m=-l}^{l} F_{lm}(r)\,Y_{lm}(\theta, \phi) \tag{7.20b}$$

so that

$$F_{lm}(r) = \int_0^{2\pi}\int_0^{\pi} Y_{lm}^*(\theta, \phi)\,F(\mathbf{r})\sin\theta\,\mathrm{d}\theta\,\mathrm{d}\phi, \tag{7.21}$$

to the same form as the second of Eqns (7.13) with $rF_{lm}(r)$ in place of $U_{i0}(r)\,\psi_{0l}^{(0)}(r)$;

$$\left[\frac{d^2}{dr^2} + k_i^2 - \frac{l(l+1)}{r^2}\right]\psi_{ilm}(r) = rF_{lm}(r). \tag{7.22}$$

Hence by analogy with (7.8),

$$\psi_{ilm}(r) \overset{r\to\infty}{\sim} -\exp(ik_i r - il\pi/2)\int_0^\infty j_l(k_i r')\,F_{lm}(r')r'^2\,dr' \tag{7.23}$$

The substitution of this result in (7.20a), with $F_{lm}(r')$ given by (7.21) implies that

$$\Psi_i(r) = f_{0i}(\theta, \phi)\exp(ik_i r)/r, \tag{7.24}$$

where

$$f_{0i}(\theta, \phi) = -\iiint \sum_{lm} (-i)^l\, Y^*_{lm}(\theta'\phi')\, Y_{lm}(\theta, \phi)\, j_l(k_i r')\, F(\mathbf{r}')r'^2 \sin\theta'\,dr'\,d\theta'\,d\phi'. \tag{7.25}$$

A more compact form may be obtained with the help of the identities (Abramowitz and Stegun, 1965)

$$\sum_{m=-l}^{l} Y^*_{lm}(\theta', \phi')\, Y_{lm}(\theta, \phi) = [(2l+1)/4\pi]\, P_l(\cos\Theta) \tag{7.26}$$

and

$$\sum_{l=0}^{\infty} (2l+1)(-i)^l j_l(kr')\, P_l(\cos\Theta) = \exp(-i\mathbf{k}_i . \mathbf{r}'), \tag{7.27}$$

where Θ is the angle between the directions (θ, ϕ) and (θ', ϕ') of $\mathbf{k}_i$ and $\mathbf{r}'$ respectively. Thus (7.25) becomes,

$$f_{0i}(\theta, \phi) = -\frac{1}{4\pi}\iiint \exp[i(\mathbf{k}_0 - \mathbf{k}_i).\mathbf{r}']\, U_{i0}(r')\, r'^2 \sin\theta'\,dr'\,d\theta'\,d\phi', \tag{7.28}$$

after the substitution for $F(r')$ in (7.19). This is the proper generalization of (7.12).

The compact forms of (7.12) and (7.28) have an obvious attraction compared with (7.10) and (7.25) respectively. Unfortunately they depend on the validity of the Born description for all partial waves. Typically however, as indicated by (7.18) the inelastic scattering is strong for small l values (and hence small impact parameters) but falls off rapidly with increasing l. Only the high impact parameter scattering therefore lies in the realm of the Born approximation and the relevant contribution to $f_{0i}(\theta)$ or $f_{0i}(\theta, \phi)$ must be obtained by curtailing the sum in (7.10) and (7.25).

7.2 The Distorted Wave Approximation

The distorted wave method differs from the Born approximation in taking explicit account of the diagonal, distortion potentials $U_{ii}(r)$ in zeroth order.

It again assumes however that $|\psi_i(r)| \ll |\psi_0(r)|$. Thus (7.13) are replaced by equations of the general form

$$\left[\frac{d^2}{dr^2} + k_0^2 - U_0(r)\right] \psi_0(r) \simeq 0 \tag{7.29a}$$

$$\left[\frac{d^2}{dr^2} + k_i^2 - U_i(r)\right] \psi_i(r) \simeq U_{i0}(r)\,\psi_0(r), \tag{7.29b}$$

where

$$U_i(r) = U_{ii}(r) + l(l+1)/r^2. \tag{7.30}$$

Note that a subscript l has been dropped in order to simplify the subsequent notation. The Born approximation by contrast allows only for distortion by the centrifugal terms $l(l+1)/r^2$. The formal appearance of the solution of (7.29) is identical however with that in (7.8)

$$\psi_0(r) = \psi_0^{(0)}(r)$$

$$\psi_i(r) = \int_0^\infty \mathscr{G}_i^{(+)}(r, r')\, U_{i0}(r')\, \psi_0^{(0)}(r')\, dr'$$

$$\overset{r\to\infty}{\sim} -k_i^{-1} \exp[ik_i r + i\delta_i] \int_0^\infty \psi_i^{(0)}(r')\, U_{i0}(r')\, \psi_0^{(0)}(r')\, dr', \tag{7.31}$$

where $\psi_0^{(0)}(r)$ is the regular solution of (7.29a) normalized in the form†

$$\psi_0^{(0)}(r) \sim \sin(k_0 r + \delta_0) \tag{7.32}$$

and $\psi_i^{(0)}(r)$ is similarly derived from $U_i(r)$. The transition amplitude is therefore given by

$$T_{0i} = \exp[i(\eta_0 + \eta_i)]\, T'_{0i}$$

$$T'_{0i} = 2(k_0 k_i)^{-\frac{1}{2}} \int_0^\infty \psi_i^{(0)}(r')\, U_{i0}(r')\, \psi_0^{(0)}(r')\, dr' \tag{7.33}$$

but the value of the integral will now depend on distortions of the $\psi_i^{(0)}(r)$ due to the $U_i(r)$ as well as on the form of the interaction function. The significance of the prime in (7.33) is that $\mathbf{T}$ is the transition matrix with respect to the hamiltonian for free motion, while $\mathbf{T}'$ is defined with respect to the distortion hamiltonian, which also includes the diagonal potential terms $V_{ii}(r)$.

One of the most important applications of the method in a molecular context, arises in the vibrational relaxation theory initiated by Jackson and

† δ_i here denotes $\eta_i - l\pi/2$.

Mott (1932). The model due to Landau and Teller is of a harmonic oscillator BC in collinear collision with A under an exponential repulsion between A and B.

$$V(r, \rho) = V' \exp(-ar_{ab}) = V' \exp\{-a[r - \lambda(\rho_0 + \rho)]\}, \tag{7.34}$$

where

$$\lambda = m_c/(m_b + m_c) \tag{7.35}$$

In the scaled coordinate system (Secrest and Johnson, 1966) such that

$$\begin{aligned} \xi &= (\rho/d), \qquad x = (r - r_0)/\lambda d \\ \mathrm{d} &= \hbar^{\frac{1}{2}}(k\mu)^{-\frac{1}{4}}, \qquad r_0 = a^{-1} \ln(2mV'/\hbar\omega) \\ \omega &= (k/\mu)^{\frac{1}{2}} \end{aligned} \tag{7.36}$$

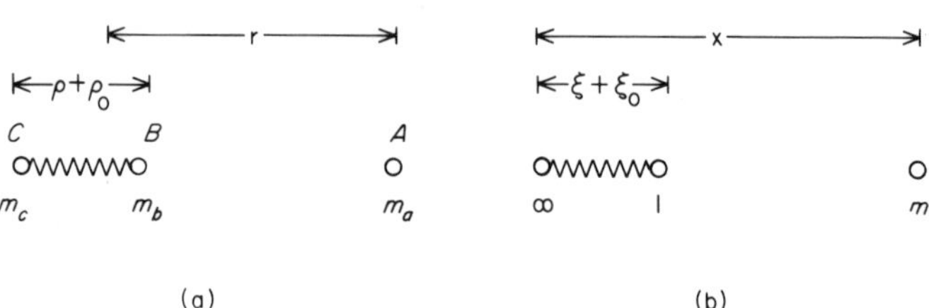

FIG. 7.1. (a) The coordinate system; (b) the equivalent system in reduced coordinates.

where μ is the reduced mass of BC, the problem is equivalent to that in Fig. 7.1(b), with a dimensionless interaction potential

$$U(x, \xi) = (\hbar\omega/2m)\ V(r, e) = \exp[-\alpha(x - \xi)], \tag{7.37}$$

the two parameters m and α being given by

$$\begin{aligned} \alpha &= \lambda a d = a[m_c/(m_b + m_c)]\hbar^{\frac{1}{2}}(\mu k)^{-\frac{1}{4}}, \\ m &= m_a m_c / M m_b. \end{aligned} \tag{7.38}$$

If the energy is measured in units of $\hbar\omega$ the internal part of the part of the Hamiltonian reduces to

$$H_{\text{int}} = -\tfrac{1}{2}\frac{\mathrm{d}^2}{\mathrm{d}\xi^2} + \tfrac{1}{2}\xi^2, \tag{7.39}$$

with eigenstates, at energies $(n + \frac{1}{2})$, (Eyring *et al.* (1944)),

$$\phi_n(\xi) = \pi^{-\frac{1}{4}}(2^n n!)^{-\frac{1}{2}} H_n(\xi) \exp(-\tfrac{1}{2}\xi^2), \tag{7.40}$$

where $H_n(\xi)$ denotes the Hermite polynomial (Abramowitz and Stegun, 1965).

Finally in this scaled form

$$k_n^2 = 2m(E - n - \tfrac{1}{2}). \tag{7.42}$$

The steps in the calculation are first to compute the distortion and interaction potentials $U_{nn'}(x)$; secondly to obtain the distorted waves $\psi_n^{(0)}(r)$ defined by (7.29) and (7.32); and finally to compute the integral in (7.33).

In the present case it follows from (7.37) and (7.40) that the potential matrix elements $U_{nn'}(x)$ may be written

$$U_{nn'}(x) = Y_{nn'}\,\mathrm{e}^{-\alpha x}, \tag{7.43}$$

where

$$\begin{aligned} Y_{nn'} &= \langle \phi_n | \mathrm{e}^{\alpha\xi} | \phi_{n'} \rangle \\ &= \langle \phi_n | 1 + \alpha\xi + \tfrac{1}{2}\alpha^2\xi^2 + \ldots 1\phi_{n'} \rangle. \\ &\simeq 1 & n' &= n \\ &\simeq \alpha[\tfrac{1}{2}(n + \tfrac{1}{2} \pm \tfrac{1}{2})]^{\frac{1}{2}}, & n' &= n \pm 1 \\ &\simeq \tfrac{1}{4}\alpha^2[(n + 1 \pm 1)(n \pm 1)]^{\frac{1}{2}}, & n' &= n \pm 2, \end{aligned} \tag{7.44}$$

the first non-vanishing terms in $Y_{nn'}$ having been evaluated with the help of standard recussion relations for $H_n(\xi)$. The validity of this approximation is based on the realistic magnitude of α; $\alpha < 0{\cdot}3$ in the cases examined by Secrest and Johnson (1966). The complications due to inclusion of higher terms are discussed below.

In the present approximation the equations for the distorted waves $\psi_n^{(0)}(x)$,

$$\left[\frac{\mathrm{d}^2}{\mathrm{d}x^2} + k_n^2 - U_{nn}(x)\right]\psi_n^{(0)}(x) = \left[\frac{\mathrm{d}^2}{\mathrm{d}x^2} + k_n^2 - \mathrm{e}^{-\alpha x}\right]\psi_n^{(0)}(x) = 0, \tag{7.45}$$

differ only in the value of k_n^2.

These equations may be cast furthermore by the substitutions

$$y = (2/\alpha)\exp(-\tfrac{1}{2}\alpha x) \tag{7.46}$$

$$q = 2k_n/\alpha$$

into the form

$$\left[\frac{\mathrm{d}^2}{\mathrm{d}y^2} + \frac{1}{y}\frac{\mathrm{d}}{\mathrm{d}y} + \frac{q^2}{y^2} - 1\right]\psi_n(y) = 0. \tag{7.47}$$

The solution of (7.47), which is regular at $y = \infty$ (corresponding according to (7.46) to $x = -\infty$) is the modified Bessel function (Abramowitz and Stegun (1965)).

$$\psi_n^{(0)}(y) = CK_{iq}(y) = \frac{i\pi C}{2}\left[\frac{I_{iq}(y) - I_{-iq}(y)}{\sinh \pi q}\right] \tag{7.48}$$

It may be verified further from the behaviour of $I_{\pm iq}(y)$ at small y (hence large x) that the choice

$$C = (q \sinh \pi q/\pi)^{\frac{1}{2}}, \tag{7.49}$$

ensures the correct normalization,

$$\psi_n^{(0)}(x) \overset{x\to\infty}{\sim} \sin(k_n x + \delta_n) \tag{7.50}$$
$$\delta_n = q \ln \alpha + \arg \Gamma(iq).$$

Finally the integral formula (Jackson and Mott (1932))

$$I = \int_0^\infty K_{iq}(y)\, K_{iq'}(y)\, y\, \mathrm{d}y = \frac{\pi^2(q^2 - q'^2)}{4(\cosh \pi q - \cosh \pi q')} \tag{7.51}$$

together with equations (7.33), (7.48), (7.49) and (7.51), and identity derived from (7.42) and (7.46)

$$(q^2 - q'^2) = 8m(n' - n)/\alpha^2, \tag{7.52}$$

yield the overall transition amplitude

$$T'_{nn'} = \frac{4\pi m(n' - n)}{\alpha^2} Y_{nn'} \frac{[\sinh \pi q_n \sinh \pi q_{n'}]^{\frac{1}{2}}}{[\cosh \pi q_n - \cosh \pi q'_n]} \tag{7.53}$$

with $Y_{nn'}$ given by (7.44). The $n = 0 \to 1$ transition probability therefore becomes

$$P_{01} = |T'_{01}|^2 = \frac{8\pi^2 m^2 \sinh \pi q_0 \sinh \pi q_1}{\alpha^2[\cosh \pi q_0 - \cosh \pi q_1]^2} \tag{7.54}$$

Now the following inequality obtains under normal molecular scattering conditions, $1 \ll \pi(q_0 - q_1) \ll \pi q_0, \pi q_1$; hence

$$P_{01} \simeq \frac{8\pi^2 m^2}{\alpha^2} \exp[-\pi(q_0 - q_1)] \simeq \frac{8\pi^2 m^2}{\alpha^2} \exp\left[\frac{\pi(q_0^2 - q_1^2)}{2\bar{q}}\right], \tag{7.55}$$

where $\bar{q} = \frac{1}{2}(q_0 + q_1)$. Expressed in terms of real quantities this means, in view of (7.36), (7.38), (7.42) and (7.52) that

$$P_{01} \simeq (8\pi^2 m^2/\alpha^2) \exp[-(v^*/v)],$$
$$v^* = 2\pi(E_1 - E_0)/\alpha\hbar. \tag{7.56}$$

The presence of the inverse velocity implies by the arguments of Section 1.3 that on thermal averaging

$$P_{01} \propto \exp[-(T^*/T)^{\frac{1}{3}}]. \tag{7.57}$$

The Schwartz *et al.* (1951) theory of vibrational relaxation expounded by Herzfeld and Litovitz (1959) is based on this result. Deviations from exact exponential behaviour of the distortion potentials, due either to the attractive nature of real intermolecular potentials or to the inclusion of centrifugal terms, are taken into account in this theory by a correction to the collision energy E. This is based on the assumption that the dominant contribution to the integral in (7.51) comes from the region of the classical turning point, and hence that the energy may be simply reduced by the value of the correction potential at this point.

Equation (7.53) also raises some points of theoretical interest. In the first place it is quite different from the Born approximation result (7.25) for an exponential interaction function, and this discrepancy may be verified to persist at the highest energies. This is a measure of the importance of the distortion terms $U_{nn}(x)$. Equation (7.50) shows in fact that the phase correction δ will never vanish in this model; hence the Born approximation cannot be valid.

For another view of the significance of the distortion potential, we might examine the form of the solution when $Y_{nn'}$ in (7.44) are evaluated exactly, so that for example

$$Y_{00} = \exp(\tfrac{1}{4}\alpha^2), \qquad Y_{11} = (1 + \tfrac{1}{2}\alpha^2)\exp(\tfrac{1}{4}\alpha^2). \tag{7.58}$$

$U_{00}(x)$ and $U_{11}(x)$ are then no longer identical, and the integral I in (7.51) must be replaced by (Mies, 1964),

$$\int_0^\infty K_{iq}(y)\, K_{iq'}(\gamma y)\, y\, \mathrm{d}y = I\gamma^{-iq} F\left(1 + \frac{i}{2}(q + q'), 1 + \frac{i}{2}(q - q'), 2, 1 - \gamma^{-2}\right), \tag{7.59}$$

where

$$\gamma = (Y_{n'n'}/Y_{nn})^{\frac{1}{2}},$$

and $F(a, b, c, z)$ is the hypergeometric function (Abramowitz and Stegun, 1965). The resulting correction to the transition probability is actually quite small for realistic cases in the harmonic model, but the discrepancy between $Y_{n'n'}$ and Y_{nn} is substantially enhanced by the introduction of vibrational anharmonicity. It is then necessary to introduce an anharmonic factor in (7.54) which may be as small as 0·03 in extreme cases (Mies, 1964).

A final comment refers to the accuracy of (7.53). Comparison with exact numerical results for the same model, except that $Y_{nn'}$ are evaluated exactly (Secrest and Johnson, 1966), shows that $T_{nn'}$ in (7.53) is in error, even when $P_{nn'}$ is small, by a factor (typically in the range 1·5 for $m \simeq 0{\cdot}1$ to 10 for $m \simeq 1{\cdot}25$) which depends strongly on m, and weakly on α, but is largely independent of collision energy, provided $P_{nn'} < 0{\cdot}1$. This discrepancy is substantially reduced by taking the distorted wave calculation to higher order

(Roberts, 1971a, 1971b) or by treating the problem in the impulse approximation (see Section 8.4). Another type of breakdown of the distorted wave method when $P_{nn'} > 0{\cdot}1$ may be partially remedied in the exponential approximation (Levine and Balint-Kurti, 1970, see also Section 7.3).

As a second example of the distorted wave method which will be useful later, we consider the coupling induced by a constant interaction term V_{12}, between wave functions $\psi_1(x)$ and $\psi_2(x)$ governed by homogeneous fields F_1 and F_2,

$$V_i(x) = E - F_i(x - a_i). \tag{7.60}$$

The origins of x and of the energy are conveniently taken at the intersection between $V_1(x)$ and $V_2(x)$, in which case

$$a_i = -(E/F_i) \tag{7.61}$$

It is assumed furthermore that $F_1 > F_2 > 0$.

In this model $\psi_1(x)$ and $\psi_2(x)$ satisfy equations of the Airy form

$$\left[\frac{d^2}{dx^2} + \beta_i^3(x - a_i)\right]\psi_i(x) \tag{7.62}$$

where

$$\beta_i = (2mF_i/\hbar^2)^{\frac{1}{3}}, \tag{7.63}$$

the proper solutions of which, chosen to vanish as $x \to -\infty$, are (Abramowitz and Stegun, 1965)

$$\psi_i^0(x) = C_i \operatorname{Ai}[-\beta_i(x - a_i)] \tag{7.64}$$

and the following properties of the Airy function may be noted

$$\operatorname{Ai}(-z) = \frac{1}{2\pi}\int_{-\infty}^{\infty} \exp\left(\frac{iu^3}{3} + izu\right) du. \tag{7.64a}$$

$$\overset{z \gg 1}{\sim} \pi^{-\frac{1}{2}} z^{-\frac{1}{4}} \sin\left(\tfrac{2}{3}z^{\frac{3}{2}} + \pi/4\right) \tag{7.64b}$$

$$\overset{z \ll -1}{\sim} \tfrac{1}{2}\pi^{-\frac{1}{2}}|z|^{-\frac{1}{4}} \exp\left(-\tfrac{2}{3}|z|^{\frac{3}{2}}\right). \tag{7.64c}$$

One small problem concerns the normalization because $\psi_i^{(0)}(x)$ in (7.64) cannot be normalized by (7.32) in the usual way since $V_i(x)$ does not vanish at infinity. However, it is the product $k_i^{-\frac{1}{2}}\psi_i^0(x)$ which occurs in the transition matrix element (7.33), and the choice of coefficient

$$C_i = \pi^{-\frac{1}{2}}\beta_i^{-\frac{1}{2}} \tag{7.65}$$

ensures, as may be verified with the help of (7.54b), that

$$\psi_i^0(x) \overset{x \to \infty}{\sim} [k_i(x)]^{-\frac{1}{2}} \sin\left[\tfrac{2}{3}\beta_i(x - a_i)\right]^{\frac{3}{2}} + \pi/4] \tag{7.66}$$

since in this problem

$$k_i^2(x) = \beta_i^3(x - a_i) \tag{7.67}$$

Now under present assumptions, $U_{12} = 2mV_{12}/\hbar^2 =$ constant; hence T_{12} in (7.33) may be derived from (7.63), (7.64a) and (7.65) in the form

$$T'_{12} = \frac{(\beta_1\beta_2)^{-\frac{1}{2}}U_{12}}{2\pi}\int_{-\infty}^{\infty}\int_{-\infty}^{\infty}\int_{-\infty}^{\infty}\exp\left[\frac{i}{3}(u^3 + v^3) + i(\beta_1 a_1 u + \beta_2 a_2 v) - i(\beta_1 u + \beta_2 v)x\right]\mathrm{d}x\,\mathrm{d}u\,\mathrm{d}v, \tag{7.68}$$

the integral over x having been extended to the range $-\infty < x < \infty$, over which $\mathrm{Ai}(x)$ is defined. However according to Dirac (1958)

$$\int_{-\infty}^{\infty}\exp\left[i(\beta_1 u + \beta_2 v)x\right]\mathrm{d}x = 2\pi\delta(\beta_1 u + \beta_2 v) \tag{7.69}$$

so that after integration with respect to u,

$$\begin{aligned} T'_{12} &= \beta_1^{-1}(\beta_1\beta_2)^{-\frac{1}{2}}U_{12}\int_{-\infty}^{\infty}\left[\frac{i}{3\beta_1^3}(\beta_1^3 - \beta_2^3)v^3 - i(a_0 - a_1)\beta_2 v\right]\mathrm{d}v \\ &= \frac{2\pi V_{12}(F_1F_2)^{\frac{1}{2}}}{\varepsilon_0(F_1 - F_2)}\mathrm{Ai}(-E/\varepsilon_0) \end{aligned} \tag{7.70}$$

where after substitution for a_i and β_i from (7.61) and (7.62),

$$\begin{aligned} (E/\varepsilon_0) &= (a_1 - a_2)\beta_1\beta_2/(\beta_1^3 - \beta_2^3) \\ \varepsilon_0 &= \left[\hbar^2F_1^2F_2^2/2m(F_1 - F_2)^2\right]^{\frac{1}{3}}. \end{aligned} \tag{7.71}$$

The same result is obtained with $|F_1F_2|^{\frac{1}{2}}$ in place of $(F_1F_2)^{\frac{1}{2}}$ if one of the forces F_i is negative.

Equation (7.70) reduces, by virtue of (7.64b) and (7.64c) for large positive and negative energies to the forms

$$T'_{12} \overset{E \gg \varepsilon_0}{\sim} \frac{2(2\pi)^{\frac{1}{2}}V_{12}}{[\hbar v(F_1 - F_2)]^{\frac{1}{2}}}\sin\left(\tfrac{2}{3}(E/\varepsilon_0)^{\frac{3}{2}} + \pi/4\right) \tag{7.72a}$$

$$\overset{E \ll -\varepsilon_0}{\sim} \frac{(2\pi)^{\frac{1}{2}}V_{12}}{[\hbar v(F_1 - F_2)]^{\frac{1}{2}}}\exp\left(-\tfrac{2}{3}|E/\varepsilon_0|^{\frac{3}{2}}\right). \tag{7.72b}$$

where v is the velocity at the crossing point.

$$v = (2E/m)^{\frac{1}{2}}. \tag{7.73}$$

The high energy form (7.72a) may also be derived in a general case (Landau and Lifshitz, 1965), by approximating $\psi_i^{(0)}(x)$ in the semi-classical form (4.27), such that

$$k_i^{-\frac{1}{2}}\psi_i^{(0)}(x) \simeq [k_i(x)]^{-\frac{1}{2}} \sin\left[\int_{a_i}^{x} k_i(x)\,\mathrm{d}x + \pi/4\right]. \tag{7.74}$$

In this case the product of sine terms which appears in the integrand in (7.33) may be expressed as a combination of cosines of the sum and difference of the two arguments, only the second of which, due to its relatively slow variation, is important. Hence

$$T'_{12} \simeq \frac{2m}{\hbar^2}\int_{-\infty}^{\infty} [k_1(x)k_2(x)]^{-\frac{1}{2}}\, V_{12}(x) \cos\left[\xi_{12} + \int_0^x (k_1(x') - k_2(x')]\,\mathrm{d}x'\right]\mathrm{d}x \tag{7.75}$$

where

$$\xi_{12} = \int_{a_1}^{0} k_1(x)\,\mathrm{d}x - \int_{a_2}^{0} k_2(x)\,\mathrm{d}x.$$

It is now argued that T'_{12} is dominated by the region around the crossing point, $x = 0$, ((7.74) being valid here if $E \gg E_0$), because the oscillation frequency of the integrand increases rapidly outside this region. Provided this range is small (see (7.80) below), the following approximations are permissible

$$V_{12}(x) \simeq V_{12}(0).$$

$$k_1(x) \simeq k_2(x) \simeq k = mv/\hbar. \tag{7.76}$$

$$k_1(x) - k_2(x) \simeq [k_1^2(x) - k_2^2(x)]/2k = (F_1 - F_2)x/\hbar v$$

in which case, the integral in (7.75) takes the standard form, (Dwight, 1961)

$$\int_{-\infty}^{\infty} \cos(\alpha + \tfrac{1}{2}\gamma x^2)\,\mathrm{d}x = \left(\frac{2\pi}{\gamma}\right)^{\frac{1}{2}} \cos(\alpha + \pi/4) \tag{7.77}$$

This means that†

$$T'_{12} = \frac{2(2\pi)^{\frac{1}{2}} V_{12}(0)}{[\hbar v(F_1 - F_2)]^{\frac{1}{2}}} \sin\left[\int_{a_2}^{0} k_2(x)\,\mathrm{d}x - \int_{a_1}^{0} k_1(x)\,\mathrm{d}x + \pi/4\right] \tag{7.78}$$

† If the curve $V_2(x)$ has negative slope, it may be verified on substituting

$$k_2^{-\frac{1}{2}}\psi_2^0(x) = [k_2(x)]^{-\frac{1}{2}} \sin\left[\int_x^{a_2} k_2(x)\mathrm{d}x + \pi/4\right]$$

(see Appendix C.2) in (7.33), that the phase term in (7.78) must be replaced by

$$\int_{a_1}^{0} k_1(x)\,\mathrm{d}x + \int_0^{a_2} k_2(x)\,\mathrm{d}x = \tfrac{2}{3}(E/\varepsilon_0)^{\frac{3}{2}}.$$

a result is in exact agreement with (7.72a) because in the present model

$$\int_{a_2}^{0} k_2(x)\,\mathrm{d}x - \int_{a_1}^{0} k_1(x)\,\mathrm{d}x = [2(2m)^{\frac{1}{2}}/3\hbar]\,[E^{\frac{3}{2}}/F_2 - E^{\frac{3}{2}}/F_1] = \tfrac{2}{3}(E/\varepsilon_0)^{\frac{3}{2}}. \tag{7.79}$$

The validity of (7.78) depends on the accuracy of (7.76) over the range

$$\gamma^{-\frac{1}{2}} = [\hbar v/(F_1 - F_2)]^{\frac{1}{2}}, \tag{7.80}$$

of slow variation in the integrand in (7.71). Clearly since $\gamma^{-1} \to \infty$ as $v \to \infty$ a breakdown at high energies must be expected for any realistic form of $V_{12}(x)$.

A model for which the proper correction terms are readily assessed is that for which $V_1(x)$ and $V_2(x)$ are given by (7.60), but

$$V_{12}(x) = V_{12}^{0}\,\exp(-\alpha x). \tag{7.81}$$

An extension of the arguments leading to (7.70) shows (Child, 1972) that

$$T'_{12} = \frac{2\pi V_{12}^{0}|F_1F_2|^{\frac{1}{2}}}{\varepsilon_0(F_1 - F_2)}\exp\left[\frac{\hbar^2\alpha^3(F_1 + F_2)}{6m(F_1 - F_2)^2}\right]\mathrm{Ai}\left[-\frac{E - \varepsilon_1}{\varepsilon_0}\right],$$

$$\underset{E \ll -\varepsilon_0}{\sim} \frac{2(2\pi)^{\frac{1}{2}}V_{12}^{0}}{[\hbar v(F_1 - F_2)]^{\frac{1}{2}}}\exp\left[\frac{\hbar^2\alpha^3(F_1 + F_2)}{6m(F_1 - F_2)^2}\right]\sin\left[\frac{2}{3}\left(\frac{E}{\varepsilon_0}\right)^{\frac{3}{2}} + \pi/4 - \frac{\alpha^2\hbar v}{(F_1 - F_2)} + \ldots\right] \tag{7.82}$$

where

$$\varepsilon_1 = [\hbar^2\alpha^2\,F_1F_2/2m(F_1 - F_2)^2].$$

The exponential term is relatively unimportant since the exponent is likely to be small in practical cases. The first correction to (7.72a) is therefore the term $-\alpha^2\hbar v/(F_1 - F_2)$ which depends as expected on the ratio of the stationary phase parameter $[\hbar v/(F_1 - F_2)]^{\frac{1}{2}}$ to the range, α^{-1}, of $V_{12}(x)$ in (7.81).

In conclusion, we note that the presence of the sine terms in (7.72a), (7.78) and (7.82) taken in conjunction with the phase shift term $\exp i(\eta_1 + \eta_2)$ in (7.33) will lead, on analysis by the methods of Section 5.1, to a number of interfering branches in the inelastic differential cross-section (6.61). Olson and Smith (1971) have shown how the resulting oscillatory cross-section may be analysed to determine the potential terms $V_{11}(r)$, $V_{22}(r)$ and $V_{12}(r)$.

7.3 The Exponential Approximation

Partly to overcome the difficulty inherent in the first order Born and Distorted wave approximations, that the unitarity of the S matrix is lost as the coupling strength increases, and partly to shed light on the inelastic collision process in the near classical limit, Levine (1971) has suggested a direct approach to **S**, in the (necessarily unitary) exponential form (6.51)

$$\mathbf{S} = \exp(i\mathbf{A}) = \mathbf{T} + i\mathbf{A} - \tfrac{1}{2}\mathbf{A}^2 - i/6\mathbf{A}^3 \ldots \tag{7.83}$$

where **A** is a hermitian matrix. The conditions on the validity of the method are either that the elements of **A** are small, in which case (7.83) reduces to the distorted wave expression (7.33), or that the de Broglie wavelengths in the interaction region are small compared with the range of the interaction potential. The latter corresponds to near classical conditions. The derivation below, based on the amplitude density method of Calogero (1967), Chapter 19, follows that of Levine (1971).

The first step is to define appropriate solutions of the distortion equations

$$\left[\frac{d^2}{dr^2} + k_i^2 - U_{ii}(r)\right]\psi_i(r) = 0, \tag{7.84}$$

modified from those defined in appendix A by the inclusion of a factor $k_i^{-\frac{1}{2}}$. Three such solutions are employed, namely

$$\psi_i^{(0)}(r) \overset{r\to\infty}{\sim} k_i^{-\frac{1}{2}} \sin(k_i r - l\pi/2 + \eta_i), \qquad \psi_i^{(0)}(0) = 0,$$

$$\psi_i^{(1)}(r) \overset{r\to\infty}{\sim} k_i^{-\frac{1}{2}} \cos(k_i r - l\pi/2 + \eta_i),$$

$$\begin{aligned} f_i^{(\pm)}(r) &= [\psi_i^{(1)}(r) \pm i\psi_i^{(0)}(r)] \\ &\overset{r\to\infty}{\sim} k_i^{-\frac{1}{2}} \exp[\pm i(kr - l\pi/2 + \eta_i)]. \end{aligned} \tag{7.85}$$

Solutions to the full set of coupled equations, rearranged in the form

$$\left[\frac{d^2}{dr^2} + k_i^2 - U_{ii}(r)\right]\psi_i(r) = \sum_j U'_{ij}(r)\psi_j(r) \tag{7.86}$$

such that

$$U'_{ij}(r) = (1 - \delta_{ij})U_{ij}(r) \tag{7.87}$$

are then written

$$\psi_i(r) = i[-f_i^{(+)}(r)\xi_i^{(+)}(r) + f_i^{(-)}(r)\xi_i^{(-)}(r)], \tag{7.88}$$

with the values of the amplitude density functions $\xi_i^{(\pm)}(r)$ at the origin chosen so that the divergent parts of $f_i^{(\pm)}(r)$ in (7.88) cancel out; thus

$$\xi_i^{(\pm)}(0) = a. \tag{7.89}$$

The subsequent analysis is most easily followed in matrix form, so that with all terms replaced by diagonal matrices, (7.88) becomes

$$\boldsymbol{\psi}(r) = i[-\mathbf{f}^{(+)}(r)\xi^{(+)}(r) + \mathbf{f}^{(-)}(r)\xi^{(-)}(r)]. \tag{7.90}$$

It follows by comparison with the definition of the S matrix (Eqn (6.34)) that

$$\begin{aligned}\mathbf{S} &= \exp(i\boldsymbol{\delta})\mathbf{S}' \exp(i\boldsymbol{\delta})\\ \mathbf{S}' &= \lim_{r\to\infty} \xi^{(+)}(r)[\xi^{(-)}(r)]^{-1},\end{aligned} \tag{7.91}$$

where the elements of the diagonal matrix $\exp(i\boldsymbol{\delta})$ are defined by (7.85). $\mathbf{S}$ here is the full matrix for the problem, made up of the distortion parts $\exp(i\boldsymbol{\delta})$ and an interaction part $\mathbf{S}'$ due to $\mathbf{U}'$ in (7.86). (7.91) suggests that if $\xi^{(\pm)}(r)$ are taken in exponential form

$$\xi^{(\pm)}(r) = \exp[\pm i\mathbf{A}^{(\pm)}(r)] \tag{7.92}$$

then, provided that $\mathbf{A}^{(\pm)}(r)$ commute, $\mathbf{S}'$ will also have the desired exponential form

$$\mathbf{S}' = \exp[\pm i\mathbf{A}]$$

where

$$\mathbf{A} = \lim_{r\to\infty}[\mathbf{A}^{(+)}(r) + \mathbf{A}^{(-)}(r)] \tag{7.93}$$

It is convenient also to define a matrix $\mathbf{S}'(r)$ at all r, by analogy with (7.91);

$$\begin{aligned}\mathbf{S}'(r) = \xi^{(+)}(r)[\xi^{(-)}(r)]^{-1} &= \exp[i\mathbf{A}(r)],\\ &\overset{r\to 0}{\sim} \mathbf{I}\\ &\overset{r\to\infty}{\sim} \mathbf{S}'\end{aligned} \tag{7.94}$$

the limiting values of which are implied by (7.89) and (7.93).

The necessary equations for the unknowns $\xi^{(+)}(r)$ are now conveniently obtained from the integral form of (7.86),

$$\boldsymbol{\psi}(r) = \boldsymbol{\psi}^{(0)}(r) + \int_0^r [\boldsymbol{\psi}^{(0)}(r)\boldsymbol{\psi}^{(1)}(r') - \boldsymbol{\psi}^{(1)}(r)\boldsymbol{\psi}^{(0)}(r')]\mathbf{U}'(r')\boldsymbol{\psi}(r')\,\mathrm{d}r', \tag{7.95}$$

chosen to ensure that $\boldsymbol{\psi}(0) = 0$, and that asymptotically $\boldsymbol{\psi}(r)$ contains equal incoming and outgoing parts as required by (7.90) and (7.91)†. This implies on comparing coefficients of $\mathbf{f}^{(+)}(r)$ in (7.90) that

$$\xi^{(\pm)}(k) = \frac{1}{2}\left[\mathbf{I} + \int_0^r f^{(\pm)}(r')\,\mathbf{U}'(r')\,\boldsymbol{\psi}(\mathbf{r}')\,\mathrm{d}r'\right]; \tag{7.96}$$

† The term $2[\psi^{(0)}(r)\psi^{(1)}(r') - \psi^{(1)}(r)\psi^{(0)}(r')]$ may be recognized as the appropriate Green's function (see Appendix B).

hence, since by (7.88) $\psi(r)$ is real, $\xi^{(\pm)}(r)$ are complex conjugates and the elements of $\mathbf{A}$ in (7.93) are real. The necessary equations for $\mathbf{A}^{(\pm)}(r)$ in (7.92) are now obtained by differentiating (7.96),

$$\frac{\mathrm{d}\xi^{(\pm)}}{\mathrm{d}r} = \frac{i}{2}\,\mathbf{f}^{(\mp)}(r)\,\mathbf{U}'(r)\,[-\,\mathbf{f}^{(+)}(\mathbf{r})\,\xi^{(+)}(r) + \mathbf{f}^{(-)}(r)\,\xi^{(-)}(r)]. \tag{7.97}$$

At this stage all equations are exact.

The first approximation is introduced with the help of (7.92) by setting

$$\frac{\mathrm{d}\xi^{(\pm)}}{\mathrm{d}r} = \pm\, i\left(\frac{\mathrm{d}\mathbf{A}^{(\pm)}}{\mathrm{d}r}\right)\xi^{(\pm)}(r) \tag{7.98}$$

on the assumption that commutators between the matrices $\mathbf{A}$ and $(\mathrm{d}\mathbf{A}/\mathrm{d}r)$ disappear,

$$\left[\mathbf{A}, \frac{\mathrm{d}\mathbf{A}}{\mathrm{d}r}\right] = 0. \tag{7.99}$$

Levine and Johnson (1970) show by comparison between the classical and quantum mechanical equations of motion that the inclusion of the proper commutators in (7.98) would introduce corrections to the distortion potentials $U_{ii}(r)$ due to the off-diagonal term $U_{ij}(r)$. It was in order to minimize such distortion effects that the $U_{ii}(r)$ were included in (7.84).

To the extent therefore that (7.93) and (7.98) are valid, it follows from (7.93), (7.94) and (7.97) that

$$\begin{aligned}\frac{\mathrm{d}\mathbf{A}}{\mathrm{d}r} &= \frac{\mathrm{d}}{\mathrm{d}r}(\mathbf{A}^{(+)}(r) + \mathbf{A}^{(-)}(r))\\ &= \tfrac{1}{2}[\mathbf{f}^{(+)}(r) - \mathbf{f}^{(-)}(r)]\,\mathbf{U}'(r)\,[\mathbf{f}^{(+)}(r) - \mathbf{f}^{(-)}(r)]\\ &\quad - \tfrac{1}{2}\{\mathbf{f}^{(-)}(r)\,\mathbf{U}'(r)\,\mathbf{f}^{(-)}(r)\,[\mathbf{I} - \mathbf{S}'^{-1}(r)] + \mathbf{f}^{(+)}(r)\,\mathbf{U}'(r)\,\mathbf{f}^{(+)}(r)\,[\mathbf{I} - \mathbf{S}'(r)]\}.\end{aligned} \tag{7.100}$$

The second approximation is to neglect the final terms in brackets { } in (7.100), and hence to obtain after substitution from (7.85)

$$\mathbf{A} = \lim_{r\to\infty}\mathbf{A}(r) = -2\int_0^\infty \boldsymbol{\psi}^{(0)}(r)\,\mathbf{U}'(r)\,\boldsymbol{\psi}^{(0)}(r)\,\mathrm{d}r. \tag{7.101}$$

In the desired exponential form (7.93), $\mathbf{S}'$ therefore depends on the very integrals which determine the transition matrix $\mathbf{T}'$ in (7.33) at the distorted wave level†. An improved approximation is therefore obtained simply by

† Recall that by the present definition (7.85), $\psi_i^{(0)}(r)$ contains an additional factor $k_i^{-\frac{1}{2}}$ over the $\psi_i^{(0)}(r)$ in (7.33).

exponentiating the distorted wave results.

$$\mathbf{S}' = \exp[i\mathbf{A}] = \exp\left[-2i\int_0^\infty \boldsymbol{\psi}^{(0)}(r)\,\mathbf{U}'(r)\,\boldsymbol{\psi}^{(0)}(r)\,dr\right]. \qquad (7.102)$$

Equally $\boldsymbol{\xi}^{(\pm)}(r)$ in (7.90) may be obtained by exponentiating $\mathbf{A}^{(\pm)}(r)$ determined by the precursors to (7.100).

The accuracy of (7.102) clearly rests on the choice of an adequate distortion potential, in order that (7.99) should be justified, and also on the dominance of the leading term in (7.100). The neglect of other terms has been justified by Levine (1971) by means of an expansion in powers of $\hbar$. It may also be supported by other arguments. At an elementary level, (7.102) is clearly valid if $\mathbf{S}(r) \simeq \mathbf{S}^{-1}(r) \sim \mathbf{I}$ for all r. This implies that $\mathbf{U}'(r)$ acts as a small perturbation and that $\mathbf{A}$ in (7.102) is small; hence by expansion to one term

$$\mathbf{S}' = \mathbf{I} - 2i\int_0^\infty \boldsymbol{\psi}^{(0)}(r)\,\mathbf{U}'(r)\,\boldsymbol{\psi}^{(0)}(r)\,dr. \qquad (7.103)$$

This is the distorted wave result (7.33), since by (6.44), $\mathbf{S}' = \mathbf{I} - i\mathbf{T}'$. At a higher level we know by (7.94) that $\mathbf{A}(0) = 0$ and $\mathbf{S}(0) = \mathbf{S}^{-1}(0) = \mathbf{I}$ which means (despite the divergence of $\mathbf{f}^{(+)}(r)$ at the origin) that the neglected term will vanish. This situation must also obtain over the entire non-classical region because the magnitude of $\mathbf{A}(r)$ is governed by the (exponentially small) regular solutions $\boldsymbol{\psi}^{(0)}(r)$. On crossing into the classical regions on the other hand, although $\mathbf{A}(r)$ may increase to the extent that $\mathbf{S}(r) \neq \mathbf{I}$, rapid oscillations at the *combination* frequency in the product terms $\mathbf{f}^{(-)}(r)\,\mathbf{U}'(r)\mathbf{f}^{(-)}(r)$ and $\mathbf{f}^{(+)}\mathbf{U}'(r)\mathbf{f}^{(+)}(r)$ can lead on integration to wholesale cancellation in the rejected terms. Nevertheless fluctuations at the difference frequency in the products $\mathbf{f}^{(+)}(r)\,\mathbf{U}'(r)\mathbf{f}^{(+)}(r)$ contained in the leading term of (7.100) may be sufficiently slow that $\mathbf{A}$ is quite large. The conclusion is that (7.102) may be valid at a non-perturbation level if the de Broglie wavelengths in the distortion channels are short compared with the range of the interaction potential $\mathbf{U}'(r)$. Levine and Balint-Kurti (1970), Balint-Kurti and Levine (1970) and Levine and Johnson (1971) certainly find by comparison with exact results for model problems that the exponential form (7.102) represents a substantial improvement on the distorted wave result (7.33). Some results obtained by Balint-Kurti and Levine (1970) are included in table (7.1). Cross (1967, 1968) has also found the Born (rather than Distorted Wave) modification of (7.102) to be useful in the theory of low angle inelastic scattering.

Levine (1971) further shows that the utility of (7.102) may be increased in practice by an alternative expansion of $\mathbf{S}$ in preference to the infinite series (7.83). This is based on the fact that the $N \times N$ hermitian matrix $\mathbf{A}$ it has N real eigenvalues, α_n, which are governed, since according to (7.87) $\mathbf{A}$ in (7.102)

TABLE 7.1. Vibrational transition probabilities

E	m	α	$i \to j$	P(distorted Wave)	P(Exponential)	P (exact)
5·4183	1/13	0·1287	$0 \to 1$	0·276	0·228	0·230
			$1 \to 2$	0·289	0·225	0·224
			$0 \to 2$	2×10^{-5}	0·018	0·018
8·4183	1/13	0·1287	$0 \to 1$	0·815	0·404	0·409
			$1 \to 2$	1·22	0·348	0·343
			$0 \to 2$	2×10^{-4}	0·141	0·150
4·0	2/3	0·3	$0 \to 1$	0·164	0·152	0·108
			$1 \to 2$	0·064	0·060	0·042
			$0 \to 2$	1×10^{-7}	0·002	0·001
8·0	2/3	0·3	$0 \to 1$	1·82	0·381	0·434
			$1 \to 2$	2·48	0·109	0·220
			$0 \to 2$	3×10^{-5}	0·351	0·291

m and α are defined by (7.38) and E is in units of $\hbar\omega$. Exact results are taken from Secrest and Johnson (1966).

is traceless, by $\sum_{n=1}^{N} \alpha_n = 0$; hence $\mathbf{S}'$ may be represented as a polynomial in $\mathbf{A}$ of degree $N - 1$,

$$\mathbf{S}' = \sum_{n=0}^{N-1} a_n \mathbf{A}^n. \tag{7.104}$$

The validity of this expansion which is based on the Cayley–Hamilton theorem, is readily verified by replacing $\mathbf{S}'$ and $\mathbf{A}$ by the unitary transforms $\tilde{\mathbf{S}}'$ and $\tilde{\mathbf{A}}$ (which must also satisfy (7.104));

$$\begin{aligned} \mathbf{S}' &= \mathbf{X}\tilde{\mathbf{S}}'\mathbf{X}^\dagger \\ \mathbf{A} &= \mathbf{X}\tilde{\mathbf{A}}\mathbf{X}^\dagger \end{aligned} \tag{7.105}$$

with $\mathbf{X}$ chosen such that $\mathbf{A}$ is diagonal; by (7.83) $\mathbf{S}'$ must be diagonal also (see also (6.53)–(6,54)).

$$\tilde{\mathbf{A}} = \begin{pmatrix} \alpha_1 & 0 & \dots & 0 \\ 0 & \alpha_2 & \dots & 0 \\ \vdots & \vdots & & \vdots \\ 0 & 0 & \dots & \alpha_n \end{pmatrix}, \qquad \tilde{\mathbf{S}}' = \begin{pmatrix} \exp i\alpha_1, & 0 & \dots & 0 \\ 0 & \exp i\alpha_2, & \dots & 0 \\ \vdots & \vdots & & \\ 0 & 0 & & \exp i\alpha_n \end{pmatrix}. \tag{7.196}$$

The coefficients a_n in (7.104) must therefore be chosen to satisfy the N equations

$$\begin{pmatrix} 1 & \alpha_1 & \dots & \alpha_1^{N-1} \\ 1 & \alpha_2 & \dots & \alpha_2^{N-1} \\ \vdots & & & \\ 1 & \alpha_N & \dots & \alpha_N^{N-1} \end{pmatrix} \begin{pmatrix} a_0 \\ a_1 \\ \vdots \\ a_{N-1} \end{pmatrix} = \begin{pmatrix} \exp i\alpha_1 \\ \exp i\alpha_2 \\ \vdots \\ \exp i\alpha_N \end{pmatrix} \tag{7.107}$$

Thus for a 2×2 problem, with $\alpha_1 = -\alpha_2 = \alpha$

$$\mathbf{S}' = \cos\alpha\,\mathbf{I} + i(\sin\alpha/\alpha)\mathbf{A}, \tag{7.108}$$

while for a 3×3 problem with eigenvalues $0, \pm\alpha$,

$$\mathbf{S}' = \mathbf{I} + i(\sin\alpha/\alpha)\,\mathbf{A} + [(\cos\alpha - 1)/\alpha^2]\,\mathbf{A}^2. \tag{7.109}$$

(7.108) and (7.109) may also be treated as the first terms in the general expansion (7.104).

It is readily verified, since $\mathbf{A}$ is real, that $\mathbf{S}'$ in (7.108) retains its unitarity if

$$\mathbf{A}^2 = \alpha^2\mathbf{I}. \tag{7.110}$$

in which case (7.109) reduces to (7.108). Strictly this requires, consistent with the traceless character of $\mathbf{A}$, that N is even and that the eigenvalues of $\mathbf{A}$ should split into N/2 pairs $\pm\alpha$. For the use of (7.108) as an approximation Levine (1972) suggests that $\sum_m |\mathbf{A}_{nm}|^2$ should depend only weakly on n, a condition derived from the diagonal elements of (7.110). This appears to be true for many distorted wave calculations if the coupling is not too strong (Lester and Bernstin, 1967). In this limited coupling regime, the selection rules for, and relative magnitudes of, the off-diagonal elements of $\mathbf{S}'$ are exactly the same as those which apply to $\mathbf{A}$. In other words the transition probabilities predicted by first-order distorted wave theory are correct in relative magnitude but require renormalization by the factor $(\sin\alpha/\alpha)$ in order to re-establish the unitarity of $\mathbf{S}'$. This means, as is readily verified from (7.83), that within the validity of (7.110), the higher terms in the exponential contribute only to the renormalization factors $\cos\alpha$ and $(\sin\alpha/\alpha)$. As (7.110) begins to break down however, the final term in (7.109) also contributes to the off-diagonal elements of $\mathbf{S}$ and transitions which are first-order forbidden in the distorted wave theory become allowed, due to the exponentiation in (7.102). Finally, in the dominant coupling regime, the eigenvalues α_1 in (7.106) will typically be large and rapidly varying with the parameters (such as the energy) of the system. Hence, on average

$$\begin{aligned}\langle |S_{ij}|^2 \rangle &= \langle \sum_{kk'} \exp i(\alpha_k - \alpha_{k'})\, X^*_{ki} X_{kj} X_{k'j} X^*_{k'j} \rangle \\ &\simeq \sum_k |X_{ki}|^2 |X_{kj}|^2,\end{aligned} \tag{7.111}$$

since

$$\langle \exp i(\alpha_k - \alpha_{k'}) \rangle \simeq \delta_{kk'}. \tag{7.112}$$

The eigenvectors of $\mathbf{A}$ appropriate to the initial and final channels, represented by the relevant columns of $\mathbf{X}$ in (7.105), therefore determine the transition probabilities.

Levine (1971) finally offers another view of the exponential approximation. We note first that (7.102) remains valid for any division of the potential $U(r, \rho)$ into distortion and interaction parts

$$U(r, \rho) = U^0(r, \rho) + U'(r, \rho), \tag{7.113}$$

provided that the matrix $U^0(r)$ is diagonal on the basis of internal states $\phi_i(r; \rho)$. Secondly under certain conditions $U^0(r, \rho)$ may be taken to determine a trajectory $r(t)$ for the relative motion; specifically anticipating the discussion in Section 8.1, this will be true if the $\psi_i^{(0)}(r)$ in (7.102) are accurately represented in the important integration region ($r > a$, where a is the turning point for the trajectory $r(t)$) by the semi-classical forms

$$\psi_i^{(0)}(r) \simeq [k_i(r)]^{-\frac{1}{2}} \sin\left(\int_a^r k_i(r)\,\mathrm{d}r + \pi/4\right), \tag{7.114}$$

and also if for the internal states in question

$$k_i(r) - k_j(r) \simeq [k_i^2(r) - k_j^2(r)]/2k(r)$$
$$= -[E_i^{(0)} + V_{ii}^{(0)}(r) - (E_j^{(0)} + V_{jj}(r))]/\hbar v, \tag{7.115}$$

where $v = (k\hbar/m) = (\mathrm{d}r/\mathrm{d}t)$. (7.115) therefore requires that the energy difference should be small compared with the available energy over the important integration region.

Now we know also that if H_0 represents the zeroth order Hamiltonian, $H_{\text{int}} + V^0(r, \rho)$, then

$$H_0\phi_i(r; \rho) = [E_i^0 + V_{ii}(r)]\phi_i(r; \rho), \tag{7.116}$$

Hence in view of (7.83) and the closure relation

$$\sum_k |\phi_k\rangle\langle\phi_k| = 1 \tag{7.117}$$

the general element of $\mathbf{S}'$ in (7.102) may be written, with the help of (7.114)–(7.117)

$$S'_{ij} = \langle\phi_i|\exp i\Delta(\rho)|\phi_j\rangle \tag{7.118}$$

where

$$\Delta(\rho) = -4\int_0^\infty \psi_i^{(0)}(r)U'(r, \rho)\psi_j^0(r)\,\mathrm{d}r$$

$$\simeq -\frac{1}{\hbar v}\left\{\int_a^\infty \exp\left[i\int_a^\infty (k_i - k_j)\,\mathrm{d}r\right]V'(r, \rho)\,\mathrm{d}r + \text{complex conjugate}\right\}$$

$$\simeq -\frac{1}{\hbar}\int_{-\infty}^\infty \exp(iH_0t/\hbar)\,V'(t, \rho)\exp(-iH_0t/\hbar)\,\mathrm{d}t. \tag{7.119}$$

This result is obtained by expanding $\psi_i^{(0)}(r)$ in exponential form, and retaining only product terms oscillating at the difference frequency, in deriving the second line of (7.119). Also, with the time origin taken at the classical turning point, dr in the second line has been replaced by $v\,\mathrm{d}t$ and the complex conjugate removed by extending the integration range to $t = -\infty$.

Expressed in a more compact form (7.119) becomes

$$\Delta(\rho) = -\frac{1}{\hbar}\int_{-\infty}^{\infty} V'(t, \rho(t))\,\mathrm{d}t. \tag{7.120}$$

where $V'(t, \rho(t))$ and $\rho(t)$ are the Heisenberg representatives

$$\begin{aligned} \rho(t) &= \exp(iH_0 t/\hbar)\rho\exp(-iH_0 t/\hbar) \\ V'(t, \rho(t)) &= \exp(iH_0 t/\hbar)\, V'(t, \rho)\exp(-iH_0 t/\hbar). \end{aligned} \tag{7.121}$$

The operator $\Delta(\rho)$ is now displayed in terms of the action integral

$$\Delta(\rho) = \frac{1}{\hbar}\int_{-\infty}^{\infty} [L(t) - L_0(t)]\,\mathrm{d}t, \tag{7.122}$$

where $L(t)$ denotes the Lagrangian (see Appendix E), appropriate to motion in which the relative coordinate follows a classical trajectory $r(t)$ governed by the mean distortion potential $\langle V(r, \rho)\rangle$ averaged over ρ, while according to (7.121) the description of the internal motion is quantum mechanically correct. The correspondence principle limit of this description offers an alternative approach (Levine, 1971) to the semi-classical S matrix, discussed in Chapter 9.

7.4 Internal Excitation Below the Inelastic Threshold

Having discussed in Sections 7.1–7.3 the problem of scattering from one open channel of the system to another, we now consider the effect of coupling to channels which are closed, in the sense that the available energy lies below the inelastic threshold.

We consider as an illustration the two-state problem represented by the equations

$$\left[\frac{\mathrm{d}^2}{\mathrm{d}r^2} + k^2 - U_{11}(r)\right]\psi_1(r) = U_{12}(r)\,\psi_2(r) \tag{7.123a}$$

$$\left[\frac{\mathrm{d}^2}{\mathrm{d}r^2} + k^2 - U_{22}(r)\right]\psi_2(r) = U_{21}(r)\,\psi_1(r), \tag{7.123b}$$

where

$$U_{11}(\infty) < k^2 < U_{22}(\infty)$$

Channel 2 is therefore closed and the problem is to determine the effect of the coupling terms, which allow for polarization of the internal motion, on the elastic scattering in channel 1.

We note first that in view of this closure, $\psi_2(r)$ must vanish at infinity, and hence that the appropriate formal solution of (7.123b) may be expressed in terms of the standing wave Green's function $\mathscr{G}_2^{(0)}(r, r')$ defined in Appendix B,

$$\psi_2(r) = \int_0^\infty \mathscr{G}_2^{(0)}(r, r')\, U_{21}(r')\, \psi_1(r')\, \mathrm{d}r'. \tag{7.124}$$

In the bilinear form (B.12),

$$\mathscr{G}_2^{(0)}(r, r') = \sum\!\!\!\!\!\!\int \frac{\psi_{2n}^{(0)}(r')\, \psi_{2n}^{(0)}(r)}{k^2 - k_{2n}^2}. \tag{7.128}$$

where the $\sum\!\int$ sign is taken to include both discrete and continuum states; $\psi_{2n}^{(0)}(r)$ is the real bounded solution of (7.123b) when $U_{21}(r) = 0$, normalized to unity for a bound state or to a delta function of k. Equally $\psi_1(r)$ may be derived from (7.123a) in the form

$$\psi_1(r) = \psi_1^{(0)}(r) + \int_0^\infty \mathscr{G}_1^{(+)}(r, r')\, U_{12}(r')\, \psi_2(r')\, \mathrm{d}r', \tag{7.126}$$

where $\mathscr{G}_1^{(+)}(r, r')$ is the familiar outgoing Green's function

$$\begin{aligned} \mathscr{G}_1^{(+)}(r, r') &= -k_1^{-1}[\psi_1^{(0)}(r)\, \psi_1^{(1)}(r') + i\psi_1^{(0)}(r)\, \psi_1^{(0)}(r')],\ r > r' \\ &\quad -k^{-1}[\psi_1^{(1)}(r)\, \psi_1^{(0)}(r') + i\psi_1^{(0)}(r)\psi_1^{(0)}(r')],\ r < r', \end{aligned} \tag{7.127}$$

expressed in terms of solutions $\psi_1^{(0)}(r)$, $\psi_1^{(1)}(r)$ of the homogeneous analogue of (7.123a), such that

$$\begin{aligned} \psi_1^{(0)}(r) &\sim \sin(k_1 r + \delta_1),\ \psi_1^{(0)}(0) = 0, \\ \psi_1^{(1)}(r) &\sim \cos(k_1 r + \delta_1), \end{aligned} \tag{7.128}$$

with

$$k_1^2 = k^2 - U_{11}(\infty). \tag{7.129}$$

Overall therefore $\psi_1(r)$ contains a term $\psi_1^{(0)}(r)$ distorted by the static potential $U_{11}(r)$ and a polarization correction;

$$\begin{aligned} \psi_1(r) &= \psi_1^{(0)}(r) + \int_0^\infty \int_0^\infty \mathscr{G}_1^{(+)}(r, r')\, U_{12}(r')\, \mathscr{G}_2^{(0)}(r', r'')\, U_{21}(r'')\, \psi_1(r'')\, \mathrm{d}r'\, \mathrm{d}r'' \\ &\overset{r\to\infty}{\sim} \psi_1^0(r) - k_1^{-1} \exp[i(k_1 r + \delta_1)] \int_0^\infty \int_0^\infty \psi_1^{(0)}(r')\, U_{12}(r')\, \mathscr{G}_2^{(0)}(r', r'')\, U_{21}(r'') \\ &\qquad \psi_1(r'')\, \mathrm{d}r'\, \mathrm{d}r''. \end{aligned} \tag{7.130}$$

This is of course merely an integral form of (7.123). In order to proceed further we require an initial approximation either to $\psi_1(r)$ or to $\psi_2(r)$. The former is readily available if the second term in (7.130) provides merely a perturbation to $\psi_1^{(0)}(r)$; this will occur if the excitation terms $k^2 - k_{2n}^2$ in the denominator of (7.126) are large compared with $U_{12}(r)$. We have then a pure polarization problem. A good approximation to $\psi_2(r)$ is available, on the other hand, if the energy, k_2, lies close to one of the bound levels k_{2n}^2. In this case $G_2^{(0)}(r, r')$ is dominated by a single term, and $\psi_2(r)$ in (7.124) is proportional to the bound function $\psi_{2n}^{(0)}(r)$. This is a resonance situation, the description of which given below, was first presented by Feshbach (1958).

In the, relatively simple, polarization case the asymptotic form of $\psi_1(r)$ in (7.130) reduces, on substituting $\psi_1^{(0)}(r)$ for $\psi_1(r)$ and using the bilinear form (7.125) of $\mathscr{G}_2^{(0)}(r', r'')$ in the integrand, to the form

$$\psi_1(r) \overset{r\to\infty}{\sim} \psi_1^{(0)}(r) - k^{-1} \exp[i(k_1 r + \delta_1)] \sum\!\!\!\!\!\!\int \frac{\langle\psi_1^{(0)}|U_{12}|\psi_{2n}^{(0)}\rangle\langle\psi_{2n}^{(0)}|U_{21}|\psi_1^{(0)}\rangle}{k^2 - k_{2n}^2} \tag{7.131}$$

This may be compared, following the derivation of (4.10), with the corresponding result

$$\psi_1(r) \overset{r\to\infty}{\sim} \psi_1^{(0)}(r) - k_1^{-1} \exp[i(k_1 r + \delta_1)] \langle\psi_1^{(0)}|W_{11}|\psi_1^{(0)}\rangle, \tag{7.132}$$

for the effect of a small additional diagonal potential term $W_{11}(r)$, on the elastic scattering due to $U_{11}(r)$. In order words the polarization terms $U_{12}(r)$, $U_{21}(r)$ in (7.123) are equivalent to a non-local potential (dependent by virtue of the energy k^2 on $\psi_1^{(0)}(r)$)

$$W_{11}(r) = \sum\!\!\!\!\!\!\int \frac{U_{12}|\psi_{2n}^{(0)}\rangle\langle\psi_{2n}^{(0)}|U_{21}}{k^2 - k_{2n}^2} \tag{7.133}$$

Now to the extent that derivatives of $U_{ij}(r)$ may be ignored compared with those of $\psi_1^{(0)}(r)$ and $\psi_{2n}^0(r)$, the denominators $k^2 - k_{2n}^2$ may be replaced by $h_1(r)$–$h_2(r)$, where

$$h_i(r) = -\frac{d^2}{dr^2} + U_{ii}(r), \tag{7.134}$$

because the $\psi_i^{(0)}(r)$ are eigenfunctions of $h_i(r)$. Hence in view of the closure property

$$\sum\!\!\!\!\!\!\int \psi_{2n}^{(0)}(r)\psi_{2n}^{(0)}(r') = \delta(r - r'), \tag{7.135}$$

we find on replacing $h_1(r) - h_2(r)$ by $U_{11}(r) - U_{22}(r)$ that

$$W_{11}(r) \simeq \frac{U_{12}(r)U_{21}(r)}{U_{11}(r) - U_{22}(r)}, \tag{7.136}$$

provided that the range of $U_{12}(r)$ is large compared with the de Broglie wavelength.

It is readily verified in a case of weak coupling to several closed channels that $W_{11}(r)$ in (7.133) and (7.136) must be replaced by a sum of analogous terms taken over the channels in question. Thorson (1963) has based the theory of non-adiabatic effects in the high energy scattering of rare gases on corrections of the above type to the adiabatic intermolecular potential. The dominant terms, which arise from terms of the type $(\hat{J}_+\hat{P}_- + \hat{J}_-\hat{P}_+)$ in (6.16), may be traced to the high angular velocities required at short range to maintain the angular momentum.

Turning to the Feshbach (1958) resonant situation, we assume that k^2 lies close to a bound level k^2_{2n}, and hence that $\psi_2(r)$ in (7.124) may be approximated, by virtue of (7.125), in the form

$$\psi_2(r) = A\psi^{(0)}_{2n}(r) \tag{7.137}$$

where

$$A = \langle\psi^{(0)}_{2n}|U_{21}|\psi_1\rangle/(k^2 - k^2_{2n}). \tag{7.138}$$

Hence, on substitution in (7.126)

$$\psi_1(r) = \psi^{(0)}_1(r) + A\int_0^\infty \mathscr{G}^{(+)}_1(r, r')\, U_{12}(r')\, \psi^{(0)}_{2n}(r')\, \mathrm{d}r'. \tag{7.139}$$

But by the above definition of A

$$(k^2 - k^2_{2n})A = \langle\psi^{(0)}_{2n}|U_{21}|\psi^{(0)}_1\rangle$$

$$+ A\int_0^\infty\int_0^\infty \psi^{(0)}_{2n}(r)\, U_{21}(r)\, \mathscr{G}^{(+)}_1(r, r')\; U_{12}(r')\, \psi^{(0)}_{2n}(r')\, \mathrm{d}r\, \mathrm{d}r' \tag{7.140}$$

from which it follows that

$$A = \frac{\langle\psi^{(0)}_{2n}|U_{21}|\psi^{(0)}_1\rangle}{k^2 - k^2_{2n} - \delta_{2n} + i\gamma_{2n}/2} \tag{7.141}$$

where

$$\delta_{2n} = \mathrm{Re}\int_0^\infty\int_0^\infty \psi^{(0)}_{2n}(r)U_{21}(r)\mathscr{G}^{(+)}_1(r, r')U_{12}(r')\psi^{(0)}_{2n}(r')\, \mathrm{d}r\, \mathrm{d}r' \tag{7.142}$$

$$\gamma_{2n} = -2\,\mathrm{Im}\int_0^\infty\int_0^\infty \psi^{(0)}_{2n}(r)\, U_{21}(r)\, \mathscr{G}^{(+)}_1(r, r')\, U_{12}(r')\, \psi^{(0)}_{2n}(r')\, \mathrm{d}r\, \mathrm{d}r'$$

$$= 2k^{-1}|\langle\psi^{(0)}_{2n}|U_{21}|\psi^{(0)}_1\rangle|^2, \tag{7.143}$$

the second line of (7.143) being obtained from the imaginary part of $\mathscr{G}^{(+)}_1(r, r')$ in (7.127); δ_{2n} cannot unfortunately be simplified in the same way. It follows on taking the asymptotic limit of (7.139) that $\psi_1(r)$ shows the characteristic resonance behaviour discussed in Section 4.3.

$$\psi_1(r) \overset{r\to\infty}{\sim} C[\exp(-ik_1r) - S_{11}\exp(ik_1r)], \tag{7.144}$$

with

$$\begin{aligned} S_{11} &= \left(\frac{k^2 - k_{2n}^2 - \delta_{2n} - i\gamma_{2n}/2}{k^2 - k_{2n}^2 - \delta_{2n} + i\gamma_{2n}/2}\right)\exp(2i\delta_1) \\ &= \left(\frac{E - E_{2n} - \Delta_{2n} - i\Gamma_{2n}/2}{E - E_{2n} - \Delta_{2n} + i\Gamma_{2n}/2}\right)\exp(2i\delta_1) \end{aligned} \tag{7.145}$$

and

$$\Delta_{2n} = \delta_{2n}\hbar^2/2m$$

$$\Gamma_{2n} = \gamma_{2n}\hbar^2/2m = \frac{4}{\hbar v_1}|\langle\psi_1^{(0)}|V_{12}|\psi_{2n}^{(0)}\rangle|^2. \tag{7.146}$$

Δ_{2n} is interpreted as a level shift due to interaction with the continuum, while as discussed in Section 4.3, Γ_{2n} determines the resonance lifetime and the level width. These results, derived on the assumption that $\mathscr{G}_2^{(0)}(r, r')$ in (7.125) is dominated by a single term may be extended to cover the full width of the resonance provided that Δ_{2n} and Γ_{2n} are small compared with the energy separation E_{2n}–E_{2m} between neighbouring levels.

For another view of the significance of the resonance terms we may apply purely outgoing boundary conditions in channel 1 by setting $\psi_1^{(0)}(r)$ in (7.139) to zero (see Section 4.3). The constant A is then undetermined, but (7.140) leads directly to the complex eigenvalue condition

$$k^2 = k_{2n}^2 + \delta_{2n} - i\gamma_{2n}/2$$

or

$$E = E_{2n} + \Delta_{2n} - i\Gamma_{2n}/2. \tag{7.147}$$

Furthermore $\psi_1(r)$ becomes asymptotically

$$\psi_1(r) \overset{r\to\infty}{\sim} -\exp i(k_1r + \delta_1)\,Ak_1^{-1}\,\langle\psi_1^{(0)}|U_{12}|\psi_{2n}^{(0)}\rangle, \tag{7.148}$$

corresponding to an outgoing flux

$$v_1|\psi_1(r)|^2 = |A|^2v_1k_1^{-2}|\langle\psi_1^{(0)}|U_{12}|\psi_{2n}^{(0)}\rangle|^2 = |A|^2\Gamma_{2n}/\hbar. \tag{7.149}$$

Since $|A|^2$ may be recognized according to (7.137) as the population of the bound state, $\psi_{2n}^{(0)}(r)$, the decay follows first order kinetics with rate constant $(\Gamma_{2n}/\hbar)$ and time constant

$$\tau_{2n} = \hbar/\Gamma_{2n} \tag{7.150}$$

in agreement with the general conclusions of Section 4.3.

On the question of the magnitude of the level width Γ_{2n} in a typical molecular case, we may note that a crossing between $U_{11}(r)$ and $U_{22}(r)$ as shown in Fig. 7.2(a) rather than 7.2(b) will lead to particularly large value because $\psi_1^{(0)}(r)$ and $\psi_{2n}^{(0)}(r)$ in the integral (7.146) will oscillate at the same frequency over the crossing region. It may be shown (Child, 1970, 1972) by extension of the derivation of (7.70) that if $U_{12}(r)$ is assumed constant and $U_{11}(r)$ and $U_{22}(r)$ assumed linear in r over the crossing region then

$$\Gamma_{2n} = \frac{4\pi\hbar\omega_2 V_{12}^2}{\hbar v^*}\{\mathrm{Ai}\,[-(v/v^*)^2]\}^2 \tag{7.151}$$

where v is the velocity at the crossing point, $\hbar\omega_2$ is the local energy spacing between the bound levels, and

$$v^* = 2(\hbar|F_1F_2|/4m^2|F_1 - F_2|)^{\frac{1}{3}}$$
$$F_i = -(\mathrm{d}V_{ii}/\mathrm{d}r) \tag{7.152}$$

This formula has been used (Child, 1970, 1972) in the analysis of the molecular predissociation linewidth pattern to determine the potential terms $V_{ij}(r)$.

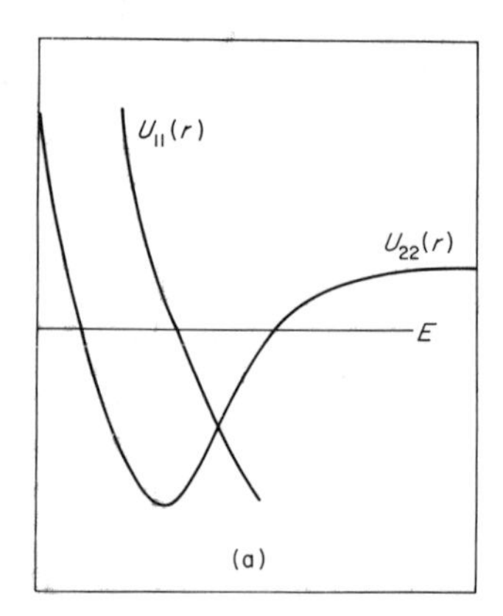

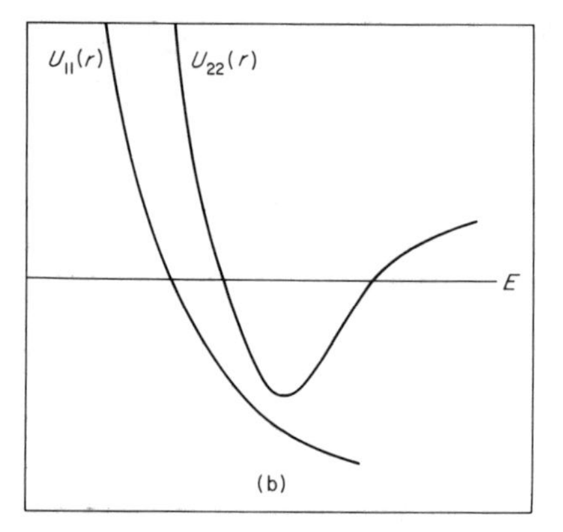

FIG. 7.2.

In cases of strong coupling or small bound state energy separations, the single level formulae (7.142)–(7.146) are no longer valid. $\psi_{2n}(r)$ in (7.139) must then be replaced by the proper sum

$$\psi_2(r) = \sum_n A_n \psi_{2n}^{(0)}(r) \tag{7.153}$$

with A_n given by (7.138). Hence on applying the outgoing boundary condition, (7.140) must be replaced by the equations

$$(k^2 - k_{2n}^2)A_n = \sum_{n'} \Lambda_{nn'}A_{n'} \tag{7.154}$$

where

$$\Lambda_{nn'} = \int_0^\infty \int_0^\infty \psi_{2n}^{(0)}(r)\, U_{21}(r)\, \mathscr{G}_1^{(+)}(r, r')\, U_{12}(r')\, \psi_{2n'}(r')\, \mathrm{d}r\, \mathrm{d}r'. \quad (7.155)$$

The resonance points are therefore given by the (complex) eigenvalues of the matrix $\mathbf{k}_2^2 + \Lambda$, where $\mathbf{k}_2^2$ is diagonal, with elements k_{2n}^2. This takes account of coupling between one continuum and several overlapping bound states in channel 2.

If on the other hand a single bound level is weakly coupled to several continua, $\psi_\nu(r)$, $\nu = 1, 3, 4\ldots$ say, $\psi_2(r)$ is again given by the single term in (7.137), except that now

$$A = \sum_\nu \langle \psi_{2n}^{(0)} | U_{2\nu} | \psi_\nu \rangle / (k^2 - k_{2n}^2) \quad (7.156)$$

where, under outgoing boundary conditions

$$\psi_\nu(r) = A \int_0^\infty \mathscr{G}_\nu^{(+)}(r, r')\, U_{\nu 2}(r')\, \psi_{2n}^{(0)}(r')\, \mathrm{d}r'. \quad (7.157)$$

Hence by elimination of $\psi_{2\nu}(r)$

$$(k^2 - k_{2n}^2)A = A \sum_\nu \int_0^\infty \int_0^\infty \psi_{2n}^{(0)}(r)\, U_{2\nu}(r)\, \mathscr{G}_\nu^{(+)}(r, r')\, U_{\nu 2}(r')\, \psi_{2n}^{(0)}(r')\, \mathrm{d}r\, \mathrm{d}r' \quad (7.158)$$

according to which the eigenvalue k^2 differs from k_{2n}^2 by a simple sum over the continuum channels, of resonance terms, $\delta_{2n}^{(\nu)}$ and $\gamma_{2n}^{(\nu)}$, defined by (7.142) and (7.143) with 1 replaced by ν throughout. In particular since the $\gamma_{2n}^{(\nu)}$ are necessarily positive the nett level width is simply the sum of the widths obtained by treating the coupling to each continuum separately.

Finally the common intervention of a bound state $\psi_{2n}^{(0)}(r)$ implies a second order coupling between the open channels. This raises the possibility of scattering from one open channel to another. In order to determine the relevant **S** or **T** matrix elements, we insert an elastically distorted wave $\psi_\mu^{(0)}(r)$ in the μth equation (7.157) as a means to allow for incident motion in this channel. The constant A is then determined by the analogue of (7.141)

$$A = \frac{\langle \psi_{2n}^{(0)} | U_{2\mu} | \psi_\mu^{(0)} \rangle}{k^2 - k_{2n}^2 - \delta_{2n} - i\gamma_{2n}/2} \quad (7.159)$$

where

$$\delta_{2n} = \sum_\nu \delta_{2n}^{(\nu)}, \qquad \gamma_{2n} = \sum_\nu \gamma_{2n}^{(\nu)} \quad (7.160)$$

It follows that

$$\psi_\nu(r) \overset{r\to\infty}{\sim} \delta_{\mu\nu} \psi_\nu^{(0)}(r) - k_\nu^{-1} \frac{\langle \psi_\nu^{(0)} | U_{\nu 2} | \psi_{2n}^{(0)} \rangle \langle \psi_{2n}^{(0)} | U_{2\mu} | \psi_\mu^{(0)} \rangle}{k^2 - k_{2n}^2 - \delta_{2n} + i\gamma_{2n}/2} \exp(ik_\nu r + i\delta_\nu) \quad (7.161)$$

and hence

$$T'_{\mu\nu} = i\left[\frac{\langle\psi_\nu^{(0)}|\,U_{\nu 2}|\psi_{2n}^{(0)}\rangle\langle\psi_{2n}^{(0)}|\,U_{2\mu}|\psi_\mu^{(0)}\rangle}{(k_\nu k_\mu)^{\frac{1}{2}}\,(k^2 - k_{2n}^2 - \delta_{2n} + i\gamma_{2n}/2)} - \delta_{\mu\nu}\right]. \tag{7.162}$$

Each element of the **T** matrix contains the same resonance denominator, but the numerators vary according to the strength of coupling between $\psi_{2n}^{(0)}(r)$ and the continuum states in question.

CHAPTER 8

Semi-Classical Models

In considering semi-classical approximations to the full quantum mechanical theory outlined in Chapter 6 it is convenient to recall the semi-classical analysis of elastic scattering in Chapters 4 and 5. The essential steps were to introduce a JWKB approximant to the (translational) wavefunction in Sections 4.2, and to use this in Sections 5.1 to obtain a differential cross-section with a close similarity to the purely classical form. The full generalization of this technique to cover inelastic events is examined in Chapter 9. At present we outline an intermediate approximation whereby the relative motion of the collision partners is treated classically, while the more tightly bound internal motion is assumed to obey the laws of quantum mechanics.

The validity of this picture is examined in detail in Section 8.6. It requires not only that the relative motion in a particular channel should be semi-classical in the sense of Sections 4.2 and 5.1 but also that this motion should lie close to a single mean trajectory for all channels in question. This implies that changes in linear and angular momentum must be small compared with their average values, over the inelastic coupling region; $\Delta p(r) \ll \bar{p}(r)$ and $\Delta l \ll l$.

Given these conditions the second order time independent Eqns (6.6), (6.56) or (6.77) may be replaced by equivalent first order time dependent forms. The general structure of the resulting theory is outlined in Sections 8.1. Perturbation and exponential forms of solution, analogous to those in Sections 7.1–7.3 are then discussed, while Section 8.5 is devoted to the important curve crossing problem.

8.1 Equations of Motion, Cross-sections and Scattering Matrices

The basic assumption of the present chapter is the existence over the inelastic coupling region of a common mean classical trajectory $\mathbf{r}(t)$ for the relative motion, governed by an appropriate average central potential $\overline{V}(r)$, as

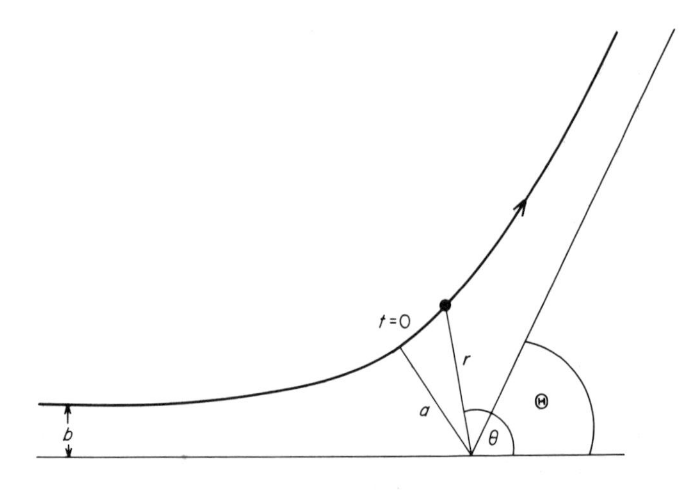

FIG. 8.1. The classical trajectory.

depicted in Fig. 8.1. The consequence of this picture, the validity of which is examined in Section 8.6, is that the relative kinetic energy operator may be omitted from the hamiltonian, to be replaced by a time dependent interaction potential $V(\mathbf{r}(t), \boldsymbol{\rho})$. Hence the time dependent Schrödinger equation

$$i\hbar \frac{\mathrm{d}\Phi}{\mathrm{d}t} = [H_{\mathrm{int}}(\boldsymbol{\rho}) + V(\mathbf{r}(t), \boldsymbol{\rho})]\, \Phi(\boldsymbol{\rho}, t), \tag{8.1}$$

together with the internal state expansion

$$\Phi(\boldsymbol{\rho}, t) = \sum_n a_n(t)\, \phi_n(\boldsymbol{\rho})\, \mathrm{e}^{-iE_n t/\hbar}, \tag{8.2}$$

where

$$H_{\mathrm{int}}(\boldsymbol{\rho})\, \phi_n(\boldsymbol{\rho}) = E_n\, \phi_n(\boldsymbol{\rho}), \tag{8.3}$$

will yield the familiar equations for the expansion coefficients,

$$\begin{aligned} i\hbar \frac{\mathrm{d}a_n}{\mathrm{d}t} &= \sum_m V_{nm}(r)\, \mathrm{e}^{i\omega_{nm} t} a_m(t) \\ &= \sum_m H'_{nm}(t)\, a_n(t), \end{aligned} \tag{8.4}$$

or more compactly

$$i\hbar \frac{\mathrm{d}\mathbf{a}}{\mathrm{d}t} = \mathbf{H}'(t)\mathbf{a}(t). \tag{8.5}$$

In these equations

$$
\begin{aligned}
V_{nm}(\mathbf{r}) &= \langle \phi_n | V(\boldsymbol{\rho}, \mathbf{r}) | \phi_m \rangle, \\
\hbar\omega_{nm} &= E_n - E_m, \\
H'_{nm}(t) &= V_{nm}(\mathbf{r})\, e^{i\omega_{nm}t}.
\end{aligned}
\tag{8.6}
$$

It must be stressed that knowledge of the trajectory, $\mathbf{r}(t)$, determined by solution of the equations

$$
\begin{aligned}
(\mathrm{d}r/\mathrm{d}t) &= \pm v[1 - b^2/r^2 - \overline{V}(r)/E]^{\frac{1}{2}}, \\
(\mathrm{d}\theta/\mathrm{d}t) &= vb/r^2, \\
\phi &= \text{constant},
\end{aligned}
\tag{8.7}
$$

is essential in integrating (6.1). The particular case of a free motion trajectory, $\overline{V}(r) = 0$, for which

$$
\begin{aligned}
r &= (b^2 + v^2t^2)^{\frac{1}{2}} \\
\theta &= \tan^{-1}(b/vt) \\
\phi &= \text{const},
\end{aligned}
\tag{8.8}
$$

provides the basis of the impact parameter method (Mott, 1931, Bates, 1962).

Alternative versions of the equations of motion (8.4) and (8.7) are given in (8.33), (8.38) and (8.43) below. We consider at present the solution of (8.4) subject to the initial boundary conditions

$$a_m(-\infty) = \delta_{nm}, \tag{8.9}$$

consistent with unit probability amplitude in entrance channel n. The required transition probabilities for the chosen trajectory are then determined in the form,

$$P_{nm} = |a_m(\infty)|^2. \tag{8.10}$$

The conversion of these probabilities into a total cross-section is equally straightforward. Since at a given energy, each trajectory is specified by an initial impact parameter, b, and azimuthal angle, ϕ, σ_{nm} is obtained simply by integration over b and ϕ,

$$\sigma_{nm} = \int_0^{2\pi} \int_0^{\infty} P_{nm}(b, \phi) b \,\mathrm{d}b\, \mathrm{d}\phi. \tag{8.11}$$

The derivation of the differential cross-section is however more complicated. On a strictly classical interpretation of the trajectory it would be given by

$$\frac{\mathrm{d}\sigma_{nm}}{\mathrm{d}\Omega} = I_{nm}(\theta, \phi) = \sum I_{\mathrm{el}}(\theta)\, P_{nm}(b, \phi), \tag{8.12}$$

with the elastic scattering term $I_{el}(\theta)$ due to the mean potential $\bar{V}(r)$ given by (2.15), and the sum taken over all trajectories passing through (θ, ϕ). The justification for Eqns (8.4) given in Section 8.5 uses however semi-classical rather than classical arguments. $I_{el}(\theta)$ in (8.12) should therefore at least be replaced by the semi-classical form derived from (5.21), but even this is not quite correct because variations in the final amplitudes $a_m(\infty)$ in (8.10), or S matrix elements below, must also be included in the stationary phase analysis used in Section 5.1. An example of the full argument is outlined in the curve-crossing section, Section 8.5 below.

The significance of the S matrix mentioned above may be readily appreciated in this formulation. It appears simply as the limit of a time evolution operator $\mathbf{U}(t, t_0)$, defined by the equation

$$\mathbf{a}(t) = \mathbf{U}(t, t_0)\,\mathbf{a}(t_0) \tag{8.13}$$

which must satisfy, according to (8.5)

$$i\hbar \frac{\mathrm{d}}{\mathrm{d}t}\,\mathbf{U}(t, t_0) = \mathbf{H}'(t)\,\mathbf{U}(t, t_0), \tag{8.14}$$

subject to the boundary condition

$$\mathbf{U}(t_0, t_0) = \mathbf{I} \tag{8.15}$$

This equation has a formal solution

$$\mathbf{U}(t, t_0) = \left\{\exp\left[-\frac{i}{\hbar}\int_{t_0}^{t} \mathbf{H}'(t)\,\mathrm{d}t\right]\right\}, \tag{8.16}$$

the meaning of which is discussed in Section 8.3. The matrix $\mathbf{S}$, which by definition connects $\mathbf{a}(\pm\infty)$, therefore becomes

$$\mathbf{S} = \mathbf{U}(\infty, -\infty) = \left\{\exp\left[-\frac{i}{\hbar}\int_{-\infty}^{\infty} \mathbf{H}'(t)\,\mathrm{d}t\right]\right\}. \tag{8.17}$$

It is convenient again to define the related transition matrix $\mathbf{T}$, such that

$$\mathbf{S} = \mathbf{I} - i\mathbf{T} \tag{8.18}$$

Certain valuable symmetry properties of $\mathbf{U}(t, t_0)$ and of $\mathbf{S}$ in this so called interaction picture (see Goldberger and Watson (1967) Section 2.5) follow from the hermitian character of $\mathbf{H}'$ and its properties under time reversal†

$$\mathbf{H}'(t) = \mathbf{H}'^{\dagger}(t)$$
$$\mathbf{H}'(t) = \mathbf{H}'^{*}(-t) \tag{8.19}$$

† Time $t = 0$ is taken at the mid-point of the necessarily symmetrical mean elastic scattering trajectory (see Fig. 2.1 and Eqn (2.3)).

In the first place a combination of the equation

$$i\hbar\, \mathbf{U}^{\dagger}(t, t_0)\left[\frac{\mathrm{d}}{\mathrm{d}t}\mathbf{U}(t, t_0)\right] = \mathbf{U}^{\dagger}(t, t_0)\,\mathbf{H}'(t)\,\mathbf{U}(t, t_0), \tag{8.20}$$

and its hermitian conjugate shows that

$$\mathbf{U}^{\dagger}(t, t_0)\left[\frac{\mathrm{d}}{\mathrm{d}t}\mathbf{U}(t, t_0)\right] + \frac{\mathrm{d}}{\mathrm{d}t}[\mathbf{U}^{\dagger}(t, t_0)]\,\mathbf{U}(t, t_0)$$

$$= \frac{\mathrm{d}}{\mathrm{d}t}[\mathbf{U}^{\dagger}(t, t_0)\,\mathbf{U}(t, t_0)] = 0. \tag{8.21}$$

In other words, in view of (8.15),

$$\mathbf{U}^{\dagger}(t, t_0)\,\mathbf{U}(t, t_0) = \mathbf{I} \tag{8.22}$$

This establishes the unitarity of $\mathbf{U}(t, t_0)$ for all t and t_0, and hence also the unitary of $\mathbf{S}$. As in the time independent formulation it has the physical consequence, according to (8.13), that the number of particles is at all times conserved, because

$$\sum |a_n(t)|^2 = \mathbf{a}^{\dagger}(t)\mathbf{a}(t) = \mathbf{a}^{\dagger}(t_0)\mathbf{U}^{\dagger}(t, t_0)\,\mathbf{U}(t, t_0)\mathbf{a}(t_0)$$

$$= \mathbf{a}^{\dagger}(t_0)\mathbf{a}(t_0) = \sum_n |a_n(t_0)|^2. \tag{8.23}$$

Equation (8.22) also establishes an identity between $\mathbf{U}(t_0, t)$ and $\mathbf{U}^{\dagger}(t, t_0)$ because by the definition (8.13),

$$\mathbf{a}(t_0) = \mathbf{U}(t_0, t)\,\mathbf{a}(t) = \mathbf{U}(t_0, t)\,\mathbf{U}(t, t_0)\,\mathbf{a}(t_0), \tag{8.24}$$

from which by comparison with (8.22)

$$\mathbf{U}(t_0, t) = \mathbf{U}^{\dagger}(t, t_0) \tag{8.25}$$

Finally it follows from (8.19) on replacing t by $-t$ in the complex conjugate of (8.14) with $t_0 = 0$, that

$$i\hbar\frac{\mathrm{d}}{\mathrm{d}t}\mathbf{U}^*(-t, 0) = \mathbf{H}'(t)\,\mathbf{U}^*(-t, 0). \tag{8.26}$$

In other words $\mathbf{U}^*(-t, 0)$ obeys the same equation as $\mathbf{U}(t, 0)$ with the same boundary conditions and hence

$$\mathbf{U}^*(-t, 0) = \mathbf{U}(t, 0). \tag{8.27}$$

In particular, with the definitions

$$\begin{aligned} \mathbf{\Omega}_+ &= \mathbf{U}(\infty, 0) \\ \mathbf{\Omega}_- &= \mathbf{U}(0, -\infty), \end{aligned} \tag{8.28}$$

it follows from (8.25) and (8.27) that

$$\mathbf{\Omega}_{\pm} = [\mathbf{\Omega}^{\dagger}_{\mp}]^* = \tilde{\mathbf{\Omega}}_{\mp} \tag{8.29}$$

where $\tilde{\mathbf{\Omega}}$ denotes the simple transpose. These matrices provide an alternative form for $\mathbf{S}$ since

$$\mathbf{a}(\infty) = \mathbf{U}(\infty, 0)\,\mathbf{a}(0) = \mathbf{U}(\infty, 0)\,U(0, -\infty)\,\mathbf{a}(-\infty) \tag{8.30}$$

with the implication that

$$\mathbf{S} = \mathbf{\Omega}_{+}\mathbf{\Omega}_{-} = \mathbf{\Omega}_{+}\tilde{\mathbf{\Omega}}_{+} = \tilde{\mathbf{\Omega}}_{-}\mathbf{\Omega}_{-}. \tag{8.31}$$

This form has the advantages over (8.17), in certain contexts, that integration is restricted to the range $0 < t < \infty$, and that the elements of $\mathbf{\Omega}_{+}$ are more simply expressed than those of $\mathbf{S}$. This completes the framework of the scattering theory based on Eqns (8.4) and (8.5).

Alternative versions of the equations of motion (8.4) are also available, with respect to each of which a modified S matrix with similar properties may be defined (see for example Bates *et al.*, 1964). In the first place a trivial modification of (8.4) is obtained by taking the diagonal terms $V_{nn}(r)$ into account in zeroth order by the substitutions

$$a_n(t) = c_n(t) \exp\left[-\frac{i}{\hbar} \int_0^t V_{nn}(\mathbf{r})\, \mathrm{d}t \right], \tag{8.32}$$

with the result for the $c_n(t)$ that

$$i\hbar \frac{\mathrm{d}c_n}{\mathrm{d}t} = \sum_{m \neq n} V_{nm}(\mathbf{r}) \exp\left\{ \frac{i}{\hbar} \int_0^t [W_n(\mathbf{r}) - W_m(\mathbf{r})]\, \mathrm{d}t \right\} c_m(t) \tag{8.33}$$

where

$$W_n(\mathbf{r}) = E_n + V_{nn}(\mathbf{r}).$$

$W_n(\mathbf{r})$ is therefore the first order perturbed energy of the state $\phi_n(\boldsymbol{\rho})$. The relation between (8.4) and (8.33) is of course exact.

An equally valid JWKB form of (8.33) may also be justified by the validity criterion, $\Delta p \ll \bar{p}$, on the classical trajectory method. The argument is most simply developed for an interaction potential $V(r, \boldsymbol{\rho})$ with spherical symmetry. It starts from the definitions

$$k_n^2(r) = 2m[E - W_n(r)]/\hbar^2$$

$$\bar{k}^2(r) = 2m[E - \bar{V}(r)]/\hbar^2 \tag{8.34}$$

in terms of which the integrand in (8.33) may be expressed

$$W_n(r) - W_m(r) = \left(\frac{\hbar^2}{2m}\right)[k_m^2(r) - k_n^2(r)]$$

$$\simeq \frac{\hbar^2}{m}[k_m(r) - k_n(r)]\bar{k}(r), \tag{8.35}$$

since by hypothesis $\Delta k(r) \ll \bar{k}(r)$. The second step is to replace t in (8.33) by the distance, s, along the trajectory, by means of the equation

$$\frac{\mathrm{d}s}{\mathrm{d}t} = v(r) = \frac{\hbar}{m}\bar{k}(r), \tag{8.36}$$

with the result, first obtained by Bates and Crothers (1970) that

$$2i\bar{k}(r)\frac{\mathrm{d}c_n}{\mathrm{d}s} = \sum_{m \neq n} U_{mn}(r) \exp\left\{i\int_0^s [k_m(r) - k_n(r)]\,\mathrm{d}r\right\} c_m(s), \tag{8.37}$$

where

$$U_{mn}(r) = 2m\, V_{mn}(r)/\hbar^2$$

and s is related to r by the equation

$$s = \int_a^r \left[\left(\frac{\mathrm{d}s}{\mathrm{d}t}\right)\bigg/\left(\frac{\mathrm{d}r}{\mathrm{d}t}\right)\right]\mathrm{d}r = \pm\int_a^r [1 - \bar{V}(r)/E]^{\frac{1}{2}}\,\mathrm{d}r, \tag{8.38}$$

upper or lower signs being taken to refer to outward or inward radial motion respectively. Bates and Crothers (1970) actually suggest the form

$$\bar{k}(r) = [k_1(r)k_2(r)]^{\frac{1}{2}} \tag{8.39}$$

in a two state problem, but Delos *et al.* (1972) find that provided $\Delta k(r) \ll \bar{k}(r)$ over the inelastic coupling region, the precise prescription for $\bar{k}(r)$ is unimportant.

A final similar form which avoids the necessity for an auxiliary variable s or t is obtained by using

$$\bar{k}_l^2(r) = 2m[E - V(r) - (l + \tfrac{1}{2})^2\hbar^2/2mr^2]/\hbar^2,$$

$$k_{nl}^2(r) = 2m[E - W_n(r) - (l + \tfrac{1}{2})^2\hbar^2/2mr^2]/\hbar^2,$$

$$\frac{\mathrm{d}r}{\mathrm{d}t} = \pm v_l(r) = \pm\frac{\hbar}{m}\bar{k}_l(r), \tag{8.40}$$

in place of (8.34) and (8.36), the difference being that $k_l(r)$ relates to the radial component of the velocity, while $\bar{k}(r)$ depends on the velocity itself. The result is a branch point in the analogue of (8.37), at the radial turning point a_l. Different equations, distinguished by upper or lower signs are therefore required for the outward or inward motion respectively.

$$\pm 2i\bar{k}_l(r)\frac{dc_n^{(\pm)}}{dr} = \sum_{m\neq n} U_{mn}(r)\exp\left\{\pm i\int_{a_l}^{r} [k_{ml}(r) - k_{nl}(r)]\,dr\right\} c_m^{(\pm)}(r). \tag{8.41}$$

The direct use of (8.41) is therefore limited to situations where coupling is unimportant at the turning point. Special provision (as discussed in Section 8.6) is also required to connect the functions $c_n^{(\pm)}(r)$ around the turning point. This form is however particularly well adapted to a general JWKB analysis (see for example Section 8.5).

8.2 Perturbation Methods

The purpose of this section is to describe simple perturbation solutions to the basic Eqns (8.4) and (8.7), and to examine the results in relation to the more accurate Born and Distorted Wave solutions in Sections 7.1 and 7.2. Similar arguments may be applied to (8.37) and (8.41). In each case different approximations may be distinguished first by the relative importance attached to the diagonal and off-diagonal terms in the basic equation

$$i\hbar\frac{da_n}{dt} = \sum_m V_{mn}(\mathbf{r})\,e^{i\omega_{nm}t}a_m(t), \tag{8.42}$$

and secondly by the choice of the assumed trajectory.

At the simplest level only the initial channel coefficient $a_0(t)$ is retained (with constant value unity) on the right-hand side of (8.42) on the grounds that there is little depletion of the initial state. The trajectory is then taken to be linear, with constant velocity v and impact parameter b as shown in

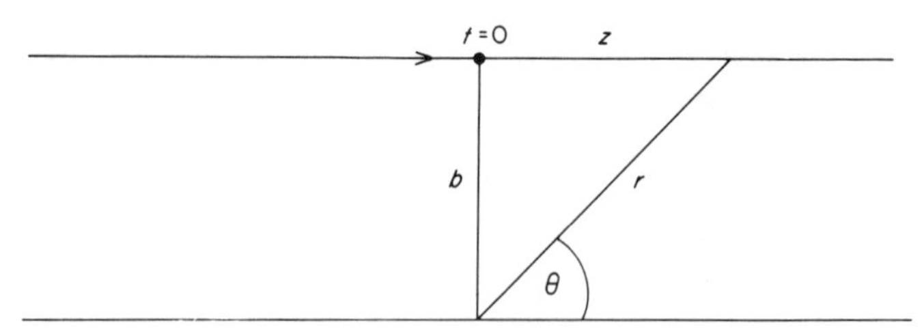

FIG. 8.2. The impact parameter model.

Fig. 8.2. The solution, subject to the boundary condition $a_n(-\infty) = 0$, gives for the T matrix element defined by (8.18),

$$T_{0n} = ia_n(\infty) = \frac{1}{\hbar}\int_{-\infty}^{\infty} V_{n0}(\mathbf{r})\,e^{i\omega_{n0}t}\,dt$$

$$= \frac{1}{\hbar v}\int_{-\infty}^{\infty} V_{n0}(\mathbf{r})\exp(i\omega_{n0}z/v)\,dz$$

$$= \frac{2}{\hbar v} \int_0^\infty V_{n0}(\mathbf{r}) \cos(\omega_{n0} z/v)\, \mathrm{d}z, \tag{8.43}$$

where the final line follows from the symmetry of $V_{n0}(\mathbf{r})$ under reversal of the signs of z or t. Since only off-diagonal interaction terms are taken into account this is the classical trajectory equivalent of the Born approximation.

The inclusion of the diagonal terms $V_{nn}(r)$ may be taken in two stages. As a first improvement we may retain the linear trajectory (impact parameter) approximation, but solve (8.33) with $c_m(t) = \delta_{m0}$ on the right-hand side, rather than (8.4) with $a_m(t) = \delta_{m0}$. This leads to a modified T matrix element

$$T'_{0n} = ic_n(\infty) = \frac{2}{\hbar v} \int_0^\infty V_{n0}(\mathbf{r}) \cos\left[\frac{1}{\hbar v} \int_0^z [W_n(\mathbf{r}) - W_0(\mathbf{r})]\, \mathrm{d}z'\right] \mathrm{d}z, \tag{8.44}$$

where

$$W_n(\mathbf{r}) = E_n + V_{nn}(\mathbf{r}) \tag{8.45}$$

which is related to T_{0n} in the form

$$T'_{0n} = \exp\left\{\frac{i}{\hbar} \int_0^\infty [V_{nn}(\mathbf{r}) + V_{00}(\mathbf{r})]\, \mathrm{d}t\right\} T_{0n}. \tag{8.46}$$

More accurately however the trajectory $\mathbf{r}(t)$ would be derived from equations (8.7), with $\bar{V}(r)$ taken as an appropriate mean of $V_{nn}(r)$ and $V_{00}(t)$. T'_{0n} is then given by

$$T'_{0n} = \frac{2}{\hbar} \int_0^\infty V_{n0}[\mathbf{r}(t)] \cos\left\{\frac{1}{\hbar} \int_0^t [W_n(\mathbf{r}(t')) - W_0(\mathbf{r}(t'))]\, \mathrm{d}t'\right\} \mathrm{d}t. \tag{8.47}$$

This is the classical trajectory variant of the distorted wave approximation (7.33). Like (8.43), it has the advantage over its quantum mechanical equivalent that the necessary integrals are more easily evaluated.

As a first application of the classical trajectory approximation, we consider the case of an exponential interaction function

$$V_{10}(r) = V \mathrm{e}^{-r/R} \tag{8.48}$$

on the constant velocity, zero impact parameter model, for which, according to (8.43)

$$T_{01} = \frac{2V}{\hbar v} \int_0^\infty \mathrm{e}^{-r/R} \cos(\omega r/v)\, \mathrm{d}r = \frac{2VR}{\hbar v(1 + \omega^2 R^2/v^2)}. \tag{8.49}$$

This may be compared with Born result (7.15) for the same model

$$T_{01} = \frac{2mVR}{\hbar^2 (k_0 k_1)^{\frac{1}{2}} [1 + (k_0 - k_1)^2 R^2]} \tag{8.50}$$

The two are seen to be identical to the extent that

$$\bar{k} = mv/\hbar = (k_0 k_1)^{\frac{1}{2}} = \tfrac{1}{2}(k_0 + k_1), \tag{8.51}$$

which is of course the condition for a valid common trajectory.

More ambitiously, (8.43) may also be applied in rotational and vibrational relaxation problems (Gray and Van Kranendonk, 1966; and Sharma, 1969) on the assumption of an orientation dependent, multipole interaction function

$$V_l(\mathbf{r}, \boldsymbol{\rho}_1, \boldsymbol{\rho}_2) = 4\pi\varepsilon_l \sum_m T_{lm}(\boldsymbol{\rho}_1, \boldsymbol{\rho}_2)\, Y_{lm}(\theta, \phi)/r^{l+1}, \tag{8.52}$$

where

$$\varepsilon_l = \frac{(-1)^{l_2}}{(2l+1)} \left\{ \frac{4\pi(2l+1)!}{(2l_1+1)!\,(2l_2+1)!} \right\},$$

$$T_{lm}(\boldsymbol{\rho}_1, \boldsymbol{\rho}_2) = \sum_{m_1, m_2} (l_1 l_2 m_1 m_2 | l_1 l_2 l m) Q_{l_1 m_1}(\boldsymbol{\rho}_1) Q_{l_2 m_2}(\boldsymbol{\rho}_2),$$

$$Q_{lm}(\boldsymbol{\rho}) = \sum_j e_j \rho_j^l Y_{lm}(\hat{\boldsymbol{\rho}}_j). \tag{8.53}$$

Given the forms of the internal matrix elements $\langle a | T_{lm}(\rho_1, \rho_2) | b \rangle$, between appropriate vibrational-rotational states, which may be obtained. and averaged by standard methods (Brink and Satchler, 1968), the problem reduces to evaluation of the integral

$$I_{lm} = \frac{1}{\hbar} \int_{-\infty}^{\infty} \mathrm{e}^{i\omega t}\, Y_{lm}(\theta, \phi) r^{-1-l}\, \mathrm{d}t. \tag{8.54}$$

It is assumed in this theory that the trajectory is governed by a hard sphere potential with radius d.

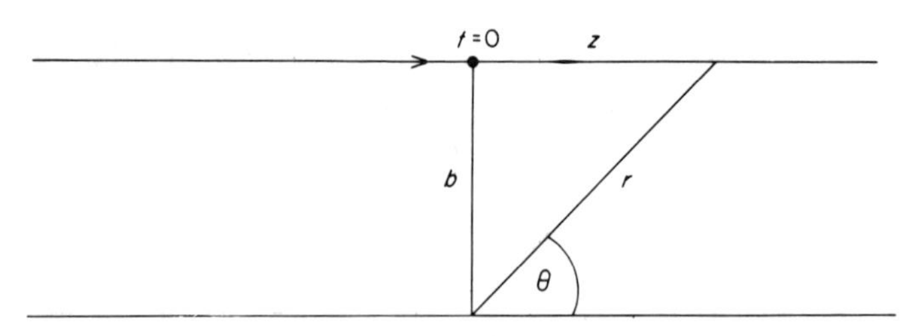

FIG. 8.3. The coordinate system.

Thus for impact parameters greater than the hard sphere radius d, for which the trajectory follows a straight line as illustrated in Fig. 8.3, we may substitute

$$r = (b^2 + v^2 t^2)^{\frac{1}{2}}$$

$$\cos\theta = vt/(b^2 + v^2t^2)^{\frac{1}{2}}$$
$$\phi = \text{constant} \tag{8.55}$$

and note that

$$Y_{lm}(\theta, \phi) = (-1)^m \left\{\frac{(2l+1)(l-m)!}{4\pi(l+m)!}\right\}^{\frac{1}{2}} P_{lm}(\cos\theta)\, e^{im\phi}. \tag{8.56}$$

It follows, for positive values of the quantity

$$x = (\omega b/v) \tag{8.57}$$

that (Gray and Van Kranendonk (1966)),

$$I_{lm}(x) = (-1)^m e^{im\phi}(\hbar v b^l)^{-1}\left\{\frac{(2l+1)(l-m)!}{4\pi(l+m)!}\right\}\int_{-\infty}^{\infty} e^{ixs}\frac{P_{lm}[s/(1+s^2)]\,ds}{(1+s^2)^{\frac{1}{2}(l+1)}}$$

$$= i^{l+m} e^{im\phi}(\hbar v b^l)^{-1}\left\{\frac{(2l+1)}{\pi(l+m)!\,(l-m)!}\right\}^{\frac{1}{2}} x^l K_m(x), \tag{8.58}$$

where $K_m(x)$ denotes the modified Bessel function (Abramowitz and Stegun, 1965). It may also be verified that

$$I_{lm}(x) = (-1)^m I_{l,-m}(x) = (-1)^m I_{lm}(-x) \tag{8.59}$$

The presence of the phase factors in (8.58) and (8.59) has in fact no effect on the transition probability because the selection rules on the matrix element $\langle a|T_{lm}|b\rangle$ ensure that only one m value can contribute to the sum in (8.52) for any given transition $a \to b$. This means in particular that the transition probability is independent of the azimuthal angle ϕ.

The case of an impact parameter $b < d$ is more complicated because the trajectory then consists of two linear segments. Sharma (1969) and Sharma and Brau (1969) suggest a quadratic interpolation with b over the range $0 < b < d$, based on Eqn (8.58) at $b = d$ and the formula

$$I_{lm}^{(0)} = \delta_{m0}[\hbar v d^l]^{-1}[(2l+1)/\pi]^{\frac{1}{2}}\int_0^{\infty}(1+s)^{-1-l}\cos ys\,ds, \tag{8.60}$$

where

$$y = (\omega d/v),$$

derived from (8.54) for $b = 0$. This is amenable to reduction (Sharma and Brau (1969)) in terms of the sine and cosine integrals Si(y) and Ci(y) (Abramowitz and Stegun, 1965).

In extending the above analysis to include models in which both diagonal and off-diagonal potential terms are important, it is convenient first to consider the zero impact parameter curve crossing problem defined by the equations

$$W_n(r) = -F_n(r - R), \quad n = 1, 2$$
$$V_{12} = \text{const}, \qquad v = \text{const.}, \tag{8.61}$$

for which, with $F_1 > F_2$, and $t(R)$ used to denote the time at the crossing point Eqn (8.44) becomes,

$$\begin{aligned}
T'_{12} &= \frac{2V_{12}}{\hbar v}\int_0^\infty \cos\left[\frac{1}{\hbar v}\int_0^r [W_2(r') - W_1(r')]\,\mathrm{d}r'\right]\mathrm{d}r \\
&= \frac{2V_{12}}{\hbar v}\int_0^\infty \cos\left\{\left(\frac{F_1 - F_2}{2\hbar v}\right)[(r - R)^2 - R^2]\right\}\mathrm{d}r \\
&\simeq \frac{2V_{12}}{\hbar v}\int_0^\infty \cos\left[\left(\frac{F_1 - F_2}{2\hbar v}\right)(x^2 - R^2)\right]\mathrm{d}x \\
&\simeq 2V_{12}\left[\frac{2\pi}{\hbar v(F_1 - F_2)}\right]^{\frac{1}{2}} \sin\left[\frac{(F_1 - F_2)R^2}{2\hbar v} + \pi/4\right] \\
&\simeq 2V_{12}\left[\frac{2\pi}{\hbar v(F_1 - F_2)}\right]^{\frac{1}{2}} \sin\left[\frac{1}{\hbar}\int_0^{t(R)} (W_1 - W_2)\,\mathrm{d}t + \pi/4\right]
\end{aligned} \tag{8.62}$$

provided that $R \gg [\hbar v/(F_1 - F_2)]^{\frac{1}{2}}$; otherwise the approximation contained in the third line is invalid. This confirms the conclusion of the previous semi-classical distorted wave description of the same problem (see Section 7.2), that the effective transition zone is localized about the crossing point R in a range governed by $[\hbar v/(F_1 - F_2)]^{\frac{1}{2}}$. The constant velocity assumption is therefore justified over this range, as evidenced by the agreement between the amplitude factors in (8.62) and the distorted wave form (7.78). The phase terms are however quite different because (8.62) makes no allowance for a decrease in velocity towards the classical turning point.

This weakness may however be eliminated by the assumption of a constant acceleration (rather than velocity) trajectory governed by a mean potential

$$\bar{V}(r) = -F(r - R) \tag{8.63}$$

along which

$$r(t) = r_0 + (F/2m)t^2, \tag{8.64}$$

where r_0 is the turning point. The result, after substitution in (8.47), is (Nikitin, 1968)

$$T'_{12} = \frac{2V_{12}}{\hbar}\int_0^\infty \cos\left[\left(\frac{F_1 - F_2}{\hbar}\right)\int_0^t (r(t') - R)\,\mathrm{d}t'\right]\mathrm{d}t$$

$$= \frac{2V_{12}}{\hbar}\int_0^\infty \cos\left[\frac{(F_1 - F_2)}{\hbar}\left(\frac{Ft^3}{6m} - (R - r_0)t\right)\right]\mathrm{d}t$$

$$= \frac{2\pi V}{[\hbar^2 F(F_1 - F_2)]^{\frac{1}{3}}}\,\mathrm{Ai}\left\{-\left[\frac{2m(F_1 - F_2)^2}{\hbar^2 F}\right]^{\frac{1}{3}}(R - r_0)\right\}. \quad (8.65)$$

A further reduction is obtained by noting that $V(r) = E$ at the turning point; hence $(R - r_0)$ may be replaced by (E/F), in which case (8.65) with $F = (F_1 F_2)^{\frac{1}{2}}$ becomes identical with the full distorted wave result (7.70). This precise equivalence at all energies is to some extent fortuitous, however, because, as shown in Section 8.5, this model happens to be one for which the classical trajectory equations are exact.

As a final example, of practical importance, we consider the semi-classical vibrational excitation probability obtained by Rapp and Sharp (1963) for the harmonic oscillator, Landau–Teller, model treated by the distorted wave method in Section 7.2. The theory is conveniently developed in the dimensionless units of the previous discussion (time, length and energy being measured in units of ω^{-1}, $\hbar^{\frac{1}{2}}(k\mu)^{-\frac{1}{4}}$ and $\hbar\omega$ respectively) with the origin of the translational variable x, chosen such that

$$\begin{aligned} V_{nn}(x) &= V_{n'n'}(x) = V(x) = \bar{E}\,\mathrm{e}^{-\alpha x} \\ V_{nn'}(x) &= Y_{nn'}\bar{E}\,\mathrm{e}^{-x}, \end{aligned} \quad (8.66)$$

where $\bar{E} = \frac{1}{2}m\bar{v}^2$ is translational energy derived from the mean velocity $\bar{v}$ in channels n and n', and $Y_{nn'}$ is defined by Eqn (7.44). The required T' matrix element is then obtained from (8.47) in the form ($\hbar = 1$ in the present units),

$$T'_{nn'} = 2\bar{E}Y_{nn'}\int_0^\infty \cos(\Delta n t)\,\mathrm{e}^{-\alpha x}\,\mathrm{d}t, \quad (8.67)$$

with t and x related by

$$\frac{\mathrm{d}t}{\mathrm{d}x} = \frac{1}{v(x)} = [2(\bar{E} - \bar{V}(x))/m]^{-\frac{1}{2}} = \bar{v}^{-1}(1 - \mathrm{e}^{-\alpha x})^{-\frac{1}{2}}. \quad (8.68)$$

The necessary integrals are conveniently evaluated by the substitution

$$\operatorname{sech}\theta = \mathrm{e}^{-\frac{1}{2}\alpha x} \quad (8.69)$$

in terms of which

$$t = \frac{1}{\bar{v}}\int_0^x (1 - \mathrm{e}^{-\alpha x})^{-\frac{1}{2}}\,\mathrm{d}x = 2\theta/\alpha\bar{v}, \quad (8.70)$$

and

$$T'_{nn'} = \frac{4\bar{E}}{\alpha\bar{v}} Y_{nn'} \int_0^\infty \cos\left(\frac{2\Delta n\theta}{\alpha\bar{v}}\right) \operatorname{sech}^2\theta \, d\theta$$
$$= \frac{2\pi m|n' - n|}{\alpha^2} Y_{nn'} \operatorname{cosech}\left(\frac{\pi|n' - n|}{\alpha\bar{v}}\right). \tag{8.71}$$

This means in the case $n' = n \pm 1$ that

$$P_{nn'} = |T'_{nn'}|^2 = \frac{2\pi^2 m^2 (n + \frac{1}{2} \pm \frac{1}{2})}{\alpha^2} \operatorname{cosech}^2 (\pi/\alpha\bar{v}). \tag{8.72}$$

Rapp and Sharp (1963) actually use the initial velocity v_n in place of $\bar{v}$, with the result that the microscopic reversibility $T_{nn'} = T_{n'n}$ is lost. Neither form is however realistic at low energies, where this difference is important because both predict a non-zero transition probability below the excitation threshold. Equation (8.71) is therefore clearly inferior to the distorted wave result (7.53). It is readily verified however that (7.53) and (8.71) tend to the same high energy limit, as expected when $\Delta v = |v_n - v_{n'}| \ll \bar{v}$ so that the assumption of a common trajectory is justified.

8.3 The Magnus and Sudden Approximations

In considering alternative forms of solution for use when the inelastic transition probability may be large, it is convenient to refer back to Eqn (8.14) for the time evolution operator $\mathbf{U}(t, t_0)$,

$$i\hbar \frac{d}{dt} \mathbf{U}(t, t_0) = \mathbf{H}'(t) \mathbf{U}(t, t_0), \tag{8.73}$$

where

$$H'_{nm}(t) = V_{nm}(\mathbf{r}(t)) \exp i\omega_{nm} t,$$

and to the formal solution given in (8.16)

$$\mathbf{U}(t, t_0) = \left\{ \exp\left[-i \int_{t_0}^t \mathbf{H}'(t) \, dt \right] \right\}. \tag{8.74}$$

The latter requires some amplification however unless the matrix $\mathbf{H}'(t)$ is independent of time.

One meaning can be given to the right-hand side by iterating the integral form of (8.73);

$$\mathbf{U}(t, t_0) = \mathbf{I} - \frac{i}{\hbar} \int_{t_0}^t \mathbf{H}'(t_1) \mathbf{U}(t_1, t_0) \, dt_1$$

$$= \mathbf{I} - \frac{i}{\hbar}\int_{t_0}^{t} \mathbf{H}'(t_1)\,\mathrm{d}t_1 + \left(\frac{i}{\hbar}\right)^2 \int_{t_0}^{t}\left[\int_{t_0}^{t_1} \mathbf{H}'(t_1)\mathbf{H}'(t_2)\,\mathrm{d}t_2\right]\mathrm{d}t_1 + \dots \tag{8.75}$$

This form, which has an analogue in the Born series derived from (7.5), has been labelled by Dyson (1949)

$$\mathbf{U}(t, t_0) = T\left[\exp\left\{-\frac{i}{\hbar}\int_{t_0}^{t} \mathbf{H}'(t_1)\,\mathrm{d}t_1\right\}\right], \tag{8.76}$$

where the chronological operator T generates the ordered products in (8.75). Like the Born series it suffers from the disadvantages that a large first term normally implies large subsequent terms and that truncation of the series destroys the unitarity of $\mathbf{U}(t, t_0)$.

A second, more useful, genuinely exponential form of solution has been obtained by Magnus (1954), Robinson (1963) and Pechukas and Light (1966) (of which the last give the simplest derivation). The result is†

$$\mathbf{U}(t, t_0) = \exp \mathbf{A}(t, t_0) = \mathbf{I} + \mathbf{A}(t, t_0) + \tfrac{1}{2}\mathbf{A}^2(t, t_0) + \dots \tag{8.77}$$

where

$$\mathbf{A}(t, t_0) = \mathbf{A}^{(1)}(t, t_0) + \mathbf{A}^{(2)}(t, t_0) + \dots$$

$$A^{(1)}(t, t_0) = -\frac{i}{\hbar}\int_{t_0}^{t} \mathbf{H}'(t_1)\,\mathrm{d}t_1$$

$$A^{(2)}(t, t_0) = \frac{1}{2}\left(\frac{i}{\hbar}\right)^2 \int_{t_0}^{t}\left[\int_{t_0}^{t_1} [\mathbf{H}'(t_1), \mathbf{H}'(t_2)]\,\mathrm{d}t_2\right]\mathrm{d}t_1$$

$$A^{(3)}(t, t_0) = -\frac{1}{6}\left(\frac{i}{\hbar}\right)^3 \int_{t_0}^{t}\left\{\int_{t_0}^{t_1}\left[\int_{t_0}^{t_2}\left[[\mathbf{H}'(t_1), \mathbf{H}'(t_2)], \mathbf{H}'(t_3)\right] + \left[[\mathbf{H}'(t_3), \mathbf{H}'(t_2)], \mathbf{H}'(t_1)\right]\mathrm{d}t_3\right]\mathrm{d}t_2\right\}\mathrm{d}t_1, \tag{8.78}$$

and $[\mathbf{H}'(t_r), \mathbf{H}'(t_s)]$ denotes the commutator. Now since $\mathbf{H}'(t)$ is an anti-hermitian matrix ($\mathbf{H}'^{\dagger} = -\mathbf{H}'$), the same is true for all commutators $[\mathbf{H}'(t_r), \mathbf{H}'(t_s)]$. Hence each term of $\mathbf{A}(t, t_0)$ is also anti-hermitian. This means that $\mathbf{U}(t, t_0)$ retains its unitarity when the series for $\mathbf{A}(t, t_0)$ is truncated at any point.

† A similar, more rapidly convergent series (the first term of which gives the time dependent analogue of the exponential solution described in Section 7.4) may be obtained for the matrix $\mathbf{U}'(t, t_0)$ which describes the time evolution of the coefficients $\mathbf{c}(t)$ in (8.33). It differs from (8.77) and (8.78) by replacing $\mathbf{H}'(t)$ by a matrix with the first order corrected elements

$$V_{nm}(\mathbf{r}(t)) \exp\left\{\frac{i}{\hbar}\int_0^t [W_n(\mathbf{r}) - W_m(\mathbf{r})]\,\mathrm{d}t\right\}.$$

The presence of the commutators in (8.78) in place of the simple products in the perturbation series (8.75) also confers advantages on the exponential solution in the case of a strong but short lived, "impulsive", perturbation.† Consider, as an illustration the effect of a perturbation matrix such that

$$\mathbf{V}(r(t)) = \mathbf{V}^0 \, \mathrm{e}^{-\alpha|t|}, \tag{8.79}$$

with a common time constant, α^{-1}, in all channels. This leads on calculating $\mathbf{\Omega}_+ = \mathbf{U}(\infty, 0)$ as a preliminary to determining the $\mathbf{S}$ matrix by means of (8.31), to the following forms for the typical elements of $\mathbf{A}^{(1)}(\infty, 0)$ and $\mathbf{A}^{(2)}(\infty, 0)$,

$$\begin{aligned} A^{(1)}_{nm}(\infty, 0) &= -\frac{i}{\hbar}\frac{V^0_{nm}}{(\alpha - i\omega_{nm})} \\ A^{(2)}_{nm}(\infty, 0) &= \frac{i}{2\hbar^2}\sum_k \frac{V^0_{nk}V^0_{km}(\omega_{nk} - \omega_{km})}{(2\alpha - i\omega_{nm})(\alpha - i\omega_{nk})(\alpha - i\omega_{km})}, \end{aligned} \tag{8.80}$$

while in the perturbation series (8.75), the first term is identical with $A^{(1)}_{nm}(\infty, 0)$ and the second becomes

$$-\frac{1}{\hbar^2}\sum_k \int_0^\infty \int_0^{t_1} H'_{nk}(t_1)H'_{km}(t_2)\,\mathrm{d}t_2\,\mathrm{d}t_1 = -\frac{1}{\hbar^2}\sum_k \frac{V^0_{nk}V^0_{km}}{(\alpha - i\omega_{nk})(\alpha - i\omega_{nm})}. \tag{8.81}$$

This means on approaching the impulsive limit by setting $\alpha \gg \omega_{nm}$ for all n, m, that the first terms in the two series are of order $(V^0/\hbar\alpha)$ which may be large, while the second terms in (8.75) and (8.78) are of order $(V^0/\hbar\alpha)^2$ and $(V^0/\hbar\alpha)^2(\omega/\alpha)$ respectively. Thus the second term in the perturbation series is also large, but that in the exponential series will be negligible if the product between the internal frequency ω and interaction time α^{-1} is sufficiently small. More general analysis in support of this conclusion is given by Pechukas and Light (1966).

This may be taken as the basis of the sudden approximation applied to rotational excitation problems by Kramer and Bernstein (1964) and Bernstein and Kramer (1966), according to which the frequency terms in $\mathbf{H}'(t)$ are set equal to zero, and only the first-order term is retained in the matrix $\mathbf{A}(t, t_0)$. In this approximation therefore

$$\mathbf{S} = \mathbf{U}(\infty, -\infty) = \exp\left[-\frac{i}{\hbar}\int_{-\infty}^{\infty} \mathbf{V}(\mathbf{r}(t))\,\mathrm{d}t\right]. \tag{8.82}$$

Equation (8.77) may also be used as a means for multistep numerical integration of (8.73) (Chan *et al.*, 1968). The result is an S matrix in the accu-

† There may also be problems of physical interest, such as that of a linearly forced oscillator (Pechukas and Light, 1966) for which the commutators lead to strict truncation of the series after a finite number of terms.

mulated product form†

$$\mathbf{S} = \mathbf{U}(-\infty, \infty) = \prod_{i=1}^{N-1} \exp\left[\mathbf{A}(t_{i+1}, t_i)\right], \tag{8.83}$$

in which for sufficiently small step lengths, $\Delta t = t_{i+1} - t_i$, the matrices $\mathbf{A}(t_{i+1}, t_i)$ may be approximated by the first term in the expansion (8.78).

8.4 The Impulse Approximation

Another quite different non-perturbative approach to the special problem of a linearly forced harmonic oscillator also has molecular applications. It is based on finding transformations

$$\begin{aligned} \xi(\rho, t) &= \rho - u(t) \\ \Psi(\rho, t) &= \Phi(\xi, t) \exp\left[\rho v(t) + w(t)\right], \end{aligned} \tag{8.84}$$

such that the original time dependent equation

$$\left[-\frac{\hbar^2}{2\mu}\frac{\partial^2}{\partial \rho^2} + \frac{1}{2}k\rho^2 - \rho F(t)\right]\Psi = i\hbar\left(\frac{\partial \Psi}{\partial t}\right)_\rho \tag{8.85}$$

where $F(t)$ is the applied force, may be cast into the "isolated" from

$$\left[-\frac{\hbar^2}{2\mu}\frac{\partial^2}{\partial \xi^2} + \tfrac{1}{2}k\xi^2\right]\Phi = i\hbar\left(\frac{\partial \Phi}{\partial t}\right)_\xi, \tag{8.86}$$

in the new time displaced variable, $\xi(\rho, t)$. It is readily verified that the necessary transformation functions $u(t)$, $v(t)$ and $w(t)$ are given by

$$\begin{aligned} &\mu\frac{\mathrm{d}^2 u}{\mathrm{d}t^2} + ku(t) = F(t), \\ &v(t) = i(\mu/\hbar)(\mathrm{d}u/\mathrm{d}t) \\ &w(t) = -\frac{i}{\hbar}\int_{-\infty}^{t}\left[\tfrac{1}{2}\mu\left(\frac{\mathrm{d}u}{\mathrm{d}t}\right)^2 - \tfrac{1}{2}ku^2\right]\mathrm{d}t. \end{aligned} \tag{8.87}$$

The important point is that the displacement function $u(t)$ must satisfy the classical equation of motion for an oscillator subject to the applied external force $F(t)$. The quantum mechanical solution derived from (8.86) is therefore centred on the classical trajectory;

$$\Phi_n(\xi, t) = \phi_n(\xi)\exp(-iE_n t/\hbar), \tag{8.88}$$

† This is not equivalent to $\exp\left[\sum_{i=0}^{N-1} \mathbf{A}(t_{i+1}, t_i)\right]$ because the matrices $\mathbf{A}(t_{i+1}, t_i)$ for different intervals will not in general commute.

where

$$\phi_n(\xi) = (\beta/\pi)^{\frac{1}{4}}(2^n - n!)^{-\frac{1}{2}} H_n(x)\, e^{-\frac{1}{2}x^2},$$

$$x = \beta^{\frac{1}{2}}\xi, \qquad \beta = (\mu k)^{\frac{1}{2}}/\hbar,$$

$$E_n = (n + \tfrac{1}{2})\hbar\omega, \qquad \omega = (k/\mu)^{\frac{1}{2}},$$

and $H_n(x)$ denotes the Hermite polynomial (Abramowitz and Stegun, 1965). Boundary conditions $u = du/dt = 0$ at $t = -\infty$ are set on equations (8.87) in order to ensure that initially $u(t) = v(t) = w(t) = 0$, and hence that the system starts from an undisplaced harmonic oscillator state,

$$\Phi_n(\rho, t) n_n^{t \to -\infty} \phi_n(\rho) \exp(-iE_n t/\hbar). \tag{8.89}$$

The probability amplitude for finding the system in the final state $\phi_m(\rho)$ $\exp(-iE_n t/\hbar)$ is therefore given by

$$S_{nm} = \lim_{t \to \infty} \left\{ \int_{-\infty}^{\infty} \phi_m(\rho)\, \Psi_n(\rho, t)\, d\rho \right\} \exp(iE_m t/\hbar)$$

$$= \lim_{t \to \infty} \left\{ \int_{-\infty}^{\infty} \phi_m(\rho)\phi_n(\rho - u) \exp\left[\rho v(t) + w(t) + \frac{i}{\hbar}(E_m - E_n)t\right] d\rho \right\}. \tag{8.90}$$

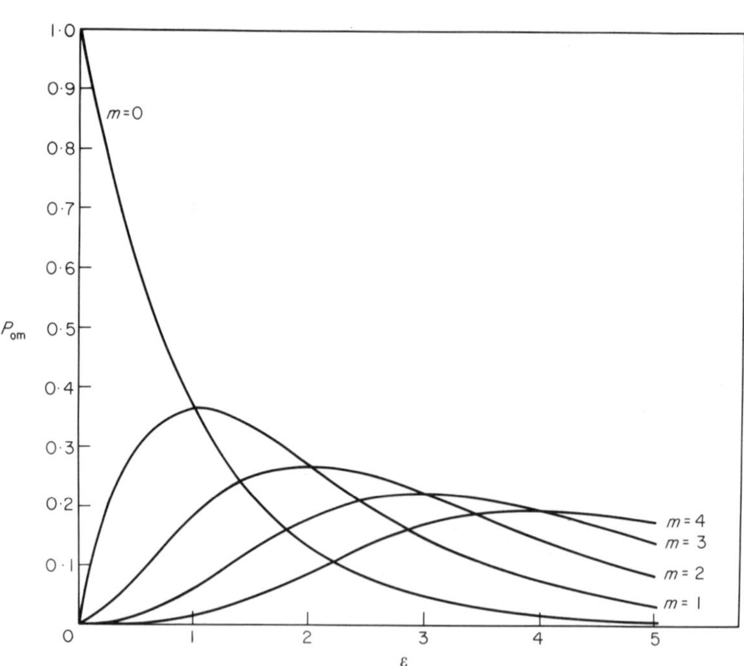

FIG. 8.4. 0 → m transition probabilities in the impulse approximation. ε denotes the mean energy transfer.

This reduces after evaluation of the necessary integrals (Kerner, 1958) to the general form

$$P_{nm} = |S_{nm}|^2 = \varepsilon^{m+n} e^{-\varepsilon} \left| \sum_{l=0}^{\min(n,m)} \frac{(-1)^l \varepsilon^{-l}}{l!\,(m-l)!\,(n-l)!} \right|^2, \tag{8.91}$$

where

$$\varepsilon = \lim_{t\to\infty} \left[\tfrac{1}{2}ku^2 + \tfrac{1}{2}\mu(\mathrm{d}u/\mathrm{d}t)^2\right]/\hbar\omega, \tag{8.92}$$

and min (n, m) denotes the lesser of m and n. ε is therefore the classical energy (expressed in units of the vibrational quantum $\hbar\omega$) imparted to the initially stationary oscillator by the transient force $F(t)$. This energy transfer is readily expressed in terms of the force $F(t)$ by integrating (8.87) to obtain

$$u(t) = (\mu\omega)^{-1} \int_{-\infty}^{t} F(t') \sin \omega(t - t')\,\mathrm{d}t', \tag{8.93}$$

from which, following (8.92)†

$$\begin{aligned} \varepsilon &= (2\mu\hbar\omega)^{-1} \left\{ \left[\int_{-\infty}^{\infty} F(t') \sin \omega t'\,\mathrm{d}t'\right]^2 + \left[\int_{-\infty}^{\infty} F(t') \cos \omega t'\,\mathrm{d}t'\right]^2 \right\} \\ &= (2\mu\hbar\omega)^{-1} \left| \int_{-\infty}^{\infty} F(t')\,e^{i\omega t'}\,\mathrm{d}t' \right|^2. \end{aligned} \tag{8.94}$$

The variation of P_{0m} given by (8.91) with the parameter ε is shown in Fig. 8.4. Note that P_{0m} reaches a maximum at the point $\varepsilon = m$, and that although the $0 \to 1$ transition is dominant at low energy transfers, this ceases to be true for $\varepsilon > 1$.

At this point the calculation is exact. In order, however, to convert a

---

† The same formula is obtained for the mean energy transferred to a non-stationary oscillator after averaging over the initial phase (Rapp and Kassal (1969)), and for the energy absorbed by a quantum oscillator averaged over all final states. The latter follows from the identity

$$\begin{aligned} \langle E - E_n \rangle &= \int_{-\infty}^{\infty} \Psi_n^*(\rho, t) \left[-\frac{\hbar^2}{2\mu}\frac{\partial^2}{\partial \rho^2} + \tfrac{1}{2}k\rho^2 - E_n \right] \Phi_n(\rho, t)\,\mathrm{d}\rho \\ &= \int_{-\infty}^{\infty} \phi_n(\rho - u)\,e^{-iv\rho} \left[-\frac{\hbar^2}{2\mu}\frac{\partial^2}{\partial \rho^2} + \tfrac{1}{2}k\rho^2 - E_n \right] \phi_n(\rho - u)\,e^{iv\rho}\,\mathrm{d}\rho \\ &= \int_{-\infty}^{\infty} \phi_n(\xi) \left[-\frac{\hbar^2}{2\mu}\frac{\partial^2}{\partial \xi^2} + \tfrac{1}{2}k\xi^2 - E_n + \tfrac{1}{2}\mu\left(\frac{\mathrm{d}u}{\mathrm{d}t}\right)^2 + \tfrac{1}{2}ku^2 - v\frac{\partial}{\partial \xi} + ku\xi \right] \phi_n(\xi)\,\mathrm{d}\xi \\ &= \tfrac{1}{2}\mu\left(\frac{\mathrm{d}u}{\mathrm{d}t}\right)^2 + \tfrac{1}{2}ku^2. \end{aligned}$$

known interaction potential $V(r, \rho)$ and collision energy into the required form $\rho F(t)$, some assumptions must be made. The obvious example for investigation is the Landau–Teller model treated by the distorted wave method in Section 7.3 and by time dependent perturbation theory in Section 8.2, which reduces in the dimensionless units of Section 7.3 to the model depicted in Fig. 8.5, with interaction potential

$$V = A \exp(-\alpha r_{ab}) \tag{8.95}$$

The simplest assumption required to convert this potential into a transient force $F(t)$ is that particle B is fixed, in which case it follows from the arguments

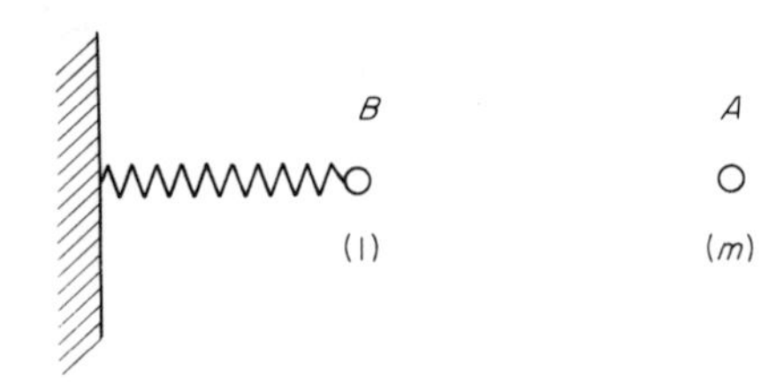

FIG. 8.5. The reduced harmonic oscillator excitation model.

of (8.69) and (8.70) that the time dependence of r_{ab} may be determined in the form

$$A \exp(-\alpha r_{ab}) = \tfrac{1}{2} m\bar{v}^2 \operatorname{sech}^2(\alpha\bar{v}t/2) \tag{8.96}$$

where $\bar{v}$ is the mean limiting velocity of A in the entrance and exit channels. Hence the force obtained from the gradient of V becomes

$$F(t) = -\left(\frac{\partial V}{\partial r_{ab}}\right) = \tfrac{1}{2}\alpha m\bar{v}^2 \operatorname{sech}^2(\alpha\bar{v}t/2) \tag{8.97}$$

An improvement to this expression has been obtained by Heidrich *et al.* (1971) by factoring out the AB centre of mass motion. The initial relative velocity is therefore again $\bar{v}$, but as a correction to the kinetic energy of relative motion, m must be replaced by the reduced mass $m/(1 + m)$. Hence in this approximation

$$F(t) = \tfrac{1}{2}\alpha\left(\frac{m}{1 + m}\right)\bar{v}^2 \operatorname{sech}^2(\alpha\bar{v}t/2). \tag{8.98}$$

The result, with this improved form for the force is according to (8.94), that

$$\varepsilon = 2\pi^2\left(\frac{m}{1 + m}\right)^2 \alpha^2 \operatorname{cosech}^2\left(\frac{\pi}{\alpha\bar{v}}\right). \tag{8.99}$$

Thus for small values of ε, ($\varepsilon \ll 1$), we have according to (8.91) for the $0 \to 1$ excitation probability,

$$P_{01} = \varepsilon\, e^{-\varepsilon} \simeq \varepsilon = 2\pi^2 \left(\frac{m}{1+m}\right)^2 \alpha^2 \operatorname{cosech}^2 \left(\frac{\pi}{\alpha \bar{v}}\right), \tag{8.100}$$

which is identical, apart from the presence of the reduced mass $[m/(1 + m)]$ in place of m, with the perturbation formula (8.72) obtained by Rapp and Sharp (1963). Equation (8.91) with ε given by (8.99) is however very much more accurate than (8.72) for large ε and for other than $\Delta n = \pm 1$ transition probabilities. Heidrich *et al.* (1971) find in fact that, with the reduced mass correction, (8.91) and (8.99) reproduce the exact numerical results of Secrest and Johnson (1966) for $\Delta n = \pm 1, \pm 2$ over a wide range of parameter values m and α. The method is however not readily adapted to other than harmonic oscillator problems.

Other problems normally classed as impulsive, which may be reduced by semi-classical assumptions to the forcing of a quantum harmonic oscillator, have been discussed by Morse and La Breeque (1971), and Levine and Bernstein (1971). Being based on Eqn (8.91), these models have the advantage that they can accommodate high transition probabilities and a wide distribution over final oscillator states.

8.5 Curve Crossing

The problem of an interaction between two states for which the potential terms $W_1(r)$ and $W_2(r)$ intersect at some point R, as illustrated in Fig. 8.6, is of considerable practical importance; it has moreover an approximate general analytical solution. First developed from a time dependent viewpoint by Landau (1932) and Zener (1932), and by phase integral methods by Stückel-

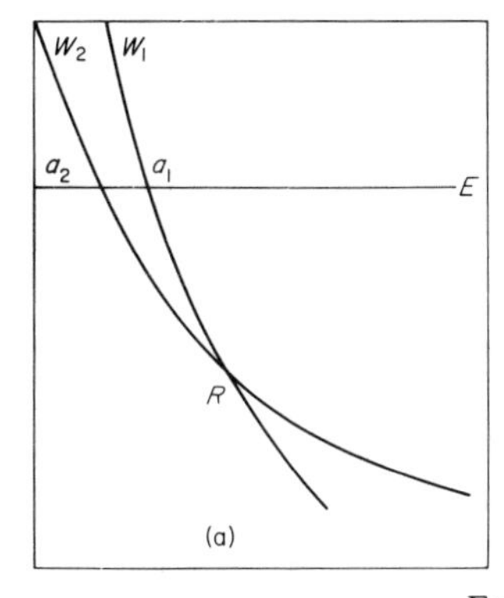

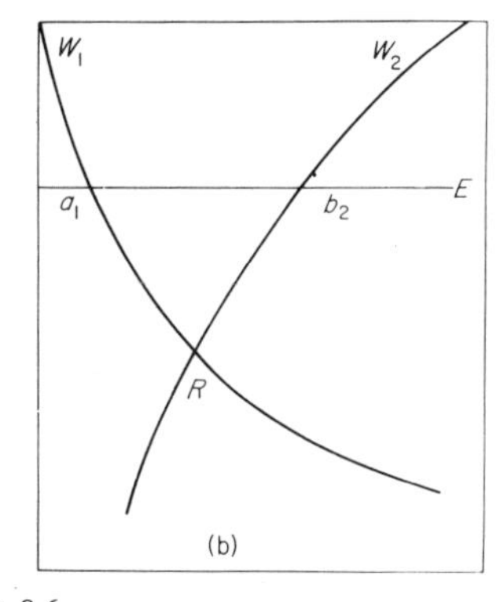

FIG. 8.6.

berg (1932) (later clarified by Crothers, 1971), the theory has also been refined by analysis in the momentum representation (Bykovskii *et al.*, 1964, Ovchinnikova, 1964), while Bates (1960) and Coulson and Zalewski (1962) have examined the practical validity of underlying model. The basic assumptions are that

$$W_i(r) = -F_i x, \qquad x = r - R \tag{8.101}$$

$$V_{12}(r) = V_{21}(r) = \text{const.} \tag{8.102}$$

$$v = \text{const.} \tag{8.103}$$

where v is the radial component of the velocity at the crossing point.† It is assumed in order to avoid subsequent ambiguities of phase that $F_1 > F_2$.

It is convenient, in order to assess the practical validity of (8.101)–(8.103) in physical terms, first to examine the problem from the time dependent viewpoint, based on (8.33). Taken together with (8.101)–(8.103), this results in the following equations:

$$\begin{aligned}\frac{\mathrm{d}b_1}{\mathrm{d}x} &= -i\frac{V_{12}}{\hbar v}\exp\left\{\frac{i}{\hbar v}\int_0^x [W_1(x') - W_2(x')]\,\mathrm{d}x'\right\} b_2(x)\\ &= -i\beta\exp\left(-\frac{i\alpha x^2}{2}\right) b_2(x)\\ \frac{\mathrm{d}b_2}{\mathrm{d}x} &= -i\frac{V_{12}}{\hbar v}\exp\left\{\frac{i}{\hbar v}\int_0^x [W_2(x') - W_1(x')\,\mathrm{d}x'\right\} b_1(x)\\ &= -i\beta\exp\left(\frac{i\alpha x^2}{2}\right) b_1(x),\end{aligned} \tag{8.104}$$

where, with t_x used to denote time at the crossing point

$$\begin{aligned}b_i(x) &= c_i(t)\exp\left[-\frac{i}{\hbar}\int_0^{t_x} W_i(t)\,\mathrm{d}t\right],\\ \alpha &= (F_1 - F_2)/\hbar v\\ \beta &= V_{12}/\hbar v.\end{aligned} \tag{8.105}$$

† Equations (8.101)–(8.103) apply to the radial equations in a partial wave formulation for a diatomic problem, with the $W_i(r)$ taken to include centrifugal terms $l(l+1)\hbar^2/2mr^2$. Alternatively in a classical trajectory development, with distance along the trajectory denoted by s, and the crossing point by S, we may write

$$W_i(s) = -F_i x, \qquad x = s - S \tag{8.101a}$$

$$V_{12}(s) = \text{const.} \tag{8.102a}$$

$$v = \text{const.} \tag{8.103a}$$

where v is now the actual velocity along the trajectory.

The next step is to eliminate $b_2(x)$ and to replace $b_1(x)$ by

$$d_1(x) = \exp(i\alpha x^2/4)b_1(x), \tag{8.106}$$

in order to reduce (8.104) to the second order equations

$$\left[\frac{d^2}{dx^2} + \beta^2 - \frac{i\alpha}{2} + \frac{\alpha^2 x^2}{4}\right]d_1 = 0, \tag{8.107}$$

details of the solution of which are given by Landau (1932), Zener (1932), and Delos and Thorson (1972). Since however (8.104)–(8.107) are shown in Section 8.6 to apply only to cases for which the signs of the F_i in (8.101a) are both positive, and since more accurate and more general results may be derived from the momentum representation, we do not choose to follow the solution of (8.107) in detail. We note merely that any approach to the problem relies on manipulating asymptotic solutions of a similar equation. The particular value of (8.107) for present purposes is that the independent variable is a physical distance. Hence we may use the general validity criterion on an asymptotic form (see Eqn (C.6)),

$$\left|\frac{d}{dx}\left[\beta^2 - \frac{i\alpha}{2} + \frac{\alpha^2 x^2}{4}\right]^{-\frac{1}{2}}\right| = \frac{2\alpha^{-1}x^{-2}}{\{1 + x^{-2}(4\beta^2/\alpha^2 - 2i/\alpha)\}^{\frac{3}{2}}} \ll 1, \tag{8.108}$$

to fix the width of the effective transition zone over which the assumptions of the model, (8.101)–(8.103), must apply. Sufficient conditions on the validity of (8.108) are that

$$\alpha x^2 \gg 2 \quad \text{and} \quad x^2 \gg |4\beta^2/\alpha^2 - 2i/\alpha|, \tag{8.109}$$

from which the width of the transition zone may be seen to be set by the larger of the two quantities

$$\Delta x_s = 2\beta/\alpha = 2V_{12}/(F_1 - F_2), \tag{8.110}$$

and

$$\Delta x_d = (2\alpha)^{-\frac{1}{2}} = [\hbar v/2(F_1 - F_2)]^{\frac{1}{2}}. \tag{8.111}$$

Δx_s will be termed the static width as distinct from the velocity dependent, dynamic width Δx_d.

Clearly Δx_s is simply the range over which V_{12} is significant compared with the energy difference $|W_1(x) - W_2(x)|$. The physical meaning of Δx_d (also termed the stationary phase region in Sections 7.2 and 8.2) on the other hand is most simply understood in the light of the uncertainty principle, by defining a "classical" momentum uncertainty over the transition zone which is related to the possibility that the motion may be governed either by $W_1(x)$ or $W_2(x)$. Thus at a point Δx from the crossing point

$$\Delta p = |p_1(x) - p_2(x)| \simeq |p_1^2(x) - p_2^2(x)|/2mv$$
$$= |W_1(x) - W_2(x)|/v = (F_1 - F_2)|\Delta x|/v \qquad (8.112)$$

according to which Δx_d given by (8.111) ensures that $\Delta x \Delta p \simeq \hbar$.

Theory based on the Landau–Zener formula (derived from (8.107)),

$$P = 1 - \exp(-2\pi\delta),$$
$$\delta = V_{12}^2/\hbar v(F_1 - F_2) \qquad (8.113)$$

for the probability of a transition between $W_1(r)$ and $W_2(r)$ on a single passage through the crossing point is therefore subject to two types of limitation. At sufficiently low velocities, Δx_s must span the classical turning point at which the constant velocity assumption (8.103) is clearly invalid, while the unlimited increase in Δx_d with increasing velocity implied by (8.111) must eventually invalidate the constant interaction, linear curve crossing assumptions (8.101) and (8.103). Modifications to the theory to meet the latter difficulty have been suggested by Coulson and Zalewski (1962), Dubrovskii (1964) and Child (1971), while weaknesses of the latter type may be handled by the methods outlined below. A final limitation to the Landau–Zener theory is the obvious one that only two internal states are included in the model. This raises difficulties, as emphasized by Bates (1960) in tackling high energy strong rotational coupling between the components of states which become degenerate in the united atom limit, as for example the coupling responsible for the $2p\sigma$–$2p\pi$ transition in the H_2^+ molecular ion (Bates and McCarroll, 1962; Knudson and Thorson, 1970; Bates and Sprevak, 1970). A blanket restriction to σ–σ interactions, as suggested by Bates (1960), which would exclude phenomena such as heterogeneous predissociation arising from an accidental crossing between states of different Λ, is however too severe.

We turn now to a detailed examination of the transition probabilities and phase changes accompanying a curve crossing under various conditions on the collision energy and on the values of V_{12}, F_1 and F_2 in (8.101)–(8.102). While the essential results are obtained in momentum space in Appendix D, thereby allowing a relaxation of the constant velocity assumption (8.103), it is convenient for the sake of practical applications to cast them into the more familiar coordinate form. This involves evaluation of the Fourier transforms

$$\psi_i(r) = (2\pi)^{-\frac{1}{2}} \int_{-\infty}^{\infty} u_i(k) \exp(ikx)\, dk, \qquad (8.114)$$

which connect the coordinate and momentum representatives $\psi_i(r)$ and $u_i(k)$ respectively, of the radial components of the time independent wavefunction given for example by (6.5).

It is first assumed that $F_1 > F_2 > 0$ in which case, after substitution for $u_i(k)$ with the help of (D.5)

$$\psi_i(r) = (2\pi)^{-\frac{1}{2}} \int_{-\infty}^{\infty} A_i(k) \exp\left\{\frac{i}{f_i}[(\varepsilon + f_i x)k - k^3/3]\right\} dk. \tag{8.115}$$

The integral may now be approximated for large positive x by the method of steepest descents, (see Jeffreys and Jeffreys, 1956, p 508; Landau and Lifshitz, 1965, Appendix B), because the saddle points of the exponent lie at

$$k = \pm(\varepsilon + f_i x)^{\frac{1}{2}} = \pm k_i(x) \text{ say,} \tag{8.116}$$

and the $A_i(k)$ are shown in appendix D to take constant values as $k \to \pm\infty$. Hence on displacing the integration contour to pass through the saddle

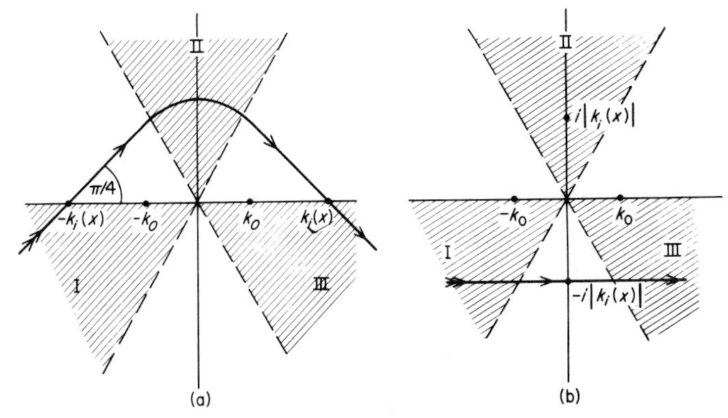

FIG. 8.7. Integration contours for x in (a) the classically accessible and (b) the classically inaccessible region. $\pm k_0$ correspond to outward and inward motion at the crossing point.

points along the lines of steepest descent, as shown in Fig. 8.7(a), $\psi_i(r)$ takes on the JWKB form

$$\psi_i(r) \overset{r\to\infty}{\sim} \frac{A_i(\infty)}{[k_i(x)]^{\frac{1}{2}}} \exp\left[i\int_{a_i}^{r} k_i(r)\,dr - i\pi/4\right] + \frac{A_i(-\infty)}{[k_i(x)]^{\frac{1}{2}}} \exp\left[-i\int_{a_i}^{r} k_i(r) + i\pi/4\right], \tag{8.117}$$

where we have made the identity

$$\int_{a_i}^{r} k_i(r)\,dr = \int_{-(\varepsilon/f_i)}^{x} (\varepsilon + f_i x)^{\frac{1}{2}}\,dx = \frac{2}{3f_i}[k_i(x)]^3. \tag{8.118}$$

It may also be verified by following the contour in Fig. 8.7(b) for $x \ll 0$, that

$\psi_i(r)$ decreases exponentially into the non-classical region, $x < -\varepsilon/f_i$, as required of the bounded solution.† It follows from (6.34) that the elements of the S matrix may be written

$$S_{ij}^{(l)} = [A_{jl}(\infty)/A_{il}(-\infty)] \exp i(\eta_{il} + \eta_{jl}), \tag{8.119}$$

where η_{il} and η_{jl} are the familiar JWKB phase shifts

$$\eta_{il} = \lim_{r\to\infty} \left\{ \int_{a_i}^{r} k_{il}(r)\,\mathrm{d}r - k_i(r) + (l + \tfrac{1}{2})\pi/2 \right\}, \tag{8.120}$$

the additional subscript l being temporarily inserted to act as a reminder that the $W_i(r)$ in (8.101) contain centrifugal terms.

Different results are obtained in Appendix D according to the values of two dimensionless parameters, given in the notation of (8.101) and (8.103) by

$$\begin{aligned} a^2 &= F(F_1 - F_2)\hbar^2/16mV_{12}^3 \\ b^2 &= E(F_1 - F_2)/2FV_{12}, \end{aligned} \tag{8.121}$$

where

$$\begin{aligned} E &= \tfrac{1}{2}mv^2 \\ F &= (F_1F_2)^{\frac{1}{2}}. \end{aligned} \tag{8.122}$$

The first case (termed sub-case (1a) in Appendix D) for which $b^2 \gg 1$ and $b^3/a \gg 1$ covers the Landau–Zener model, because although there is no explicit constant velocity assumption, these inequalities ensure that

$$\Delta x_s = 2V_{12}/(F_1 - F_2) \ll E/F$$

and

$$\Delta x_d = [\hbar v/2(F_1 - F_2)]^{\frac{1}{2}} \ll E/F \tag{8.123}$$

and the point $-E/F$ may be identified, according to the discussion at the end of Section 8.6, with the turning point on the proper mean trajectory. There is therefore no question of overlap between the transition zone and this turning point. The result obtained from (8.119) and (D.40) under these conditions is

$$\mathbf{S} = \begin{pmatrix} S'_{11}\,\mathrm{e}^{2i\eta_1}, & S'_{12}\,\mathrm{e}^{i(\eta_1+\eta_2)} \\ S'_{21}\,\mathrm{e}^{i(\eta_1+\eta_2)}, & S'_{22}\,\mathrm{e}^{2i\eta_2} \end{pmatrix}, \tag{8.124}$$

† Solutions $\psi_i(r)$ valid only in the classical region may also be derived from integration contours between areas I and II or II and III in Fig. 8.7(a), provided they do not pass through the transition zones around the points $\pm k_0$. Such contours provide expressions for $\psi_i(r)$ at short and long range in terms of the coefficients X_i and $A_i(\pm\infty)$ in (D.20) respectively. It is then possible with the help of (D.36) and (D.39) to follow changes in the $\psi_i(r)$ due to a single passage through the crossing point. Such knowledge may be valuable in multiple curve crossing problems (see Woolley, 1971).

where

$$\begin{aligned} S'_{11} &= S'^{*}_{22} = e^{-2\pi\delta} + (1 - e^{-2\pi\delta})\, e^{-2i\phi} \\ S'_{12} &= S'_{21} = 2i\, e^{-\pi\delta}(1 - e^{-2\pi\delta})^{\frac{1}{2}} \sin\phi \end{aligned} \tag{8.125}$$

and

$$\begin{aligned} \delta &= 1/8ab = V_{12}^2/\hbar v(F_1 - F_2), \\ \phi &= -\tfrac{2}{3}(b^3/a) - 2\delta \ln(2b/a) + \delta \ln 4\delta + \arg\Gamma(i\delta) + \pi/4. \end{aligned} \tag{8.126}$$

Hence to the extent that $\sin\phi^2$ may be replaced by its average value of one half, the overall transition probability for a double passage problem becomes

$$\overline{P_{12}} = |\overline{S_{12}}|^2 = 2\, e^{-2\pi\delta}(1 - e^{-2\pi\delta}), \tag{8.127}$$

as may be deduced from (8.113). This means that $P_{12} = 0$ for both $\delta = 0$ and $\delta \to \infty$, in the former case because the system remains in one of the diabatic internal states $\phi_i(\rho)$, either because $V_{12} = 0$ or because infinite velocity allows insufficient time for a transition to occur. At the opposite limit $\delta \to \infty$ on the other hand, the necessary low velocity or strong interaction, causes a very slow variation with distance in the forms of the adiabatic states

$$\begin{pmatrix} \phi_+(\rho;r) \\ \phi_-(\rho;r) \end{pmatrix} = \begin{pmatrix} \cos\theta, & \sin\theta \\ -\sin\theta, & \cos\theta \end{pmatrix} \begin{pmatrix} \phi_1(\rho) \\ \phi_2(\rho) \end{pmatrix} \tag{8.128}$$

with

$$\theta(r) = \tfrac{1}{2}\tan^{-1}\left\{\frac{2V_{12}(r)}{W_1(r) - W_2(r)}\right\}$$

and hence allows an adiabatic (Born–Oppenheimer) relaxation of the internal state, with the result that the motion is governed by one or other of the adiabatic terms

$$W_\pm(r) = \tfrac{1}{2}[W_1(r) + W_2(r)] \pm \tfrac{1}{2}\{[W_1(r) - W_2(r)]^2 + 4V_{12}^2\}^{\frac{1}{2}}, \tag{8.129}$$

rather than by $W_1(r)$ or $W_2(r)$.

Circumstances may also arise where oscillations in the S matrix elements, first predicted by Stuckelberg (1932), have important effects, notably in the interpretation of the high resolution differential cross-section, and of detailed resonance linewidth patterns discussed below. Bandrauk and Child (1970) have shown that the quantity ϕ, defined by (8.126), responsible for these oscillations may also be expressed as a phase integral correction appropriate to a switch from $W_1(r)$ to $W_-(r)$ or $W_2(r)$ to $W_+(r)$ in crossing the transition zone,†

† Differences between (8.130) and equations (28) and (29) of Bandrauk and Child (1970) arise from the assumption $F_2 > F_1$ employed by the latter.

$$\phi = \int_{a_1}^{r_0} k_1(r)\,\mathrm{d}r - \int_{a_-}^{r_0} k_-(r) + \chi$$

or

$$\phi = \int_{a_+}^{r_0} k_+(r)\,\mathrm{d}r - \int_{a_2}^{r_0} k_2(r) + \chi, \tag{8.130}$$

where the point r_0 is taken sufficiently far to the right of the crossing point that $W_-(r) \simeq W_1(r)$ and $W_+(r) \simeq W_2(r)$ for $r > r_0$. The additional correction term

$$\chi = \arg\Gamma(i\delta) + \delta - \delta\ln\delta + \pi/4 \tag{8.131}$$

is typically small; $\chi = -\pi/4$ for $\delta = 0$ and $\chi \sim -1/12\delta$ for $\delta \gg 1$.

Other cases amenable to analysis in closed form arise when $|\delta| \ll 1$ and when the energy is so low that $b^2 \ll -1$, $|b^3/a| \gg 1$. In the former we have, according to (8.119) and the perturbation formula (D.60)

$$S'_{12} = S_{12}\,\mathrm{e}^{-i(\eta_1+\eta_2)} = -i\pi a^{\frac{2}{3}}\,\mathrm{Ai}(-b^2a^{-\frac{2}{3}})$$

$$\overset{b^2 \gg a^{\frac{2}{3}}}{\sim} -4i\pi\delta\sin\left(\tfrac{2}{3}b^3/a + \pi/4\right) = 4i\pi\delta\sin\left(\int_{a_2}^{R} k_2(r)\,\mathrm{d}r - \int_{a_1}^{R} k_1(r)\,\mathrm{d}r + \pi/4\right)$$

$$\overset{b^2 \ll -a^{\frac{2}{3}}}{\sim} -\ 2i\pi\delta\exp\left(-\tfrac{2}{3}|b^3/a|\right), \tag{8.132}$$

a result also given by (7.70) and (8.65). In the latter limit, $b^2 \ll -1$, $|b^3/a| \gg 1$, on the other hand it follows from (D.56) that

$$P_{12} = |S_{12}|^2 = B(|\delta|)\exp(-\tfrac{2}{3}|b^3/a|), \tag{8.133}$$

where

$$\begin{aligned} B(|\delta|) &= 2\pi^{-1}|\delta|\sin^2\pi|\delta|[\Gamma(|\delta|)]^2|\delta|^{-2|\delta|}\exp[2|\delta|] \\ &\simeq 2\pi|\delta|, \qquad |\delta| \ll 0{\cdot}1 \\ &\simeq 4\sin^2|\delta|, \qquad |\delta| \gg 1. \end{aligned} \tag{8.134}$$

Numerical values of $B(|\delta|)$ in the intermediate range are given in Appendix D.

Analogues of the above results are also available when the slopes of $W_1(r)$ and $W_2(r)$ have opposite signs, $F_1 > 0$ and $F_2 < 0$, in which case the parameters a and b defined by (8.121) may be retained except that

$$F = |F_1F_2|^{\frac{1}{2}}. \tag{8.135}$$

The equivalent of (8.117) is then (Child (1969))

$$\psi_1(r) \overset{r \gg a_1}{\sim} \frac{A_1(\infty)}{[k_1(r)]^{\frac{1}{2}}}\exp\left(i\int_{a_1}^{r} k_1(r)\,\mathrm{d}r - i\pi/4\right)$$

$$+\frac{A_1(-\infty)}{[k_1(r)]^{\frac{1}{2}}}\exp\left(-i\int_{a_1}^{r}k_1(r)+i\pi/4\right)$$

$$\psi_2(r)\overset{r\ll b_2}{\sim}\frac{A_2(\infty)}{[k_2(r)]^{\frac{1}{2}}}\exp\left(i\int_{b_2}^{r}k_1(r)\,\mathrm{d}r+i\pi/4\right)$$

$$+\frac{A_2(-\infty)}{[k_2(r)]^{\frac{1}{2}}}\exp\left(-i\int_{b_2}^{r}k_2(r)-i\pi/4\right), \qquad (8.136)$$

where b_2 is the (right-hand) turning point on $W_2(r)$ as shown in Fig. 8.6(b). Note that since classical motions governed by $W_1(r)$ and $W_2(r)$ lie to the right of a_1 and to the left of b_2 respectively we must expect a unitary connection between $(A_1(-\infty), A_2(\infty))$ and $(A_1(\infty), A_2(-\infty))$, rather than between $(A_1(-\infty), A_2(-\infty))$ and $(A_1(\infty), A_2(\infty))$ as in the previous case. This point is amplified by the discussed after equation (D.65).

This means that under the Landau–Zener conditions, $b^2 \gg 1$, $(b^3/a) \gg 1$, the matrix $\mathbf{P}$ which connects $\mathbf{A}(\infty)$ and $\mathbf{A}(-\infty)$ by the equation

$$\mathbf{A}(\infty)=\mathbf{PA}(-\infty) \qquad (8.137)$$

is as given by (D.70) no longer unitary;

$$\mathbf{P}=\begin{pmatrix}\mathrm{e}^{2\pi\delta}+(\mathrm{e}^{2\pi\delta}-1)\,\mathrm{e}^{-2i\tilde{\phi}}, & -2i\,\mathrm{e}^{\pi\delta}(\mathrm{e}^{2\pi\delta}-1)^{\frac{1}{2}}\cos\tilde{\phi}\\ 2i\,\mathrm{e}^{\pi\delta}(\mathrm{e}^{2\pi\delta}-1)^{\frac{1}{2}}\cos\tilde{\phi}, & \mathrm{e}^{2\pi\delta}+(\mathrm{e}^{2\pi\delta}-1)\,\mathrm{e}^{2i\tilde{\phi}}\end{pmatrix}. \qquad (8.138)$$

The parameter δ is however again given by (8.126), while (Child, 1969),

$$\tilde{\phi}=\tfrac{2}{3}(b^3/a)+\arg\Gamma(i\delta)+\delta\ln 4\delta-2\delta\ln(2b/a)+\pi/4$$

$$\simeq\int_{a_+}^{b_+}k_+(r)\,\mathrm{d}r+\chi$$

$$\simeq\int_{a_1}^{r_1}k_1(r)\,\mathrm{d}r+\int_{r_2}^{b_2}k_2(r)\,\mathrm{d}r-\int_{r_2}^{r_1}k_-(r)\,\mathrm{d}r+\chi, \qquad (8.139)$$

with χ defined by (8.131) and r_1 and r_2 chosen such that $W_1(r)\simeq W_-(r)$ for $r>r_1$ and $W_2(r)\simeq W_-(r)$ for $r<r_2$.

A perturbation formula similar to that given by (8.132) may also be derived when $|\delta|\ll 1$, namely

$$[A_1(\infty)/A_2(-\infty)]=[A_2(\infty)/A_1(-\infty)]^*=-i\pi a^{-\frac{2}{3}}\,\mathrm{Ai}\,(-b^2/a^{\frac{4}{3}})$$

$$\overset{b^2\gg a^{\frac{2}{3}}}{\sim}-4i\pi\delta\sin(\tfrac{2}{3}b^3/a+\pi/4)=-4\pi i\delta\sin\left(\int_{a_1}^{R}k_1(r)\,\mathrm{d}r+\int_{R}^{b_2}k_2(r)\,\mathrm{d}r+\pi/4\right),$$

$$\overset{b^2\ll -a^{\frac{2}{3}}}{<}-2i\pi|\delta|\exp(-\tfrac{2}{3}|b^3/a|). \qquad (8.140)$$

Finally at energies well below the crossing point, the probability of penetrating from $r > a_1$ to the region $r < b_2$ is given according to (D.75) by

$$T = \tilde{B}(|\delta|) \exp\left(-2\left|\int_{a_-}^{b_-} k_-(r)\,dr\right|\right), \tag{8.141}$$

where

$$\begin{aligned} \tilde{B}(|\delta|) &= 1, \qquad &|\delta| \gg 1 \\ &\simeq 2\pi|\delta|, \qquad &|\delta| \ll 0{\cdot}1. \end{aligned} \tag{8.142}$$

The exponential term in (8.141) may be recognized as the normal semi-classical tunnelling coefficient for the lower adiabatic potential barrier $W_-(r)$ given by (8.129) (see Landau and Lifshitz, 1965). Hence $B(|\delta|)$ has the role of a non-adiabatic correction factor to account for the influence of the impenetrable term $W_1(r)$.

Two examples of the application of the above results may now be considered. The first concerns the effect of the oscillations in S_{12} given by (8.125) or (8.132) on the form of the inelastic scattering amplitude

$$f_{12}(\theta) = \frac{1}{2i[k_1 k_2]^{\frac{1}{2}}} \sum_{l=0}^{\infty} (2l + 1) S_{12}^{(l)} P_l(\cos\theta). \tag{8.143}$$

It is assumed that semi-classical conditions apply in the sense that $f_{12}(\theta)$ may be evaluated by the stationary phase methods of Section 5.1. This requires in view of its oscillatory character, that the $S_{12}^{(l)}$ must be divided into two branches, with the phases given by

$$\begin{aligned} \nu_+ &= \eta_1 + \eta_2 + \phi = \eta_1 + \eta_+ + \chi \\ \nu_- &= \eta_1 + \eta_2 - \phi = \eta_2 - \eta_- - \chi, \end{aligned} \tag{8.144}$$

where, according to (8.130)

$$\eta_+ = \eta_2 + \int_{a_+}^{r_0} k_+(r)\,dr - \int_{a_2}^{r_0} k_2(r)\,dr = \lim_{r\to\infty}\left\{\int_{a_+}^{r} k_+(r)\,dr - k_2 r + (l+\tfrac{1}{2})\pi/2\right\},$$

because $k_+(r) \simeq k_2(r)$ for $r > r_0$. Similarly

$$\eta_- = \eta_1 - \int_{a_1}^{r_0} k_1(r)\,dr + \int_{a_-}^{r} k_-(r)\,dr = \lim_{r\to\infty}\left\{\int_{a_-}^{r} k_-(r)\,dr - k_1 r + (l+\tfrac{1}{2})\pi/2\right\}. \tag{8.145}$$

Hence the phase ν_+ is appropriate to incoming motion governed by $W_1(r)$ followed by a switch to $W_+(r)$, while ν_- arises from a switch from $W_-(r)$ to $W_2(r)$ (see Fig. 8.6(a)). Oscillations in the differential cross-section for the process

$$\text{He} + \text{Ne}(2p^6) \rightarrow \text{He} + \text{Ne}(2p^5 3s)$$

have been attributed to interference between these two types of trajectory; furthermore Olson and Smith (1971) have been able, with the help of (8.132), to deduce the forms of $W_1(r)$ and $W_2(r)$ and the value of the interaction constant V_{12}.

As a second example we consider the pattern of linewidths for the Feshbach resonances (see Section 6.4) supported by a potential $W_2(r)$ of which the curves in Figs 8.6(a) or 8.6(b) are the repulsive or attractive limbs respectively, as shown in Fig. 8.8. The presence of the second turning point b_2 in the inner

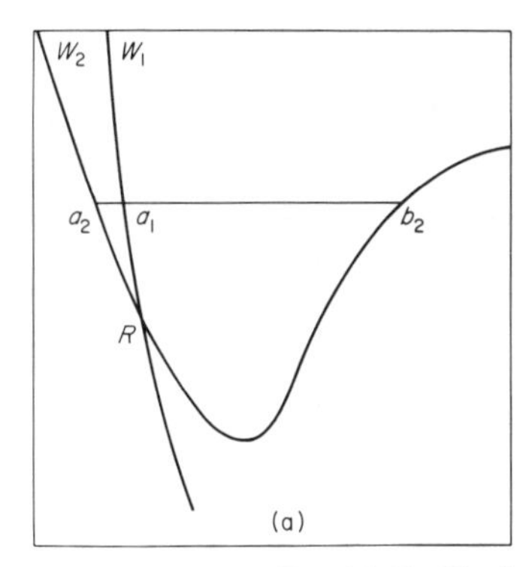

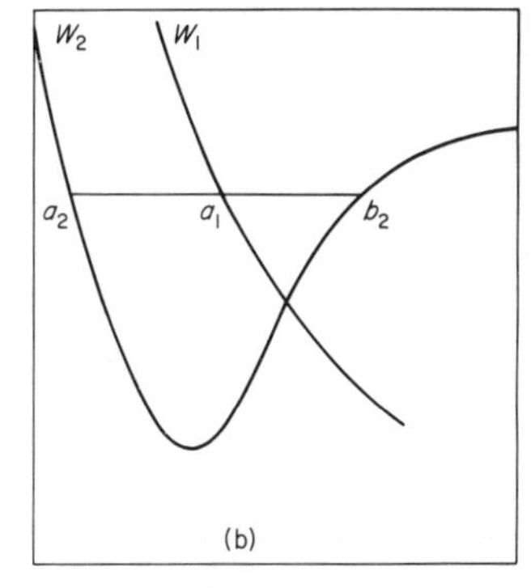

FIG. 8.8. Feshbach resonance situations.

crossing case (Fig. 8.8(a)) for example, sets another boundary condition, namely that $\psi_2(r)$ given by (8.121) must behave according to (C.33) as,

$$\psi_2(r) \overset{r \ll b_2}{\sim} C[k_2(r)]^{-\frac{1}{2}} \sin\left[\int_r^{b_2} k_2(r)\,\mathrm{d}r - \pi/4\right], \tag{8.146}$$

This requires by comparison with (8.117) that

$$[A_2(\infty)/A_2(-\infty)] = -\exp(-2i\beta) \tag{8.147}$$

where

$$\beta = \int_{a_2}^{b_2} k_2(r)\,\mathrm{d}r, \tag{8.148}$$

and hence by elimination of $A_2(\pm\infty)$ from (D.40) that

$$S_{11} = \left[\frac{\cos\beta + u\exp(-i\phi)\cos(\beta+\phi)}{\cos\beta + u\exp(i\phi)\cos(\beta+\phi)}\right]\exp 2i\eta_1, \tag{8.149}$$

with

$$u = \exp(2\pi\delta) - 1. \tag{8.150}$$

It follows from the general arguments of Section 4.4 that the complex resonance points

$$E = E_n - i\Gamma_n/2, \tag{8.151}$$

corresponding to poles in the S matrix, lie at the roots of the equation

$$\cos\beta + u\exp(i\phi)\cos(\beta + \phi) = 0. \tag{8.152}$$

Analytical expressions for E_n and Γ_n are available when $u \ll 1$ or $u \gg 1$. In the former weak interaction limit, the first term is dominant except in regions around the energies E_n^0 at which

$$\beta = \int_{a_2}^{b_2} k_2(r)\,\mathrm{d}r = (n + \tfrac{1}{2})\pi. \tag{8.153}$$

Such energies coincide, by the Bohr quantization condition, with the bound states supported by $W_2(r)$. A first order expansion of $\beta(E)$ around such points leads to the resonance conditions

$$\begin{aligned} E_n &= E_n^{(0)} - \delta_n \sin 2\phi_n \,.\, \hbar\omega_n \\ \Gamma_n &= 4\delta_n \sin^2\phi_n \,.\, \hbar\omega_n, \end{aligned} \tag{8.154}$$

where $\hbar\omega_n$ denotes the local energy spacing

$$\hbar\omega_n = \left(\frac{\partial E}{\partial n}\right) = \left(\frac{\partial E}{\partial \beta}\right)\left(\frac{\partial \beta}{\partial n}\right) = \pi\left(\frac{\partial E}{\partial \beta}\right), \tag{8.155}$$

and the subscript n signifies a value taken at the unperturbed energy $E_n^{(0)}$. Oscillatory variations in E_n and Γ_n are therefore anticipated according to the local value of ϕ_n given by (8.130), which reduces in the weak interaction limit to the form

$$\phi \simeq -\left[\int_{a_2}^{R} k_2(r)\,\mathrm{d}r - \int_{a_1}^{R} k_1(r)\,\mathrm{d}r + \pi/4\right]. \tag{8.156}$$

A general theory for inversion of such linewidth oscillations (obtained for example from spectroscopic predissociation measurements) to yield the forms of $W_1(r)$, $W_2(r)$ and V_{12} has been given by Child (1970, 1972) and Ramsay and Child (1971).

Similar theory applies to the strong interaction limit, $u \gg 1$, except that the roots of (8.152) lie close to the $E_{n'}^{(0)}$ at which

$$\begin{aligned} \beta + \phi &= \int_{a_2}^{b_2} k_2(r)\,\mathrm{d}r + \int_{a_+}^{r_0} k_+(r)\,\mathrm{d}r - \int_{a_2}^{r_0} k_2(r) + \chi \\ &\simeq \int_{a_+}^{b_+} k_+(r)\,\mathrm{d}r = (n' + \tfrac{1}{2})\pi. \end{aligned} \tag{8.157}$$

These coincide with the bound levels supported by the upper adiabatic curve $W_+(r)$. The equivalent formulae to (8.154) may be shown to be

$$E_{n'} = E_{n'}^{(0)} + (2\pi)^{-1} \exp(-2\pi\delta_{n'}) \sin 2\beta_{n'} \,.\, \hbar\omega_{n'}$$
$$\Gamma_{n'} = (2/\pi) \exp(-2\pi\delta_{n'}) \cos^2 \beta_{n'} \,.\, \hbar\omega_{n'}. \qquad (8.158)$$

Analogous results to cover the outer curve crossing case (Fig. 8.10(b)) have been given by Bandrauk and Child (1970), while Child (1969) has investigated the effects of such resonances on the elastic scattering cross-sections in Channel 1.

As a final indication of the scope of the Landau–Zener theory, we consider an extension described by Nikitin (1968) to cover the possibility of a transition from one two dimensional surface to another in the neighbourhood of a conical intersection, or an avoided conical intersection as shown in Fig. 8.9.

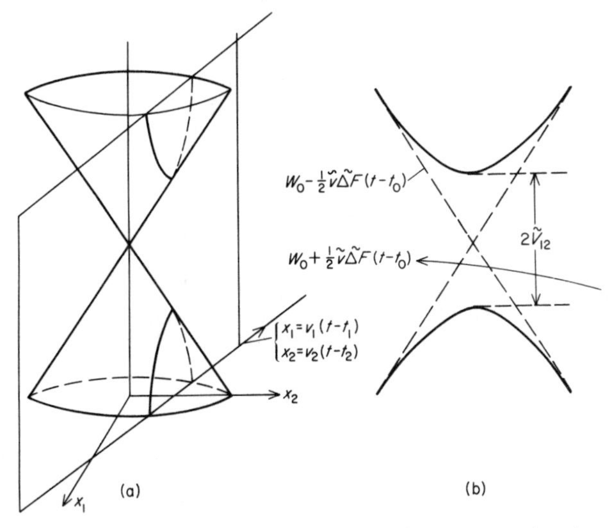

FIG. 8.9. (a) A section through the double cone cut by the trajectory $x_1 = v_1(t-t_2)$, $x_2 = v_2(t-t_2)$; (b) the corresponding curves.

The model relies on a linear expansion of the potential matrix close to the intersection point

$$V(x_1\, x_2) = \begin{pmatrix} W_0 + \frac{1}{2}\Delta F_1 x_1, & iA + \frac{1}{2}\Delta F_2 x_2 \\ -iA + \frac{1}{2}\Delta F_2 x_2, & W_0 - \frac{1}{2}\Delta F_1 x_1 \end{pmatrix}, \qquad (8.159)$$

together with the assumption of a linear trajectory

$$x_1 = v_1(t - t_1)$$
$$x_2 = v_2(t - t_2) \qquad (8.160)$$

across the surface. The constant term A in (8.159) is supposed to arise from spin–orbit coupling, while other terms are electrostatic in origin. The results of the theory outlined by Nikitin (1968) are readily understood from the forms of the adiabatic eigensurfaces derived from $V(x_1, x_2)$ in (8.159);

$$V_{\pm}(x_1, x_2) = W_0 \pm \tfrac{1}{2}[4A^2 + \Delta F_1^2 x_1^2 + \Delta F_2^2 x_2^2]^{\frac{1}{2}} \tag{8.161}$$

as illustrated in Fig. 8.9.†

The important point is that the linear trajectory (8.160) cuts a section through these surfaces, along which the one dimensional forms

$$V_{\pm}(t) = W_0 \pm \tfrac{1}{2}[4A^2 + \alpha_1^2(t - t_1)^2 + \alpha_2^2(t - t_2)^2]^{\frac{1}{2}}, \tag{8.162}$$

with $\alpha_i = v_i \Delta F_i$, are identical with the adiabatic terms derived from a Landau–Zener model based on the two linear diagonal terms,

$$W = W_0 \pm \tfrac{1}{2}\tilde{v}\widetilde{\Delta F}(t - t_0), \tag{8.163}$$

where

$$\tilde{v}\widetilde{\Delta F} = (\alpha_1^2 + \alpha_2^2)^{\frac{1}{2}} = (v_1^2 \Delta F_1^2 + v_2^2 \Delta F_2^2)^{\frac{1}{2}}$$
$$t_0 = (\alpha_1^2 t_1 + \alpha_2^2 t_2)/(\alpha_1^2 + \alpha_2^2), \tag{8.164}$$

which intersect at t_0, together with a constant interaction term

$$\tilde{V}_{12} = [A^2 + \alpha_1^2\alpha_2^2(t_1 - t_2)^2/(\alpha_1^2 + \alpha_2^2)]^{\frac{1}{2}}. \tag{8.165}$$

Hence the probability of a transition from one surface to another arising from a single traverse past the intersection point, $(x_1, x_2) = (0, 0)$, along the defined trajectory, is given by the analogue of (8.113);

$$P = \exp(-2\pi \tilde{V}_{12}^2/\hbar\tilde{v}\widetilde{\Delta F}). \tag{8.166}$$

8.6 Validity of the Classical Trajectory Approximation

In assessing the validity of the classical trajectory approximation employed in Sections 8.2 and 8.3, it is sufficient for illustrative purposes to consider systems with spherical symmetry. Our purpose is to examine the relation between the exact time independent radial Eqns (6.56),

$$\left[\frac{d^2}{dr^2} + k_{nl}^2(r)\right]\psi_{nl}(r) = \sum_{m \neq n} U_{nm}(r)\,\psi_{ml}(r) \tag{8.167}$$

where‡

† Note that the existence of A destroys the possibility of a true conical intersection (Herzberg, 1967 p 442) between $V_{\pm}(x_1, x_2)$.

‡ This assumes the Langer correction (see Section 4.2).

$$k_{nl}^2(r) = 2m(E - W_n(r) - (l + \tfrac{1}{2})^2\hbar^2/2mr^2)/\hbar^2,$$
$$U_{nm}(r) = 2mV_{nm}(r)/\hbar^2, \tag{8.168}$$

and the classical trajectory equations, most conveniently given by (8.33)

$$i\hbar\frac{dc_n}{dt} = \sum_{m \neq n} V_{nm}(r) \exp\left\{\frac{i}{\hbar}\int_0^t [W_n(r) - W_m(r)]\, dt\right\} c_m(t), \tag{8.169}$$

which are to be integrated along a trajectory $\mathbf{r}(t)$ with the same conserved angular momentum $(l + \frac{1}{2})\hbar$.

Any connection between (8.167) and (8.169) must clearly depend on the properties of the assumed trajectory, which is governed by the same energy E, and the angular momentum $(l + \frac{1}{2})\hbar$, and an appropriate mean central potential $\overline{W}(r)$. Of particular importance is the relation between position and time given by (2.3);

$$\begin{aligned}\frac{dr}{dt} &= \pm v\left[1 - \frac{\overline{W}(r)}{E} - \frac{b^2}{r^2}\right]^{\frac{1}{2}} \\ &= \pm\frac{\hbar}{m}[k^2 - \overline{U}(r) - (l + \tfrac{1}{2})^2/r^2]^{\frac{1}{2}} \\ &= \pm\frac{\hbar}{m}\bar{k}_l(r),\end{aligned} \tag{8.170}$$

where

$$mvb = (l + \tfrac{1}{2})\hbar, \qquad E = \tfrac{1}{2}mv^2,$$
$$k^2 = 2mE/\hbar^2, \qquad \overline{U} = 2m\overline{W}(r)/\hbar^2; \tag{8.171}$$

the upper and lower signs in (8.170) refer to outward and inward motion respectively. We may also recognize the existence of an equivalent semi-classical elastically scattered wave $\psi_l(r)$,

$$\psi_l(r) = f_l^{(+)}(r) - f_l^{(-)}(r), \tag{8.172}$$

where

$$f_l^{(\pm)}(r) = [\bar{v}_l(r)]^{-\frac{1}{2}} \exp\left[\pm i\int_{a_l}^r \bar{k}_l(r)\, dr \pm i\pi/4\right], \tag{8.173}$$

subject to the validity criterion

$$[d \ln \bar{k}_l(r)/dr] \ll \bar{k}_l(r), \tag{8.174}$$

the normalization in (8.173) being taken to ensure unit incoming and outgoing flux, and the phase terms $\pm i\pi/4$ being included to ensure a proper connection at the classical turning point a_l where (8.174) becomes invalid.

The general method, initiated by Cross (1969), Bates and Crothers (1970) and Delos *et al.* (1972), is now to employ the functions $f_l^{(\pm)}(r)$ as carrier waves in the time independent wavefunction. Thus Eqns (8.167) are rewritten,

$$\left[\frac{\mathrm{d}^2}{\mathrm{d}r^2} + \bar{k}_l^2(r)\right]\psi_{nl}(r) = [\bar{k}_l^2 - \bar{k}_{nl}^2]\psi_{nl}(r) + \sum_{m \neq n} U_{nm}(r)\,\psi_{ml}(r), \quad (8.175)$$

and solutions are sought in the form

$$\psi_{nl}^{(\pm)}(r) = \alpha_n^{(\pm)}(r) f_l^{(\pm)}(r), \quad (8.176)$$

with the $\alpha_n^{(\pm)}(r)$ taken as slowly varying functions of r in a sense to be made more precise below. The full solution is then written

$$\psi_{nl}(r) = \psi_{nl}^{(+)}(r) + \psi_{nl}^{(-)}(r), \quad (8.177)$$

with

$$\alpha_n^{(+)}(a_l) + \alpha_n^{(-)}(a_l) = 0, \quad (8.178)$$

in order to ensure a correct description at the classical turning point. Note that the $(\pm)$ separation implied by (8.176) is inherent in any trajectory formulation since inelastic coupling between $\psi_{nl}^{(+)}(r)$ and $\psi_{ml}^{(-)}(r)$ would represent a reversal of radial motion other than at the classical turning point.

The consequence of (8.176), after neglect of small terms, implied by (8.174) (on the assumption that inelastic coupling is negligible to close to the turning point) is that

$$\left[\frac{\mathrm{d}^2\alpha_n^{(\pm)}}{\mathrm{d}r^2} \pm 2i\bar{k}_l(r)\frac{\mathrm{d}\alpha_n^{(\pm)}}{\mathrm{d}r}\right] = [\bar{k}_l^2(r) - k_{nl}^2(r)]\alpha_n^{(\pm)}(r) + \sum_{m \neq n} U_{nm}(r)\,\alpha_m^{(\pm)}(r). \quad (8.179)$$

Hence

$$\pm\, 2i\bar{k}_l(r)\frac{\mathrm{d}\alpha_n^{(\pm)}}{\mathrm{d}r} \simeq [\bar{k}_l^2(r) - k_{nl}^2(r)]\alpha_n^{(\pm)}(r) + \sum_{m \neq n} U_{nm}(r)\,\alpha_m^{(\pm)}(r), \quad (8.180)$$

to the extent that, in view of the assumed slow variation of the $\alpha_n^{(\pm)}(r)$,

$$\left|\frac{\mathrm{d}^2\alpha_n^{(\pm)}}{\mathrm{d}r^2}\right| \ll \left|2\bar{k}_l(r)\frac{\mathrm{d}\alpha_n^{(\pm)}}{\mathrm{d}r}\right|. \quad (8.181)$$

Now since the $\alpha_n^{(\pm)}(r)$ may be assumed under semi-classical conditions to vary rapidly compared with $k_l^2(r)$ and $U_{mn}(r)$, (8.180) implies that

$$\pm 2i\bar{k}_l(r)\frac{\mathrm{d}^2\alpha_n^{(\pm)}}{\mathrm{d}r^2} \simeq [\bar{k}_l^2(r) - k_{nl}^2(r)]\frac{\mathrm{d}\alpha_n^{(\pm)}}{\mathrm{d}r} + \sum_{m \neq n} U_{nm}(r)\frac{\mathrm{d}\alpha_m^{(\pm)}}{\mathrm{d}r}. \quad (8.182)$$

Compatibility between (8.180) and (8.181) therefore requires that

$$|\bar{k}_l^2(r) - k_{nl}^2(r)| \ll 4\bar{k}_l^2(r)$$
$$U_{nm}(r) \ll 4k_l^2(r). \tag{8.183}$$

The physical significance of these conditions is discussed below.

The derivation of (8.169) from (8.180) is now a matter of rearrangement. The $\alpha_n^{(\pm)}(r)$ are first replaced by

$$\beta_n^{(\pm)}(r) = \exp\left\{\pm i \int_{a_l}^{r} \frac{[\bar{k}_l^2(r) - k_{nl}^2(r)]}{2\bar{k}_l(r)} \mathrm{d}r\right\} \alpha_n^{(\pm)}(r), \tag{8.184}$$

subject therefore to the equations

$$\pm 2i\bar{k}_l(r)\frac{\mathrm{d}\beta_n^{(\pm)}}{\mathrm{d}r} = \sum_{m\neq n} U_{mn}(r) \exp\left\{\pm i \int_{a_l}^{\infty} \frac{[k_{ml}^2(r) - k_{nl}^2(r)]}{2\bar{k}_l(r)} \mathrm{d}r\right\}. \tag{8.185}$$

Time variations are then introduced in place of variations with respect to r by means of (8.170), and a single function

$$\begin{aligned} c_n(t) &= \beta^{(+)}(r), \qquad t > 0 \text{ (outward motion)} \\ &= -\beta^{(-)}(r), \qquad t < 0 \text{ (inward motion)} \end{aligned} \tag{8.186}$$

is substituted for $\beta^{(\pm)}(r)$. Overall therefore, after substituting for $k_{nl}^2(r)$ from (8.168),

$$i\hbar\frac{\mathrm{d}c_n}{\mathrm{d}t} = \sum_{m\neq n} V_{nm}(r) \exp\left\{\frac{i}{\hbar}\int_0^t [W_n(r) - W_m(r)]\,\mathrm{d}t\right\} c_m(t), \tag{8.187}$$

with time and position related, according to (8.170), by†

$$t = \pm\frac{1}{v}\int_{a_l}^{r}\left[1 - \frac{W(r)}{E} - \frac{b^2}{r^2}\right]^{-\frac{1}{2}} \mathrm{d}r. \tag{8.188}$$

Thus the inequalities (8.174) and (8.183) are seen to justify the form of the classical trajectory Eqns (8.169), although nothing in the argument should be taken to support the substance of a truly classical picture in which the system could be observed to move along a defined trajectory. Of the above restrictions, (8.174) which would require zero inelastic coupling at the classical turning point, $t = 0$, for the assumed mean trajectory, is shown below to be open to some relaxation. The essential conditions are therefore expressed by equation (8.183), the first of which $\bar{k}_l^2(r) - k_{nl}^2(r) \ll 4k_l^2(r)$ or equivalently $|\bar{k}_l(r) - k_{nl}(r)| \ll 2k_l(r)$, establishes that the radial momenta in all coupled diabatic channels must lie close to the assumed mean; the second restriction, $U_{nm}(r) \ll 4k_l^2(r)$, ensures that kinetic energy differences between the diabatic and adiabatic channels are too small to upset the above criterion. Hence,

† A further form due to Bates and Crothers (1970), equivalent to both (8.184) and (8.186) but expressed in terms of distance s along the classical trajectory is given by (8.38).

since angular momentum is conserved, (8.102) imply, as might be expected, that for a valid trajectory formulation the unperturbed trajectories in all coupled channels must lie close the assumed mean. This criterion would imply a further angular momentum constraint $\Delta l \ll l$ in situations with non-spherical symmetry.

On the question of the importance of coupling at the classical turning point, Bates and Crothers (1970), and Knudson and Thorson (1970) have found remarkably close agreement with exact numerical values for the total cross-section and the results derived from (8.37) and (8.33) respectively, despite the existence of strong short range coupling. One explanation might be that radial turning points which appear in (8.167) no matter how high the actual velocity along the trajectory, could be removed without affecting the argument by a different choice of coordinate system.

We may also recognize however that there is at least one model system containing genuine turning points, in the sense that the kinetic energy vanishes, for which (8.33) are exact, namely the Landau–Zener model with

$$\begin{aligned} W_1(x) &= -F_1 x \\ W_2(x) &= -F_2 x \\ V_{12}(x) &= V_{12} = \text{const.} \end{aligned} \tag{8.189}$$

with $F_1 > 0$, $F_2 > 0$. In this case the exact equations in the momentum representation

$$\left[\frac{1}{2m}p^2 - E - i\hbar F_i \frac{p}{\partial p}\right] u_i(p) = -V_{ij} u_j(p) \tag{8.190}$$

may be rearranged by the substitution

$$u_i(p) = F_i^{-\frac{1}{2}} \exp\left[-\frac{i}{\hbar F_2}\int_0^p (\tfrac{1}{2}mp^2 - E)\,\mathrm{d}p\right] c_i(p) \tag{8.191}$$

to the same general form as (8.33)

$$i\hbar F \frac{\mathrm{d}c_i}{\mathrm{d}p} = V_{ij} \exp\left\{\frac{i}{\hbar}\int_0^p \left[\frac{1}{F_i}\left(\frac{1}{2m}p^2 - E\right) - \frac{1}{F_j}\left(\frac{1}{2m}p^2 - E\right)\right] \mathrm{d}p\, c_j(p)\right\} \tag{8.192}$$

where

$$F = (F_1 F_2)^{\frac{1}{2}}. \tag{8.193}$$

Hence if F is identified as the force acting along a mean trajectory governed by

$$\overline{W} = -Fx, \tag{8.194}$$

so that

$$F = \frac{\mathrm{d}p}{\mathrm{d}t} \tag{8.195}$$

and

$$\frac{1}{2m}p^2 = E - \overline{W} = E + Fx, \tag{8.196}$$

we find on combining (8.192)–(8.196) that

$$\begin{aligned} i\hbar\frac{\mathrm{d}c_i}{\mathrm{d}t} &= V_{ij}\exp\left\{\frac{i}{\hbar}\int_0^t (-F_i x + F_j x)\,\mathrm{d}t\right\} c_j(t) \\ &= V_{ij}\exp\left\{\frac{i}{\hbar}\int_0^t [W_i(x) - W_j(x)]\,\mathrm{d}t\right\} c_j(t) \end{aligned} \tag{8.197}$$

in exact agreement with (8.33). This result has led Delos (1972) to examine the validity of the classical trajectory equations in the momentum representation. His conclusions support the validity of (8.33), despite the existence of strong coupling at the classical turning point, provided that the forces in all coupled channels have the same sign, and provided that a mean force can be chosen such that

$$|F_i - \overline{F}| \ll \overline{F} \tag{8.198}$$

Like (8.193), this ensures that all unperturbed trajectories lie close to the assumed mean.

CHAPTER 9

The Semi-Classical S Matrix

The discussion in previous chapters has centred on reduction of the scattering problem to a set of ordinary differential or integral equations, with the intention of forcing the general multidimensional problem into a one dimensional form. In many situations the resulting one dimensional motion may be treated by semi-classical methods, but there is still a severe constraint on the types of system for which analytical or even numerical solution is practicable. A quite different potentially powerful approach, which we shall term the semi-classical S matrix method, has therefore been developed by Marcus (1970, 1971, 1972a, b), Connor and Marcus (1971), and Connor (1973a, b) on one hand, and by Pechukas (1969a, b), Miller (1970a, b), Miller and George (1972a, b) and Doll and Miller (1972) on the other.

The first important result of this theory is that the S matrix for a transition $a \rightarrow b$ may be reduced under semi-classical (short wavelength) conditions to a sum over classical trajectories between a and b;

$$S_{ab} = \sum P_{ab}^{\frac{1}{2}} \exp(i\Delta_{ab}), \tag{9.1}$$

where P_{ab} is the corresponding classical probability and Δ_{ab} depends on the classical trajectory in question. This idea, which generalizes the semi-classical theory of elastic scattering outlined in Section 5.1 to include inelastic and reactive events, means that quantum mechanically accurate transition probabilities for important heavy particle problems may be obtained by solution of the classical equations of motion. Paradoxically this result may even be extended to include processes which are forbidden by ordinary classical mechanics, because the normal classical solutions are obtained by treating time as a real variable. The methods developed below show that the classically forbidden events are covered by analytical continuation of the real classical solutions into the complex time and complex coordinate domains.

There are two distinct but equivalent approaches to the theory. The first originated by Pechukas (1969a, b) and Miller (1970a, b) starts from the path integral approach to quantum mechanics described by Feynman and Hibbs (1965), and follows arguments similar to those which establish the eikonal relation between physical and geometrical optics (see Landau and Lifshitz, 1960). The second method, due to Marcus (1970), which relies on the semi-classical relation between the Schrödinger equation and the Hamilton–Jacobi equation of classical mechanics (see Goldstein, 1959 and Appendix E.4) is a direct extension of the JWKB treatment of motion in one dimension. The following account takes the latter viewpoint, as being more in keeping with the development of the book. A brief outline of the classical propagator approach is included in Section 9.4. Reference may also be made to related work by Chen and Watson (1968), Percival and Richards (1970a, b), Levine and Johnson (1970a, b, 1971), Pattengill, Curtiss and Bernstein (1971), Freed (1972) and Eu (1972).

9.1 Angle Action Variables and the Semi-Classical Wave Function

The first step in the theory is to obtain a tractable semi-classical wave-function. This requires the use of angle action variables (w, N) (see Goldstein, 1958 and Appendix E.4) rather than conventional cartesian coordinates and momenta (q, p) for the internal degrees of freedom, in order to avoid the introduction of semi-classical singularities at the classical turning points.†

The important difference between the two canonical systems is that the conventional classical Hamiltonian $H(p, q)$, for an isolated internal system, depends on both coordinates and momenta. Hence the semi-classical eigenfunction of the quantum mechanical analogue $\hat{H}(\hat{p}, q)$ with $\hat{p} = -i\hbar\partial/\partial q$,

$$\phi_n(q) = C[p(q)]^{-\frac{1}{2}} \sin\left(\frac{1}{\hbar}\int_a^q p(q)\,\mathrm{d}q + \pi/4\right) \tag{9.2}$$

diverges at the zeros of $p(q)$. The classical Hamiltonian $H(N)$ in the angle action system, on the other hand, is necessarily independent of the angle variable w (see Appendix E.4). Hence the quantum mechanical operator, $\hat{H}(\hat{N} + \delta h)$ which contains an additional small zero-point energy parameter δ‡ (Dirac, 1926, 1927), commutes with the action operator, $\hat{N} = -i\hbar\partial/\partial w$. It follows that the angle action equivalent of $\phi_n(q)$ in (9.2) is

$$\phi_n(w) = \exp(2\pi i n w), \qquad n = 0, 1, 2, \ldots \tag{9.3}$$

(Motion is assumed periodic in w over the range $w = (0, 1)$). Being devoid of singularities this form is uniformly accurate for all phases of the internal motion. The price to be paid for this advantage is that the ability to transform

† Singularities at the turning points of the radial motion are of less significance because they are well separated from the asymptotic region where physical measurements are made.

‡ $\delta = 0$ for a plane rotor, $\frac{1}{2}$ for an oscillator and $\frac{1}{2}$ for orbital motion in the semi-classical limit.

from the angle to the action representation (the analogue of a transform from coordinate to momentum space) may be retained only in a stationary phase (or semi-classical) sense (see Leaf (1969)). It must also be recognized that there are serious problems in determining the appropriate angle-action variables for non-separable systems (see Born (1960), Keller (1958), Gutzwiller (1967, 1969–1971), Miller (1972) and Marcus (1973)). Attention is therefore largely restricted below to the case of a single internal degree of freedom and to a single radial or translational coordinate r, with conjugate momentum p.

It is convenient to discuss the corresponding classical motion before turning to the quantum mechanical case because the classical and semi-classical descriptions are closely related. This involves solution of the following equations for the classical trajectory:

$$H(N, p, w, r) = H_0(N) + \frac{1}{2m}p^2 + V(w, r) = E \tag{9.4}$$

$$\dot{N} = -(\partial H/\partial w), \qquad \dot{w} = (\partial H/\partial N)$$
$$\dot{p} = -(\partial H/\partial r), \qquad \dot{r} = (\partial H/\partial p) \tag{9.5}$$

subject to five stated boundary conditions. Four of these will be used to fix the energy, E, the initial action N, and particle separation r_1, and the particle separation of interest. The final condition will be taken to specify either the initial angle w_1 or the current angle w. Variation of w_1 (or w) over the interval (0, 1) then generates a family of trajectories for the motion of interest. It is important for the following discussion to recognize that different trajectories lead to different divisions of energy between the internal and external motion. Hence if N_1 and w_1 are used to fix the trajectory the current angle and action must be regarded as strictly dependent quantities, $w(N_1, w_1)$ and $N(N_1, w_1)$ respectively. Similarly if w (rather than w_1) is taken as the independent variable, the dependent quantities are $w_1(N_1, w)$ and $N(N_1, w)$.

We turn now to determination of the semi-classical wavefunction with initial asymptotic form

$$\psi(w, r) \sim v_1^{-\frac{1}{2}} \exp(2\pi i n_1 w - i k_1 r), \tag{9.6}$$

where

$$\tfrac{1}{2}mv_1^2 = E - E_{\text{int}}(n_1)$$
$$k_1 = mv_1/\hbar, \tag{9.7}$$

corresponding to an incoming wave in the n_1th internal state. A semi-classical solution to the appropriate Schrödinger equation

$$\left[H_0\left(-i\hbar\frac{\partial}{\partial w} + h\delta\right) - \frac{\hbar^2}{2m}\frac{\partial^2}{\partial r^2} + V(w, r)\right]\psi = E\psi, \tag{9.8}$$

is obtained in the usual way by means of an expansion in powers of $\hbar$,

$$\psi(w, r) = A(w, r) \exp\left\{\frac{i}{\hbar}[W(w, r) - (h\delta)w]\right\}. \tag{9.9}$$

($A(w, r)$ may be regarded as the logarithm of the second term of an expansion of the exponent). It follows on ignoring terms in $\hbar^2$, that $W(w, r)$ must satisfy an equation of the Hamilton–Jacobi form (see (E.39)).

$$H\left(\frac{\partial W}{\partial w}, \frac{\partial W}{\partial r}, w, r\right) = H_0\left(\frac{\partial W}{\partial w}\right) + \frac{1}{2m}\left(\frac{\partial W}{\partial r}\right)^2 + V(w, r) = E. \tag{9.10}$$

A solution is readily found by comparison with equation (9.4), according to which we may write

$$\frac{\partial W}{\partial w} = N, \qquad \frac{\partial W}{\partial r} = p, \tag{9.11}$$

where the momenta N and p are governed by the classical equations of motion (9.5). Hence equations (9.11) must be integrated *along a classical trajectory* to yield

$$W(w, r) = \int_{w_1}^{w} N\,\mathrm{d}w + w_1 N_1 + \int_{r_1}^{r} p\,\mathrm{d}r + r_1 p_1 \qquad (-\pi/2) \tag{9.12}$$

the terms $w_1 N_1$ being included to eliminate contributions to $(\partial W/\partial w)$ from the lower integration limit (recall that as stated above w_1 depends on w). The term $r_1 p_1$ is included for later convenience, and the factor $(-\pi/2)$ in (9.12) is included to account by the usual semi-classical arguments (see Appendix C.2) for the influence of a radial turning point. It is readily verified by setting $N = N_1 = (n_1 + \delta)h$, and $p = p_1 = -k_1\hbar$, that this exponent connects with that in the initial asymptotic form in (9.6). This line integral for $W(w_1 r)$ may be recognized as the classical action for a trajectory from (N_1, p_1, w_1, r_1) to (N, p, w, r), and the family of such integrals constitutes Hamilton's characteristic function (see Goldstein (1958)).

The equation for the pre-exponential factor $A(w, r)$ in (9.9) is slightly more complicated. Let us suppose that the internal classical hamiltonian varies as

$$H_0(N) = C_s N^s; \tag{9.13}$$

the term in $\hbar$ in the Schrödinger equation (9.8) then yields

$$-\frac{1}{2A}\left\{\frac{\partial}{\partial w}\left[sA^s C_s\left(\frac{\partial W}{\partial w}\right)^{s-1}\right] + \frac{\partial}{\partial r}\left[\frac{A^2}{m}\frac{\partial W}{\partial r}\right]\right\} = 0. \tag{9.14}$$

It follows moreover by use of (9.4), (9.5), (9.11) and (9.13) that

$$sC_s(\partial W/\partial w)^{s-1} = sC_s N^{s-1} = \partial H/\partial N = \dot{w}$$

$$\frac{1}{m}\frac{\partial W}{\partial r} = \frac{1}{m}p = \frac{\partial H}{\partial p} = \dot{r} \tag{9.15}$$

Hence Eqn (9.14) requires that A should satisfy the flux conservation condition

$$\frac{\partial}{\partial w}(A^2\dot{w}) + \frac{\partial}{\partial r}(A^2\dot{r}) = 0. \tag{9.16}$$

Written more generally this becomes

$$\nabla . \mathbf{i} = 0, \tag{9.17}$$

where $\mathbf{i}$ is the current vector $A^2\dot{\mathbf{q}}$. This means that the current has zero divergence and hence by Gauss' theorem that the net outflow from any region of (w, r) space is zero. When applied to the area between neighbouring trajectories bounded by the lines $r = r_1$ and $r = r$ (Fig. 9.1), this means that

$$A_1^2\dot{r}_1\, \mathrm{d}w_1 = A^2\dot{r}\, \mathrm{d}w \tag{9.18}$$

because there is no flow across the sides of the tube. Hence

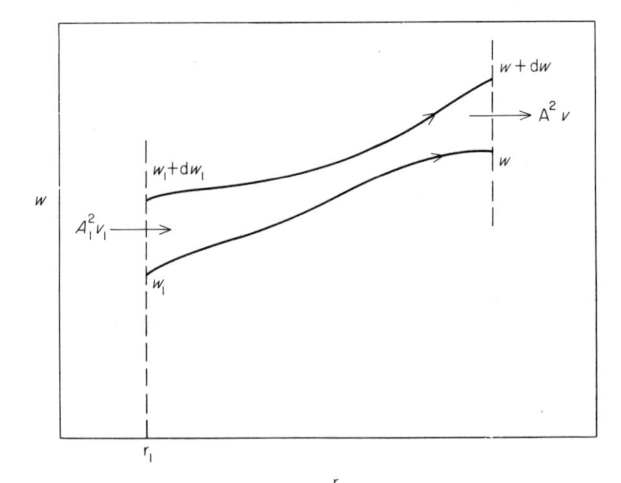

FIG. 9.1. Neighbouring trajectories in the (w, r) representation. $A_1^2v_1$ and A^2v represent the radial components of the current at r_1 and r respectively.

$$A(w, r) = A_1 \left[\frac{v_1}{v}\left(\frac{\partial w_1}{\partial w}\right)\right]^{\frac{1}{2}}. \tag{9.19}$$

Extension of the argument to many internal dimensions shows that in general

$$A(\mathbf{w}, r) = A_1 \left[\frac{v_1}{v} \frac{\partial(w_{a_1}, w_{b_1} \ldots w_{n_1})}{\partial(w_a, w_b \ldots w_n)}\right]^{\frac{1}{2}} \tag{9.20}$$

where $\partial(w_{a_1}, w_{b_1} \ldots w_{n_1})/\partial(w_a, w_b \ldots w_n)$ denotes the Jacobian of the transformation from $\mathbf{w}$ to $\mathbf{w}_1$.

The value of A_1 implied by (9.6) is clearly

$$A_1 = v_1^{-\frac{1}{2}} \tag{9.21}$$

Hence, to order $\hbar^{-1}$ in Eqn (9.8) the asymptotic form of the outgoing part of the overall wavefunction may be written

$$\psi^{(+)}_{n_1 E}(w, r) = v^{-\frac{1}{2}}(\partial w_1/\partial w)^{\frac{1}{2}} \exp\left[i\chi(w, r)\right], \tag{9.22}$$

where

$$\chi(w, r) = 2\pi \int_{w_1}^{w} n \, \mathrm{d}w + 2\pi n_1 w_1 + \int_{r_1}^{r} k \, \mathrm{d}r - k_1 r_1 - \pi/2 \tag{9.23}$$

and

$$nh = N - h\delta. \tag{9.24}$$

The integration in (9.23) is taken along a classical trajectory specified by E, n_1, r_1, r and w. It is important to recognize however that although E, n_1, r_1 and r may be chosen to be fixed in a given calculation, variation of w (as required to generate the wavefunction $\psi_{n, E}(w_1 r)$ at given r) implies variation of the trajectory. In other words the wavefunction specified by (9.22) contains information from *all* classical trajectories with initial energy E and internal action $(n_1 + \delta)h$.

The validity of Eqn (9.22) requires, as in the one dimensional theory (see Appendix Section C.1) that the pre-exponential factor should remain finite. It will be recalled that the term $\pi/2$ in (9.23) is introduced to overcome a singularity of $A(w, r)$ at the radial turning point. A less serious type of singularity in the final asymptotic region has also been observed in a study of harmonic oscillator excitation by Wong and Marcus (1971). This arises from an apparently non-monotomic variation of $w_1(w)$ if $(\partial w_1/\partial w)$ is evaluated at a fixed value of r, because both w and r are functions of time, and the time variations of both w and r may vary non-monotonically from one trajectory to another. This difficulty may be overcome by making comparisons at a fixed time because (with w_1 now regarded as the independent variable)

$$\left(\frac{\partial w}{\partial w_1}\right)_\tau = \left(\frac{\partial w}{\partial w_1}\right)_r - \left(\frac{\partial w}{\partial \tau}\right)_{w_1}\left(\frac{\partial \tau}{\partial w_1}\right)_r$$

$$= \left[\frac{\partial}{\partial w_1}(w - v\tau)\right]_r = \left(\frac{\partial \bar{w}}{\partial w_1}\right)_r, \tag{9.25}$$

where

$$\bar{w} = w - v\tau = w - vrp/m, \tag{9.26}$$

and v is the frequency of the internal motion. The analogous initial variable

$$\bar{w}_1 = w_1 - vr_1p_1/m, \tag{9.27}$$

differs from w_1 only by a constant, because p_1 takes the same value for all trajectories. Hence (9.25) may also be written

$$\left(\frac{\partial w}{\partial w_1}\right)_\tau = \left(\frac{\partial \bar{w}}{\partial \bar{w}_1}\right)_r. \tag{9.28}$$

The precise point r at which the latter derivative is taken is immaterial because $\bar{w}$ defined by (9.26) is constant along the final asymptotic part of any given trajectory in asymptotic region.

This change from r to τ as the independent variable in the asymptotic region also necessitates a unitary transformation of the wavefunction to the τ representation. The appropriate transformation formula takes the typical form (Van Vleck, 1928)

$$\tilde{\psi}(\tau) = \langle\tau|\psi\rangle = \int_{-\infty}^{\infty} \langle\tau|r\rangle\langle r|\psi\rangle \, dr, \tag{9.29}$$

where

$$\langle r|\tau\rangle = \left[\frac{i}{2\pi\hbar}\frac{\partial^2 F_1}{\partial r \partial \tau}\right]^{\frac{1}{2}} \exp\left[-\frac{i}{\hbar}F_1(r, t)\right], \tag{9.30}$$

and $F_1(r, \tau)$ is the generator for a classical canonical transformation from the (p, r) to the (E_τ, τ) representation. Hence according to Appendix E.3, $F_1(r, \tau)$ must satisfy

$$p = \frac{\partial F_1}{\partial r}, \qquad E_\tau = -\frac{\partial F_1}{\partial \tau}. \tag{9.31}$$

The second of these equations defines the momentum E_τ conjugate to τ. An appropriate generator in the asymptotic (free motion) region is

$$F_1(r, \tau) = (mr^2/2\tau), \tag{9.32}$$

in which case according to (9.31)

$$p = mr/\tau,$$
$$E_\tau = mr^2/2\tau^2 = \tfrac{1}{2}mp^2. \tag{9.33}$$

In other words E_τ is the translational energy. It is readily verified by application of (9.29)–(9.32) that the transform of a typical plane wave state

$$\psi(r) = \langle r|\psi\rangle = v^{-\frac{1}{2}} \exp(ikr), \tag{9.34}$$

is

$$\tilde{\psi}(\tau) = \exp(iE_\tau \tau/\hbar), \tag{9.35}$$

where

$$E_\tau = k^2\hbar^2/2m = \tfrac{1}{2}mv^2. \tag{9.36}$$

It only remains to note that $\psi^{(+)}_{n_1 E}(w, r)$ defined by (9.22) reduces asymptotically to the same plane wave form, multiplied by a factor

$$C = (\partial \bar{w}_1/\partial \bar{w})_r \exp\{i[\chi(w, r) - kr]\}, \tag{9.37}$$

which is independent of r. Hence the required transform of $\psi^{(+)}_{n_1 E}(w, r)$ may be written

$$\begin{aligned}\tilde{\psi}^{(+)}_{n_1 E}(\bar{w}, \tau) &= C\exp(iE_\tau \tau/\hbar) = C\exp\{i[E - E(n)]\tau/\hbar\}, \\ &= \left(\frac{\partial \bar{w}_1}{\partial \bar{w}}\right)_r \exp[i\bar{\chi}(n_1, \bar{w}) + iE\tau/\hbar],\end{aligned} \tag{9.38}$$

where

$$\bar{\chi}(n_1, \bar{w}) = 2\pi n\bar{w} - 2\pi\int_{n_1}^{n} w\, dn - \int_{k_1}^{k} r\, dk. \tag{9.39}$$

Here we have used the fact that

$$2\pi nw - E_n\tau/\hbar = 2\pi n(w - v\tau) = 2\pi n\bar{w}, \tag{9.40}$$

for E_n linearly dependent on n. The same result has been justified on more general grounds by Marcus (1971).

9.2 The S Matrix: Derivation and Approximations

The derivation of the S matrix is readily achieved by comparison between Eqn (9.38) and the corresponding transform of the defining equation

$$\psi^{(+)}_{n_1 E}(w, r) \sim \sum_{n_2} v_2^{-\frac{1}{2}} S_{n_1 n_2} \exp[2\pi i n_2 w + ik_2 r], \tag{9.41}$$

namely

$$\tilde{\psi}^{(+)}_{n_1 E}(\bar{w}, \tau) = \sum_{n_2} S_{n_1 n_2} \exp\left[2\pi i n_2 \bar{w} + iE\tau/\hbar\right]. \quad (9.42)$$

Hence on projecting out the n_2th term from $\tilde{\psi}^{(+)}_{n_1 E}(w, \tau)$ in (9.38) and introducing $\bar{w}_1$ as the independent variable

$$S_{n_1 n_2} = \int_0^1 \left(\frac{\partial \bar{w}}{\partial w}\right)^{\frac{1}{2}} \exp\left[i\Delta_{n_1 n_2}(\bar{w}_1)\right] d\bar{w}_1 \quad (9.43)$$

where

$$\Delta_{n_1 n_2}(\bar{w}_1) = 2\pi(n - n_2)\,\bar{w} - 2\pi \int_{n_1}^{n} w\, dn - \int_{k_1}^{k} r\, dk. \quad (9.44)$$

This form is due to Miller (1970b) (see Eqn (9.111)) and Marcus (1971). It differs from the result obtained by comparison between (9.22) and (9.41) by the appearance of $\bar{w}$ in place of w, and by the absence of a weak but strictly inadmissable r dependence in $S_{n_1 n_2}$ (see Marcus, 1970, 1971).

The first point to be made in connection with Eqn (9.43) is that it allows in principle for a direct determination of $S_{n_1 n_2}$ by means of a quadrature based on the results of a series of classical trajectory calculations, at appropriate initial phase angles $\bar{w}_1$. Paradoxically this calculation can even by extended to final quantum numbers n_2 which lie outside the classically accessible domain; we shall return to this point below. Results obtained by this purely numerical method (Miller, 1970b; Wong and Marcus, 1971) are however seen from Tables 9.2 and 9.3 below, to be in good agreement with the exact values only for the classically accessible or nearly accessible transitions $n_1 \to n_2$. The reason is that the integral is not symmetrical† in n_1 and n_2, because the chosen trajectories all have initial action $(n_1 + \delta)h$, but at best only a discrete number can also have final action $(n_2 + \delta)h$ (see Fig. 9.2); indeed there are by definition no such real trajectories if n_2 is classically accessible. Fortunately, as discussed below, the major contributions to the integral come from the neighbourhood of the special trajectories $n_1 \to n_2$, and this explains the level of agreement with exact results in the classically accessible case.

Attention is now directed towards finding analytical approximations to $S_{n_1 n_2}$ defined by (9.43). The argument, which is a generalization of that applied to the theory of elastic scattering in Sections 5.2–5.3 is facilitated by considering the typical variation of $n(n_1, \bar{w}_1)$ with $\bar{w}_1$ as shown in Fig. 9.2. This diagram shows the expected periodic behaviour of $n(n_1, \bar{w}_1)$, and the division of n

† Marcus (1971) has suggested a more complicated symmetrical version of (9.43) to meet this difficulty in numerical work. It has however no advantage over the formula given, at the level of the stationary phase and uniform approximations discussed below.

values into classical, $n_{min} < n < n_{max}$, and non-classical regions. It may be seen that there are typically two special $n_1 \rightarrow n_2$ trajectories, characterized by initial phases $\bar{w}_1^{(a)}$ and $\bar{w}_2^{(b)}$, for classically accessible values of n_2. We may

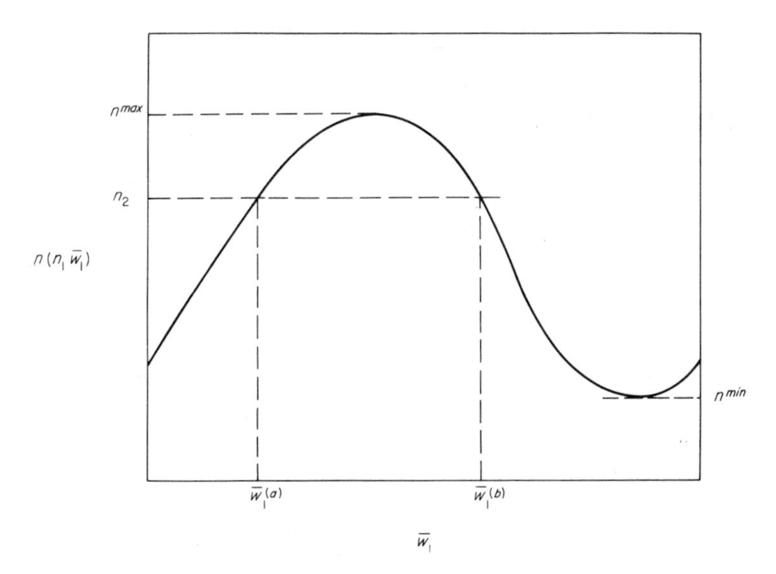

FIG. 9.2. Variation of the final quantum number $n(n_1, \bar{w}_1)$, as a function of the initial phase angle $\bar{w}_1$ at fixed n_1. n^{max} and n^{min} indicate the upper and lower classical thresholds respectively. $\bar{w}_1^{(a)}$ and $\bar{w}_2^{(b)}$ are the initial angles for which the trajectory has final action $(n_2 + \delta)h$. The classical (stationary phase) transition probabilities P_a and P_b in (9.56) depend inversely on the slopes of tangents at $w_1^{(a)}$ and $w_1^{(b)}$ respectively.

also anticipate (from the presence of the turning points in Fig. 9.2) two complex roots of the equation

$$n(n_1, \bar{w}_1) = n_2 \tag{9.45}$$

when $n_2 > n_{max}$ or $n_2 < n_{min}$. These will be found to correspond to "complex" trajectories from n_1 to classically forbidden points n_2.

The first analytical approximation to be considered is the stationary phase approximation, applicable for $n_{min} < n_2 < n_{max}$. This is included as a separate exercise, although the results are contained within those of the more powerful uniform approximation which follows, because the argument indicates most clearly the central significance of the special $n_1 \rightarrow n_2$ trajectories. This approximation rests on the postulate that the phase term, $\Delta_{n_1 n_2}(\bar{w}_1)$ is large and strongly dependent on $\bar{w}_1$. Hence rapid oscillations in the integrand in (9.43) may be expected to lead to widespread cancellation except near the stationary points of $\Delta_{n_1n_2}(\bar{w}_1)$. It is further assumed that all such points are separated by regions of rapid phase variation, and hence that they can be treated independently, and that the integration range around each point may be extended to the interval $(-\infty, \infty)$. These assumptions imply that $S_{n_1n_2}$ may be approximated as a sum of stationary phase contributions

$$S_{n_1n_2} \simeq \sum_{v} S^{(v)}_{n_1 n_2}, \tag{9.46}$$

with

$$S^{(v)}_{n_1 n_2} \simeq \int_{-\infty}^{\infty} [\partial \bar{w}/\partial \bar{w}_1]^{\frac{1}{2}} \exp [i\Delta_{n_1n_2}(\bar{w}_1)] \, d\bar{w}_1, \tag{9.47}$$

and

$$\Delta_{n_1n_2}(\bar{w}_1) = 2\pi[n - n_2]\bar{w}_1 - 2\pi \int_{n_1}^{n} w \, dn - \int_{k_1}^{k} r \, dk. \tag{9.48}$$

The dependence of $\Delta_{n_1n_2}(\bar{w}_1)$ on $\bar{w}_1$ in this equation comes from the variation of the functions $n(n_1, \bar{w}_1)$, $w(n_1, \bar{w}_1)$ and $k(n_1, \bar{w}_1)$, the dependence of k being implied by that of n because

$$k(n_1, \bar{w}_1) = \{2m[E - (n + \delta)h\nu]\}^{\frac{1}{2}}/\hbar. \tag{9.49}$$

We now proceed to use the first and second derivatives of $\Delta_{n_1, n_2}(\bar{w}_1)$ to perform a quadratic approximation about the stationary phase point $\bar{w}_1^{(v)}$. Thus $\bar{w}_1^{(v)}$ is first determined by the stationary phase condition

$$\begin{aligned}(\partial \Delta/\partial \bar{w}_1) &= 2\pi[n(n_1, \bar{w}_1) - n_2](\partial \bar{w}/\partial \bar{w}_1) + 2\pi(\partial n/\partial \bar{w}_1)\bar{w}(n_1, \bar{w}_1) \\ &\quad - 2\pi[w(n_1, \bar{w}_1) - r(\partial k/\partial n)](\partial n/\partial \bar{w}_1) \\ &= 2\pi[n(n_1, \bar{w}_1) - n_2](\partial \bar{w}/\partial \bar{w}_1) = 0,\end{aligned} \tag{9.50}$$

where Eqns (9.26) and (9.49) have been used to deduce that

$$w(n_1, \bar{w}_1) - r(\partial k/\partial n) = \bar{w}_1(n_1, \bar{w}_1). \tag{9.51}$$

Equation (9.50) requires that $\bar{w}_1^{(v)}$ should be chosen such that the trajectory from $(n_1, \bar{w}_1^{(v)})$ should lead to the final quantum number n_2. In other words $\bar{w}_1^{(v)}$ is one of the points $\bar{w}_1^{(a)}$ or $\bar{w}_2^{(b)}$ in Fig. 9.2.

The second derivative of $\Delta_{n_1n_2}(\bar{w}_1)$ is given at $\bar{w}_1 = \bar{w}_1^{(v)}$ by

$$\left(\frac{\partial^2 \Delta}{\partial \bar{w}_1}\right)_v = 2\pi\left[\left(\frac{\partial n}{\partial \bar{w}_1}\right)\left(\frac{\partial \bar{w}}{\partial \bar{w}_1}\right)\right]_v, \tag{9.52}$$

with the label v used to denote the point in question. Hence according to Eqn (9.47)

$$S^{(v)}_{n_1 n_2} \simeq \left[\frac{\partial \overline{w}}{\partial \overline{w}_1}\right]_v \int_{-\infty}^{\infty} \exp\left\{i\Delta_{n_1 n_2}(\overline{w}_1^{(v)}) + i\pi\left[\left(\frac{\partial n}{\partial \overline{w}_1}\right)\left(\frac{\partial \overline{w}}{\partial \overline{w}_1}\right)(\overline{w}_1 - \overline{w}_1^{(v)})^2\right\} \mathrm{d}\overline{w}_1$$

$$\simeq P_v^{\frac{1}{2}} \exp[i\Delta_v(n_2, n_1)] \qquad (9.53)$$

where

$$P_v = |\partial n/\partial \overline{w}_1|_v^{-1}$$

$$\Delta_v(n_2, n_1) = \Delta_{n_1 n_2}(\overline{w}_1^{(v)}) \pm \pi/4 = -\int_{n_1}^{n_2} w\, \mathrm{d}n - \int_{-k_1}^{k} r\, \mathrm{d}k \pm \pi/4, \qquad (9.54)$$

the sign of the term $\pm\pi/4$ being that of $(\partial n/\partial \overline{w}_1)_v$. It follows that, in the case illustrated in Fig. 9.2,

$$S_{n_1 n_2} = P_a^{\frac{1}{2}} \exp[i\Delta_a(n_2, n_1) + i\pi/4] + P_b^{\frac{1}{2}} \exp[i\Delta_b(n_2, n_1) - i\pi/4]. \qquad (9.55)$$

The corresponding transition probability is

$$P_{n_1 n_2} = |S_{n_1 n_2}|^2 = P_a + P_b + 2(P_a P_b)^{\frac{1}{2}} \sin(\Delta_a - \Delta_b). \qquad (9.56)$$

The terms in this equation have the following significance. P_a and P_b are the "classical" transition probabilities derived from the density of trajectories per unit quantum number at the stationary phase points $\overline{w}_1^{(a)}$ and $\overline{w}_1^{(b)}$, and the phases Δ_a and Δ_b depend on the classical actions for the two trajectories from n_1 to n_2. Hence the overall transition probability appears as a sum of classical contributions together with a semi-classical interference term governed by the difference in classical actions along the contributing paths.

The validity of this description requires that the two stationary phase approximations should not overlap. Problems are therefore encountered as n_2 approaches the classical threshold values n_{min} and n_{max}. A similar difficulty, encountered in connection with rainbow scattering in Section 5.2, was overcome by means of a cubic expansion to the equivalent of $\Delta_{n_1 n_2}(\overline{w}_1)$. The argument will not be repeated here, but a similar more powerful "uniform" approximation will be introduced instead. This is specifically designed to handle coalescing saddle points. It is based on a general method due to Chester *et al.* (1957), which may be extended to more complicated situations in more than one dimension. Relevant examples have been discussed by Connor and Marcus (1971), Marcus (1972) and Connor (1973a, b).

The general technique in one dimension, given an integral of the form

$$S(n) = \int_{-\infty}^{\infty} g(w) \exp[i\Delta(n, w)]\, \mathrm{d}w, \qquad (9.57)$$

where n is a parameter, is to map the function $S(n)$ onto an appropriate standard function by means of a variable transformation in the integrand. (The notation here is chosen to evoke that of the previous discussion, for which $n = n_2$ and $w = \bar{w}_1$.) A suitable transformation in the present case of two coalescing stationary phase points is defined by the identity

$$\Delta(n, w) = \tfrac{1}{3}y^3 - \zeta(n)\, y + A(n), \tag{9.58}$$

where $\zeta(n)$ and $A(n)$ are constants. (Note that this is not an approximation.) The condition for a valid mapping of this sort is that there should be one-to-one correspondence between the stationary points on either side of (9.58);

$$\begin{aligned} w_a &\leftrightarrow y_a = \zeta^{\frac{1}{2}} \\ w_b &\leftrightarrow y_b = -\zeta^{\frac{1}{2}}. \end{aligned} \tag{9.59}$$

Hence ζ and A may be expressed in terms of $\Delta(n, w_a)$ and $\Delta(n, w_b)$ by substitution in (9.58);

$$\begin{aligned} \Delta_a &= \Delta(n, w_a) = A - \tfrac{2}{3}\zeta^{\frac{3}{2}} \\ \Delta_b &= \Delta(n, w_b) = A + \tfrac{2}{3}\zeta^{\frac{3}{2}}. \end{aligned} \tag{9.60}$$

Similarly the pre-exponential term $g\,(\mathrm{d}w/\mathrm{d}y)$ which appears after the change of integration variable may be expanded as

$$g(\mathrm{d}w/\mathrm{d}y) = p + qy, \tag{9.61}$$

with p and q determined by solving the simultaneous equations†

$$\begin{aligned} g_a(\mathrm{d}w/\mathrm{d}y)_a &= p + qy_a = p + q\zeta^{\frac{1}{2}} \\ g_b(\mathrm{d}w/\mathrm{d}y)_b &= p + qy_b = p - q\zeta^{\frac{1}{2}}. \end{aligned} \tag{9.62}$$

Finally the derivatives $(\mathrm{d}w/\mathrm{d}y)$ may be expressed in terms of ζ and derivatives of $\Delta(n, w)$ by taking the second derivatives on both sides of the transformation equation (9.58)

$$\left(\frac{\partial^2 \Delta}{\partial w^2}\right)\left(\frac{\partial w}{\mathrm{d}y}\right) + \left(\frac{\partial \Delta}{\partial w}\right)\left(\frac{\mathrm{d}^2 w}{\mathrm{d}y^2}\right) = 2y. \tag{9.63}$$

It follows since $(\partial\Delta/\partial w) = 0$ at w_2 that

$$\begin{aligned} (\mathrm{d}w/\mathrm{d}y)_a &= (2\zeta^{\frac{1}{2}}/\Delta_a'') \\ (\mathrm{d}w/\mathrm{d}y)_b &= (-2\zeta^{\frac{1}{2}}/\Delta_b''). \end{aligned} \tag{9.64}$$

The effect of these transformations is to reduce $S(n)$ to the form

$$\begin{aligned} S(n) &= \int_{-\infty}^{\infty} (p + qy) \exp\left[iA + \tfrac{1}{3}y^3 - \zeta y\right] \mathrm{d}y \\ &= 2\pi\, \mathrm{e}^{iA}[p\,\mathrm{Ai}\,(-\zeta) - iq\,\mathrm{Ai}'\,(-\zeta)] \end{aligned}$$

† The theory clearly allows one term in (9.61) for each stationary phase point.

$$= \pi^{\frac{1}{2}} e^{iA}[(P_a^{\frac{1}{2}} + P_b^{\frac{1}{2}})\zeta^{\frac{1}{4}} \mathrm{Ai}(-\zeta) - i(P_a^{\frac{1}{2}} - P_b^{\frac{1}{2}})\zeta^{-\frac{1}{4}} \mathrm{Ai}'(-\zeta)], \tag{9.65}$$

where

$$P_a = 2\pi g_a^2/\Delta_a'', \qquad P_b = -2\pi g_b^2/\Delta_b'', \tag{9.66}$$

and Ai $(-\zeta)$ and Ai′ $(-\zeta)$ denote the Airy function and its first derivative respectively (Abramowitz and Stegun (1965)). It is readily verified that P_a and P_b defined by (9.66) correspond in the previous notation to the values given by (9.54).

Equations (9.60), (9.65) and (9.66) give the required uniform approximation, the validity of which depends only on that of the linear expansion given by (9.61). Furthermore it is readily verified that $S(n)$ given by (9.65), passes smoothly into the classically forbidden region ($\zeta < 0$) where the stationary points are complex, because the divergent terms $P_a^{\frac{1}{2}}$ and $P_b^{\frac{1}{2}}$ exactly cancel in the second term and the divergence of the sum $(P_a^{\frac{1}{2}} + P_b^{\frac{1}{2}})$ is balanced by the disappearance of $\zeta^{\frac{1}{4}}$ at the classical threshold, ($\zeta = 0$). Discussion of the physical significance of complex values of w (corresponding to complex initial phases $\bar{w}_1$ in the previous discussion), is postponed until Section 9.4. We note at present merely that according to (9.59) real and imaginary values of y_a and y_b correspond to positive and negative values of the real parameter ζ respectively.

It follows that the asymptotic expressions

$$\begin{aligned}
\mathrm{Ai}(-\zeta) &\overset{\zeta \gg 1}{\sim} \pi^{-\frac{1}{2}}\zeta^{-\frac{1}{4}} \sin\left(\tfrac{2}{3}\zeta^{\frac{3}{2}} + \pi/4\right) \\
&\overset{\zeta \ll -1}{\sim} \tfrac{1}{2}\pi^{-\frac{1}{2}}|\zeta|^{-\frac{1}{4}} \exp\left(-\tfrac{2}{3}|\zeta|^{\frac{3}{2}}\right), \\
\mathrm{Ai}'(-\zeta) &\overset{\zeta \gg 1}{\sim} -\pi^{-\frac{1}{2}}\zeta^{-\frac{1}{4}} \cos\left(\tfrac{2}{3}\zeta^{\frac{3}{2}} + \pi/4\right) \\
&\overset{\zeta \ll -1}{\sim} -\tfrac{1}{2}\pi^{-\frac{1}{2}}|\zeta|^{\frac{1}{4}} \exp\left(-\tfrac{2}{3}|\zeta|^{\frac{3}{2}}\right),
\end{aligned} \tag{9.67}$$

may be used to examine the limiting form of $S(n)$ in the two limiting regions.

The result obtained on substituting the forms for $\zeta \gg 1$ in Eqn (9.65) and using (9.60) is that $S(n)$ reduces to the form identical with that obtained by the stationary phase approximation (see Eqn (9.65))

$$S(n) \sim P_a^{\frac{1}{2}} \exp(i\Delta_a + i\pi/4) + P_b^{\frac{1}{2}} \exp(i\Delta_b - i\pi/4). \tag{9.68}$$

This covers the case when n lies in the classically accessible region with P_a and P_b real if w_a and w_b are chosen such that $\Delta_a'' > 0$ and $\Delta_b'' < 0$.

In the opposite "n classically inaccessible" case, we find on choosing the branch $\zeta = |\zeta| e^{i\pi}$ that

$$S(n) \sim P_a^{\frac{1}{2}} \exp\left[i\left(A + \frac{2i}{3}|\zeta|^{\frac{3}{2}}\right) + i\pi/4\right]$$

$$= P_a^{\frac{1}{2}} \exp(i\Delta_a + i\pi/4). \qquad (9.69)$$

Hence only one stationary phase trajectory contributes in this case, namely that for which the exponent terms Δ_a has a *positive* imaginary part. This leads, as usual for classically forbidden phenomena, to an exponentially small transition probability.

One defect in this particular uniform approximation is that it ignores the necessary periodicity of the function $\Delta(w)$ in a case such as that illustrated in Fig. 9.2. Hence the influence of the minimum in $(\partial\Delta/\partial w)$ is not taken into account in dealing with values of $n_2 \simeq n_{max}$, and vice versa. This leads to significant inaccuracies in cases of weak interaction such that the total classical range $n_{max} - n_{min}$ is small. Stine and Marcus (1974) have therefore suggested an alternative "uniform Bessel" approximation to overcome this difficulty, based on the mapping

$$\Delta(n, w) = A(n) - \zeta(n) \cos 2\pi y - 2\pi m y \qquad (9.70)$$

with $A(n)$ and $\zeta(n)$ given by

$$\begin{aligned} \Delta_a &= A - \frac{m\pi}{2} - (\zeta^2 - m^2)^{\frac{1}{2}} + m \arccos(m/\zeta) \\ \Delta_b &= A - \frac{m\pi}{2} + (\zeta^2 - m^2)^{\frac{1}{2}} - m \arccos(m/\zeta), \end{aligned} \qquad (9.71)$$

and with $g(\mathrm{d}x/\mathrm{d}y)$ expanded in the form

$$g(\mathrm{d}x/\mathrm{d}y) = p \cos 2\pi y + q \sin 2\pi y. \qquad (9.72)$$

A similar approach has been applied to the theory of glory scattering (see Section 5.4) by Berry (1969). The parameter m (which must be integral in order to preserve the periodicity of the problem) is conveniently set equal to $(n_2 - n_1)$ in practical applications. The final result may be written

$$\begin{aligned} S_m(n) = \left(\frac{\pi}{2}\right)^{\frac{1}{2}} \mathrm{e}^{iA}[(P_a^{\frac{1}{2}} + P_b^{\frac{1}{2}})(\zeta^2 - m^2)^{\frac{1}{4}} J_m(\zeta) \\ - i(P_a^{\frac{1}{2}} - P_b^{\frac{1}{2}}) \zeta^{\frac{1}{2}}(\zeta^2 - m^2)^{-\frac{1}{4}} J_m'(\zeta)], \end{aligned} \qquad (9.73)$$

where $J_m(\zeta)$ and $J_m'(\zeta)$ denote the mth order Bessel function and its first derivative (see Abramowitz and Stegun, 1965), and P_a and P_b are defined by (9.66). It may be verified that this formula reduces to those given by (9.68) and (9.69) in the stationary phase regions $\zeta \gg m$ and $\zeta \ll m$ respectively.

A similar, but less sophisticated formula

$$P_{n_1 n_2} = |S_{n_1 n_2}|^2 = |J_{n_1 - n_2}(N)|^2, \qquad (9.74)$$

based on making the approximations

$$n_2(\bar{w}_1) = n_1 + N \cos(2\pi\bar{w} + \text{const}),$$
$$\bar{w}_2 = \bar{w}_1 + \text{const}, \tag{9.75}$$

in Eqn (9.43) has also been suggested by Miller (1970b) for use in situations where $n_{\max} - n_{\min} < 1$.

We defer discussion of the numerical accuracy of these various approximations until Section 9.4.

9.3 The Semi-classical Propagator

Before considering the practical utilization of the above results it is pertinent to examine an alternative approach initiated in the molecular context by Pechukas (1969a, b) and Miller (1970a, b). This starts from the path integral approach to quantum mechanics developed by Feynman and Hibbs (1965).

The central quantity in the theory is the kernel or propagator, $K(q_2t_2; q_1t_1)$, which relates the wavefunction at a given set of coordinates q_1 and time t_1 to its value at q_2 and t_2 by the equation

$$\Psi(q_2t_2) = \int_{-\infty}^{\infty} K(q_2t_2; q_1t_1)\,\Psi(q_1t_1)\,\mathrm{d}q_1. \tag{9.76}$$

This propagator is given in the path integral formulation by

$$K(q_2t_2; q_1t_1) = \int_{\text{all paths}} \exp\left[i\Omega(q_2t_2; q_1t_1)/\hbar\right] \mathrm{d}\,\text{path} \tag{9.77}$$

where $\Omega(q_2t_2; q_1t_1)$ is the classical action (see Appendix E) along a particular path between $q_1(t_1)$ and $q_2(t_2)$.

$$\begin{aligned}\Omega(q_2t_2; q_1t_1) &= \int_{t_1}^{t_2} L[q(t), \dot{q}(t)]\,\mathrm{d}t \\ &= \int_{t_1}^{t_2} \left[\sum p\dot{q} - H(p, q)\right] \mathrm{d}t \\ &= \sum \int_{q_1}^{q_2} p\,\mathrm{d}q - E(t_2 - t_1),\end{aligned} \tag{9.78}$$

the final line of Eqn (9.78) being based on the assumption of a time independent Hamiltonian. It follows in this case that Eqn (9.76) may be reduced to the form

$$\psi(q_2) = \int_{-\infty}^{\infty} \kappa(q_2, q_1)\,\psi(q_1)\,\mathrm{d}q_1, \tag{9.79}$$

by the substitutions

$$\Psi(q_i t_i) = \psi(q_i) \exp[-iEt_i/\hbar]$$

$$\kappa(q_2, q_1) = K(q_2 t_2; q_1 t_1) \exp[iE(t_2 - t_1)/\hbar]. \quad (9.80)$$

Note that the integral in (9.79) is to be taken over all coordinates q, and that although the explicit dependence of q_1 and q_2 on time has been suppressed for later notational convenience, the values of these coordinates depend implicitly on t_1 and t_2.

The appearance of the path integral in the definition of $K(q_2 t_2; q_1 t_1)$ implies that in contrast to the classical situation where a particle is constrained to follow a particular trajectory for which, by Hamilton's principle, $\Omega(q_2 t_2; q_1 t_1)$ is stationary, a quantum mechanical particle can find its way from one point to another by an infinity of paths each with an appropriate phase. The probability amplitude for propagation from $q_1(t_1)$ to $q_2(t_2)$ is then obtained by considering the interference between these paths. The equivalence between this approach and the Schrödinger theory is established on pp 76–79 of Feynman and Hibbs (1965). Particular features exploited by Pechukas (1969a, b) and Miller (1970a, b) are the simple form for the reduced propagator $\kappa(q_2, q_1)$ in the near classical limit, and the direct relation between this propagator and the S matrix.

The reduction of $\kappa(q_2, q_1)$ relies on the semi-classical condition that the action $\Omega(q_2 t_2; q_1 t_1)$ should be large compared with $\hbar$ (see Section 2.4). Hence a small variation in the path will lead in general to a large phase change $\Delta\Omega/\hbar$, and consequently to strong cancellation in the path integral. An exception occurs however near the classical path $q_{\mathrm{cl}}(t)$ governed by Hamilton's equations, since by Hamilton's principle, $\Omega(q_2 t_2; q_1 t_1)$ then takes a stationary value. Hence in view of the destructive interference between all non-classical paths, the propagator may be reduced to the form†

$$\kappa'_{\mathrm{cl}}(q_2, q_1) = A'(q_2, q_1) \exp[i\phi'(q_2, q_1)/\hbar] \quad (9.81)$$

where

$$\begin{aligned} \phi'(q_2, q_1) &= \Omega_{\mathrm{cl}}(q_2 t_2; q_1 t_1) + E(t_2 - t_1) \\ &= \sum \int_{q_1}^{q_2} p \, \mathrm{d}q. \end{aligned} \quad (9.82)$$

Thus the semi-classical propagator, like the semi-classical wavefunction given in Section 9.1 depends on the classical action (a line integral) along the trajectory from q_1 to q_2. The normalization function $A'(q_2, q_1)$ remains to be determined later.

† It will be convenient later to apply small modifications to $\kappa'(q_2, q_1)$, $A'(q_2, q_1)$ and $\phi'(q_2, q_2)$.

We turn now to the relation between $\kappa'_{cl}(q_2, q_1)$ and the S matrix. The obvious connection according to Eqn (9.79), is that $K(q_2, q_1)$ governs the evolution of the system from $q_1(t_1)$ to $q_2(t_2)$. A simple illustration of the sense of this evolution is provided by application to the one dimensional JWKB expression

$$\psi(q_1) = p'(q_1) \exp\left[\frac{i}{\hbar}\int_a^{q_1} p'(q)\,dq\right], \tag{9.83}$$

in which the momentum $p'(q)$ may or may not coincide with the trajectory momentum $p(q)$ which appears in the phase term $\phi'(q_2, q_1)$ given by (9.82). The combination of (9.79), (9.81) and (9.83) shows that,

$$\begin{aligned}\psi(q_2) &= \int_{-\infty}^{\infty} A'(q_2, q_1)\,[p'(q)]^{-\frac{1}{2}} \exp\left[\frac{i}{\hbar}\int_a^{q_1} p'(q)\,dq + \frac{i}{\hbar}\int_{q_1}^{q_2} p(q)\,dq\right] dq_1 \\ &= \exp\left[\frac{i}{\hbar}\int_a^{q_2} p(q)\,dq\right]\int_{-\infty}^{\infty} A'(q_2, q_1)\,[p'(q_1)]^{-\frac{1}{2}} \\ &\quad\times \exp\left[\frac{i}{\hbar}\int_{q_1}^{q_2} (p' - p_1)\,dq\right] dq_1\end{aligned} \tag{9.84}$$

But by the now familiar semi-classical arguments, rapid oscillations in the integrand must destroy the integral unless $p(q) = p'(q)$. It follows that $A'(q_2, q_1)$ must contain a factor $[p(q_1)/p(q_2)]^{\frac{1}{2}}$, together with a suitable constant such that

$$\psi(q_2) = \delta(p - p')[p(q_2)]^{-\frac{1}{2}} \exp\left[\frac{i}{\hbar}\int_a^{q_2} p'(q)\,dq\right]. \tag{9.85}$$

The implication is that $\kappa'(q_2, q_1)$ will propagate the correct JWKB form, for which $p(q) = p'(q)$, but will otherwise destroy the wavefunction.

Extension of the theory to the multidimensional case requires a distinction between the translational variables (r, p) and the internal variables henceforth denotes (w, N) as in Sections 9.1 and 9.2 above. Repetition of the above argument applied to the initial asymptotic wavefunction

$$\psi(w_1\, r_1) = k_1^{-\frac{1}{2}} \exp[ir_1 p_1/\hbar]\,\xi_a(w_1) \tag{9.86}$$

shows that

$$\psi(w_2\, r_2) \overset{r_2\to\infty}{\sim} \delta(p_1 - p_1')\,k_2^{-\frac{1}{2}} \exp[ir_2 p_2/\hbar] \int_0^1 \kappa(w_2, w_1)\,\xi_a(w_1)\,dw_1. \tag{9.87}$$

It is also desirable as in the previous theory to transform from the (w, r) to the $(\overline{w}, \tau)$ representation by use of Eqns (9.27) and (9.29)–(9.35) to obtain

$$\tilde{\psi}(\overline{w}_2, \tau) \overset{\tau\to\infty}{\sim} \exp(iE\tau/\hbar) \int_0^1 \kappa(\overline{w}_2, \overline{w}_1)\,\xi_a(\overline{w}_1)\,d\overline{w}_1 \tag{9.88}$$

where

$$\kappa(\bar{w}_2, \bar{w}_1) = A(\bar{w}_2, \bar{w}_1) \exp\left[i\phi(\bar{w}_2, \bar{w}_1)/\hbar\right], \tag{9.89}$$

$$\phi(\bar{w}_2, \bar{w}_1) = \sum N_2 \bar{w}_2 - N_1 \bar{w}_1 - \int_{N_1}^{N_2} w \, \mathrm{d}N - \int_{p_1}^{p_2} r \, \mathrm{d}p. \tag{9.90}$$

The relation between this propagator $\kappa(w_2, w_1)$ and the S matrix is readily established by comparison between (9.88) and the defining equation

$$\tilde{\psi}(\bar{w}_2, \tau) \overset{\tau \to \infty}{\sim} \sum_b S_{ab} \exp(iE\tau/\hbar) \, \xi_b(\bar{w}_2). \tag{9.91}$$

Thus

$$S_{ab} = \int_0^1 \int_0^1 \xi_b^*(\bar{w}_2) \, \kappa(w_2, w_1) \, \xi_a(\bar{w}_1) \, \mathrm{d}\bar{w}_1 \, \mathrm{d}\bar{w}_2. \tag{9.92}$$

It follows by analogy with the Dirac (1958) form

$$S_{ab} = \int_0^1 \int_0^1 \langle \xi_b | \bar{w}_2 \rangle \langle \bar{w}_2 | S | \bar{w}_1 \rangle \langle \bar{w}_1 | \xi_a \rangle \, \mathrm{d}\bar{w}_1 \, \mathrm{d}\bar{w}_2, \tag{9.92a}$$

that the limit of the radially corrected propagator as $t_1 \to -\infty, t_2 \to \infty$ may be identified with the coordinate representation of the S matrix.

This offers a solution to the problem of the normalization of $\kappa(\bar{w}_2, \bar{w}_1)$, at least in cases for which there is only one classical trajectory between all pairs of points $\bar{w}_1$ and $\bar{w}_2$, (we assume a single variable $\bar{w}$ for simplicity) because the unitarity condition

$$\int_0^1 \langle \bar{w}_2 | S | \bar{w}_1 \rangle \langle \bar{w}_2 | S | \bar{w}_1' \rangle^* \, \mathrm{d}\bar{w}_2 = \delta(\bar{w}_1 - \bar{w}_1'), \tag{9.93}$$

may be used to determine the function $A(\bar{w}_2, \bar{w}_1)$ in the near classical limit, $\hbar \to 0$. This is achieved by substituting $\kappa(\bar{w}_2, \bar{w}_1)$ for $\langle \bar{w}_2 | \mathrm{S} | \bar{w}_1 \rangle$ and noting that rapid oscillations in the integrand on the left of the resulting equation,

$$\int_0^1 A(\bar{w}_2, \bar{w}_1) A^*(\bar{w}_2, \bar{w}_1') \exp\{i[\phi(\bar{w}_2, \bar{w}_1) - \phi(\bar{w}_2, \bar{w}_1')]/\hbar\} \, \mathrm{d}\bar{w}_2$$
$$= \delta(\bar{w}_1 - \bar{w}_1'), \tag{9.94}$$

will lead to zero unless $\bar{w}_1 \simeq \bar{w}_1'$. Hence it is justified to extend the integration to the range $(-\infty, \infty)$ and to expand $\phi(\bar{w}_2, \bar{w}_1)$ about the point $\bar{w}_1' = \bar{w}_1$;

$$\phi(\bar{w}_2, \bar{w}_1') = \phi(\bar{w}_2, \bar{w}_1) + (\bar{w}_1' - \bar{w}_1) x(\bar{w}_2, \bar{w}_1)/\hbar + \ldots \tag{9.95}$$

where

$$x(\bar{w}_2, \bar{w}_1) = \hbar^{-1} (\partial \phi / \partial \bar{w}_1)_{\tau w_2}. \tag{9.96}$$

It follows on introducing x in place of w_2 as the integration variable, on the assumption that x is monotomic in w_2, that

$$\int_{-\infty}^{\infty} |A(\bar{w}_2, \bar{w}_1)|^2 \exp\left[ix(\bar{w}_1 - \bar{w}_1')\right](\partial x/\partial \bar{w}_2)\,\mathrm{d}x = \delta(\bar{w}_1 - \bar{w}_1'). \quad (9.97)$$

This may be compared with the identity Dirac (1958)

$$\frac{1}{2\pi}\int_{-\infty}^{\infty} \exp\left[ix(\bar{w}_1 - \bar{w}_1')\right]\mathrm{d}x = \delta(\bar{w}_1 - \bar{w}_1'), \quad (9.98)$$

from which it is seen that the unitarity of $\langle \bar{w}_2 | S | \bar{w}_1 \rangle$ requires that

$$\begin{aligned} |A(\bar{w}_2, \bar{w}_1)| &= (2\pi)^{-\frac{1}{2}} |\partial x/\partial \bar{w}_2|^{\frac{1}{2}} \\ &= \left[\frac{1}{2\pi\hbar}\left|\frac{\partial^2 \phi}{\partial \bar{w}_1 \partial \bar{w}_2}\right|\right]^{\frac{1}{2}}. \end{aligned} \quad (9.99)$$

Extension of the argument to n internal dimensions shows that in general

$$|A(\bar{w}_2, \bar{w}_1)| = \left[\left(\frac{1}{2\pi\hbar}\right)^n \left|\frac{\partial^2 \phi}{\partial \bar{w}_{1i} \partial \bar{w}_{2j}}\right|\right]^{\frac{1}{2}}, \quad (9.100)$$

where $|\partial^2 \phi / \partial \bar{w}_{1i} \partial \bar{w}_{2j}|$ denotes the Van Vleck (1928) determinant of second derivatives.

The validity of this normalization rests on the assumed monotomic behaviour of the function $x(\bar{w}_2, \bar{w}_1)$. This requires that the initial action $N_1(\bar{w}_2, \bar{w}_1)$ appropriate to a trajectory from $\bar{w}_1$ to $\bar{w}_2$ should vary monotonically with $\bar{w}_2$ since according to (9.90) and (9.96)

$$x(\bar{w}_2, \bar{w}_1) = \hbar^{-1}(\partial \phi / \partial \bar{w}_1)_{\bar{w}_2} = -N_1(\bar{w}_2, \bar{w}_1). \quad (9.101)$$

(Note that derivatives of $\phi(\bar{w}_2, \bar{w}_1)$ arising from $N_1(\bar{w}_2, \bar{w}_1)$ and $N_2(\bar{w}_2, \bar{w}_1)$ cancel out by an argument similar to that applied in equation (9.50).) In other words $\bar{w}_2$ must be accessible from $\bar{w}_1$ by a single trajectory characterized by the unique initial momentum $N_1(\bar{w}_2, \bar{w}_1)$. In this case the coordinate representation of S may be written

$$\begin{aligned} \langle \bar{w}_2 | S | \bar{w}_1 \rangle &= \left[\frac{1}{2\pi i\hbar}\left(\frac{\partial^2 \phi}{\partial \bar{w}_1 \partial \bar{w}_2}\right)\right]^{\frac{1}{2}} \exp\left[i\phi(\bar{w}_2, \bar{w}_1)/\hbar\right] \\ &= \left[\frac{i}{2\pi\hbar}\left(\frac{\partial N_1}{\partial \bar{w}_2}\right)\right]^{\frac{1}{2}} \exp\left[i\phi(\bar{w}_2, \bar{w}_1)/\hbar\right] \\ &= \left[\frac{1}{2\pi i\hbar}\left(\frac{\partial N_2}{\partial \bar{w}_1}\right)\right]^{\frac{1}{2}} \exp\left[i\phi(\bar{w}_2, \bar{w}_1)/\hbar\right]. \end{aligned} \quad (9.102)$$

It remains to substitute $\langle \bar{w}_2 | S | \bar{w}_1 \rangle$ in (9.92) in order to determine the S

matrix elements, which may also be considered to form the momentum (or action) representation of S if the states of interest are the momentum (or action) eigenstates. The necessary unitary transformation is defined in the semi-classical limit by the generator

$$\langle \bar{w} | N \rangle = (2\pi i \hbar)^{-\frac{1}{2}} \exp(iN\bar{w}/\hbar) \tag{9.103}$$

which may be seen to differ from the normalized eigenfunction only by a factor $(2\pi i\hbar)^{-\frac{1}{2}}$. It is important in following the argument, which will be developed in stages, to distinguish between values of N_1 and N_2 of interest, and the dependent quantities $N_1(\bar{w}_2, \bar{w}_1)$ and $N_2(\bar{w}_2, \bar{w}_1)$ appropriate to the end point on a trajectory from $\bar{w}_1$ to $\bar{w}_2$.

The first step is to perform the integral over $\bar{w}_1$ in (9.92) in order to obtain the mixed representative

$$\begin{aligned}\langle \bar{w}_2 | S | N_1 \rangle &= \int_0^1 \langle \bar{w}_2 | S | \bar{w}_1 \rangle \langle \bar{w}_1 | N \rangle \, d\bar{w}_1 \\ &= \frac{1}{2\pi\hbar} \int_0^1 \left| \frac{\partial N_1}{\partial \bar{w}_2} \right|^{\frac{1}{2}} \exp\{i[\phi(\bar{w}_2, \bar{w}_1) + N_1 \bar{w}_1]/\hbar\} \, d\bar{w}_1 . \end{aligned} \tag{9.104}$$

This integral may be approximated by the stationary phase method in the near classical limit. Thus the stationary phase condition

$$\frac{\partial}{\partial \bar{w}_1} [\phi(\bar{w}_2, \bar{w}_1) + N_1 \bar{w}_1] = -N_1(\bar{w}_1, \bar{w}_2) + N_1 = 0, \tag{9.105}$$

shows that the point of stationary phase is the value of $\bar{w}_1$, say $\bar{w}_1(\bar{w}_2, N_1)$, appropriate to a trajectory with initial action N_1 and final position $\bar{w}_2$. The second derivative of the exponent required for a quadratic expansion about this point is clearly

$$\frac{\partial^2}{\partial \bar{w}_1^2} (\phi(\bar{w}_2, \bar{w}_1) + N_1 \bar{w}_1) = -(\partial N_1 / \partial \bar{w}_1)_{w_2} . \tag{9.106}$$

It follows on extending the integration to the range $(-\infty, \infty)$ that

$$\langle \bar{w}_2 | S | N_1 \rangle \simeq \left[\frac{1}{2\pi i \hbar} \left(\frac{\partial N_1}{\partial \bar{w}_2} \right)_{\bar{w}_1} \Big/ \left(\frac{\partial N_1}{\partial \bar{w}_1} \right)_{\bar{w}_2} \right] \exp[i\chi(\bar{w}_2, N_1)], \tag{9.107}$$

where

$$\begin{aligned}\chi(\bar{w}_2, N_1) &= \phi[\bar{w}_2, \bar{w}_1(\bar{w}_2, N_1)] + N_1 \bar{w}_1 \\ &= N_2 \bar{w}_2 - \int_{N_1}^{N_2} w \, dN - \int_{p_1}^{p_2} r \, dp. \end{aligned} \tag{9.108}$$

The product of partial derivatives in (9.107) is readily contracted by use of the Jacobian notation (see Margenan and Murphy, 1956) to yield,

$$\left(\frac{\partial N_1}{\partial \bar{w}_2}\right)_{\bar{w}_1} \bigg/ \left(\frac{\partial N_1}{\partial \bar{w}_1}\right)_{\bar{w}_2} = \frac{\partial(N_1, \bar{w}_1)}{\partial(\bar{w}_2, \bar{w}_1)} \cdot \frac{\partial(\bar{w}_1, \bar{w}_2)}{\partial(N_1, \bar{w}_2)} = -\frac{\partial(N_1, \bar{w}_1)}{\partial(N_1, \bar{w}_2)}$$

$$= -\left(\frac{\partial \bar{w}_1}{\partial \bar{w}_2}\right)_{N_1} = \frac{\partial^2 \chi}{\partial \bar{w}_2 \partial N_1}. \tag{9.109}$$

Thus $\langle \bar{w}_2 | S | N_1 \rangle$ reduces in the near classical limit to the general unitary form (Van Vleck, 1928),

$$\langle \bar{w}_2 | S | N_1 \rangle \simeq \left[\frac{1}{2\pi i\hbar} \frac{\partial^2 \chi}{\partial \bar{w}_2 \partial N_1}\right]^{\frac{1}{2}} \exp\left[i\chi(\bar{w}_2, N_1)/\hbar\right], \tag{9.110}$$

similar to that of $\langle \bar{w}_2 | S | \bar{w}_1 \rangle$ in (9.102). A further close similarity between $\langle \bar{w}_2 | S | N_1 \rangle$ and the semi-classical wavefunction given by (9.39) should cause no surprise, because the product $\langle \bar{w}_2 | S | \bar{w}_1 \rangle \langle \bar{w}_1 | N_1 \rangle$ in (9.104) is very close to the $\kappa(w_2, w_1)\xi_a(w_1)$ in (9.88).

It follows by repeating the same general argument in evaluating the integral over $\bar{w}_2$ in (9.92), that in the stationary phase limit

$$\langle N_2 | S | N_1 \rangle = \left[\frac{1}{2\pi i\hbar}\left(\frac{\partial^2 \Delta}{\partial N_1 \partial N_2}\right)\right]^{\frac{1}{2}} \exp\left[i\Delta(N_2, N_1)/\hbar\right], \tag{9.111}$$

where

$$\begin{aligned} \Delta(N_2, N_1) &= \chi(\bar{w}_2, N_1) - N_2\bar{w}_2 \\ &= \phi(\bar{w}_2, \bar{w}_1) - N_2\bar{w}_2 + N_1\bar{w}_1 \\ &= -\int_{N_1}^{N_2} w\, dN - \int_{p_1}^{p_2} r\, dp, \end{aligned} \tag{9.112}$$

and $\bar{w}_1(N_2, N_1)$ and $\bar{w}_2(N_2, N_1)$ are the coordinate end-points on a supposedly unique trajectory specified by N_1 and N_2. This may be recognized by comparison with (9.53) as a typical stationary phase approximant to the previous integral representation of S given by (9.43).

Finally an even closer connection between $\langle N_2 | S | N_1 \rangle$ and the previous integral representation for $S_{n_1 n_2}$ given by (9.43) may be obtained by omitting the final integration step and introducing $\bar{w}_1$ as the integration variable instead. The result

$$\langle N_2 | S | N_1 \rangle = \frac{1}{2\pi\hbar} \int_0^1 \left(\frac{\partial \bar{w}_2}{\partial \bar{w}_1}\right)_{N_1} \exp\{i[\chi(\bar{w}_1, N_1) - N_2\bar{w}_2]/\hbar\}\, d\bar{w}_1, \tag{9.113}$$

may be verified to differ from $S_{n_1 n_2}$ in (9.43) only by a factor $(2\pi h)^{-1}$ which arises from the normalization difference between the generator $\langle \bar{w} | N \rangle$ given by (9.103) and the normalized action eigenstate $\xi_N(\bar{w})$.

This "derivation" of the S matrix in an integral form must however be viewed with some caution because the normalization given by (9.99) rests on the assumption of a unique trajectory from $\bar{w}_1$ to $\bar{w}_2$, whereas the most interesting semi-classical effects arise from interference between multiple trajectories linking the initial and final states of interest.

The strength of the propagator approach lies in its consistent use of the stationary phase approximation. Thus it provides a short cut to what we term below the "primitive semi-classical" or "stationary phase" results represented by Eqns (9.55), (9.68) and (9.69), which are obtained from the semi-classical wavefunction only by considerable labour. Another feature of the propagator method is the facility it provides for transforming from one representation to another. Further transformations to other systems of canonical variables are also readily performed by the same general technique (see Van Vleck, 1928, Miller (1970a)).

9.4 Classically Forbidden Events

One of the most remarkable aspects of the theory, shown most clearly in Section 9.2 is the ability to predict probabilities for classically forbidden† events by solution of the classical equations of motion. Thus these equations apparently contain information about processes such as elastic scattering beyond the rainbow angle, and quantum mechanical tunnelling which are inaccessible by normal classical methods.

The key to this knowledge is to be found in the analytical topology of the problem. This is most clearly seen in the methods used to derive a uniform approximation to the S matrix (see Eqns (9.57)–(9.69)). It will be recalled that the stationary phase points $\bar{w}_1^{(v)}$ of the integrand in (9.43) were shown to correspond to the required initial phases $w_1^{(v)}(n_2, n_1)$ appropriate to a trajectory from n_1 to n_2. These phases are of course real quantities if the passage from n_1 to n_2 is allowed by classical mechanics. Furthermore these stationary phase points can be analytically continued into the complex $\bar{w}_1$ plane in order to account for the classically forbidden events. It is this introduction of a complex angle $\bar{w}$ which marks the break from classical mechanics; but it causes no difficulty in quantum mechanics, because the only necessarily real quantities are the physical observables: in this context the actions N or quantum numbers n.

The only question therefore is how best to determine these complex initial angles required to cover the classically inaccessible range. One relatively crude method is to employ an appropriate power series or other analytical

† "Forbidden" is intended here in the weak sense that no solution of the classical equations leads to the relevant transition, rather than in a strong sense requiring violation of a conservation law.

fit to the final actions $n_2(n_1, \bar{w}_1)$ and phases $\bar{w}_2(n_1, \bar{w}_1)$ predicted by normal classical methods, and then to continue these functions into the complex $\bar{w}_1$ plane in order to predict the necessary parameters $A(n_2)$, $S(n_2)$, P_a and P_b for substitution in for example Eqn (9.65) when n_2 lies outside the classical range. The formula given by Eqn (9.74) is an example of this approach.

The possibility of a more fundamental and more powerful approach may however be recognized, by turning back to the derivation of the semi-classical wavefunction given by (9.22) and (9.38) or of the propagator given by (9.89). The argument is seen in both cases to require that the motion should be governed by Hamilton's equations but not that the solutions should be constrained in any other classical sense. Thus by analytical continuation of the classical solutions into the complex coordinate plane (Miller and George, 1972; Stine and Marcus, 1972) we may find what are termed "complex" or "hidden" classical trajectories between otherwise mutually inaccessible points n_1 and n_2. These are classical in the sense that they satisfy the laws of classical mechanics, but hidden in the complex plane; they are found also to involve motion in complex time. A simple trajectory of this type is examined below. Questions of numerical stability in finding such trajectories are discussed by Miller and George (1972) and Stine and Marcus (1972).

One final general point concerns the relative contributions of the hidden trajectories to the transition probability, because each complex solution of the real classical equations will be matched by a complex conjugate solution. Thus the imaginary parts of the phases Δ_v in (9.60) corresponding to the two trajectories will be positive and negative respectively. The first treated alone would yield an acceptable exponentially small transition probability, but the second would predict a probability which is exponentially large. The uniform approximation technique shows, by Eqn (9.68), that the two contributions are essentially equivalent in the allowed classical regions, but that the relative contribution of the exponentially small solution increases through the classical threshold to become completely dominant, deep in the classically inaccessible region, as shown by Eqn (9.69).

We consider as an example of this behaviour the problem of passage through a quadratic potential barrier (Miller and George, 1972)

$$V(q) = -\tfrac{1}{2}\lambda q_2 \tag{9.114}$$

at energy $-\Delta E$, subject to the boundary condition $p < 0$ for $t < 0$, the time zero being conveniently taken at the classical turning point at which the momentum $n = 0$. The solution of the classical equations is readily shown to be

$$q = -(2\Delta E/\lambda)^{\frac{1}{2}} \cosh \omega^* t$$

$$p = -(2\Delta E/m)^{\frac{1}{2}} \sinh \omega^* t$$
$$\omega^* = (\lambda/m)^{\frac{1}{2}} \tag{9.115}$$

Thus the coordinate q remains negative at all real times, while the momentum p changes sign at $t = 0$. In other words the particle is simply reflected by the barrier.

Suppose however that the time acquires an imaginary increment $\pm i\pi/\omega^*$ during the motion, so that finally

$$t_2 = t_2' \pm i\pi/\omega^* \tag{9.116}$$

It follows from (9.115) that

$$\begin{aligned} q_2 &= -(2\Delta E/\lambda)^{\frac{1}{2}} \cosh(\omega^* t_2' \pm i\pi) = (2\Delta E/\lambda)^{\frac{1}{2}} \cosh \omega^* t_2' \\ p_2 &= -(2\Delta E/m)^{\frac{1}{2}} \sinh(\omega^* t_2' \pm i\pi) = (2\Delta E/m)^{\frac{1}{2}} \sinh \omega^* t_2' \end{aligned} \tag{9.117}$$

The particle has now passed through the barrier moving in the positive q direction.

Since we are interested in a coordinate change from $q_1 < 0$ to $q_2 > 0$, the barrier penetration probability is governed by the phase term

$$\phi(q_2, q_1) = \int_{t_1}^{t_2} p\dot{q}\, dt, \tag{9.118}$$

in the coordinate representation $\langle q_2|S|q_1\rangle$ analogous to $\langle \bar{w}_1|S|\bar{w}_1\rangle$ in (9.102). It is immaterial how the integration path in (9.118) is taken, but it is convenient for physical reasons to choose that q should remain real. This involves integration along the real time axis to $t = 0$, followed by an imaginary increment to the point $t = \pm i\pi/\omega^*$ after which Im t remains constant. Hence over the imaginary region, $t = it''$, and

$$\begin{aligned} q &= -(2\Delta E/\lambda)^{\frac{1}{2}} \cosh i\omega^* t'' = -(2\Delta E/\lambda)^{\frac{1}{2}} \cos \omega^* t'' \\ p &= -(2\Delta E/m)^{\frac{1}{2}} \sinh i\omega^* t'' = -2i(\Delta E/m)^{\frac{1}{2}} \sin \omega^* t''. \end{aligned} \tag{9.119}$$

Note that q passes smoothly through the barrier from $q = -(2\Delta E/\lambda)^{\frac{1}{2}}$ to $q = +(2\Delta E/\lambda)^{\frac{1}{2}}$ as t'' varies from 0 to $\pm\pi$, while p remains purely imaginary. The required imaginary part of the exponent in $\langle q_2|S|q_1\rangle$ is therefore given by

$$\begin{aligned} \text{Im}\,[\phi(q_2, q_1)/\hbar] &= \frac{1}{\hbar}\int_0^{\pm i\pi/\omega^*} p\dot{q}\, dt = 2\Delta E/\hbar \int_0^{\pm i\pi/\omega^*} \sinh^2 \omega t\, dt \\ &= \pm i\pi\Delta E/\hbar\omega^*. \end{aligned} \tag{9.120}$$

This would be reflected in the theory by the appearance of a factor $\exp(\pm\pi\Delta E/\hbar\omega^*)$ in the JWKB wavefunction to the right of the barrier. The above arguments show that the trajectory leading to the exponential enhancement makes no contribution at energies well below the barrier. The remaining factor $\exp(-\pi\Delta E/\hbar\omega^*)$ is identical with that obtained by more conventional methods in Section 4.2 and Appendix C.4.

9.5 Summary and Numerical Applications

The semi-classical nature of the theory is clearly evident from the foregoing analysis. It is classical in the sense that the transition probabilities are derived from the solutions of the classical equations of motion, but quantum mechanical in allowing for interference effects and for extension to cover classically forbidden phenomena. The characteristic variation of the transition probability, $P_{n_1n_2}$, with quantum number n_2 is schematically illustrated in Fig. 9.3. $P_{n_1n_2}$ is seen to oscillate around the classical probability $(dn_2/d\bar{w}_1)^{-1}$; over the classically accessible region and to decrease exponentially for $n_2 > n_{max}$ and $n_2 < n_{min}$.

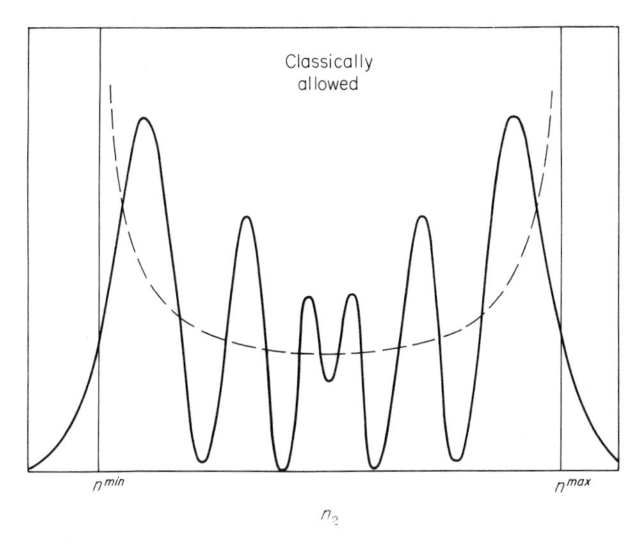

FIG. 9.3. Schematic variation of the transition probability $P_{n_1n_2}$ as a function of the continuous variable n_2. The dashed line follows the classical probability, $(dn_2/dw_1)^{-1}$, which diverges at the classical thresholds n^{max} and n^{min}.

For a quantitative explanation of this pattern, three sub-divisions into "allowed primitive semi-classical", "classical threshold", and "forbidden primitive semi-classical" categories of event may be recognized. The first, which is covered by (9.65), yields for the transition probability a sum of classical contributions, modulated by a semi-classical interference term, dependent on the difference in classical actions between the contributing trajectories. The third category is somewhat similar in that an isolated trajectory approximation, (represented by (9.69)) is applicable, with the phase term Δ_a again governed by classical action. The difference between this and the former category is that this phase is derived from a "complex" or "hidden" trajectory and hence is itself complex; this leads to an exponentially small transition probability. These two primitive semi-classical categories are ideally suited to treatment by the classical propagator approach because of its heavy reliance on the stationary phase approximation.† The intermediate, "classical threshold", category on the other hand is characterized by a confluence of two or more stationary phase regions and hence requires

TABLE 9.1. Harmonic oscillator excitation probabilities: $m = 2/3$, $\alpha = 3/10$, $E = 10$

n_1	n_2	Semi-classical[a]	Uniform Airy[a]	exact[b]
*0	0	—	0·058	(0·060)
0	1	0·422	0·211	0·218
0	2	0·416	0·381	0·366
0	3	0·359	0·266	0·267
*0	4	—	0·075	0·089
1	1	0·290	0·287	(0·286)
1	2	0·009	0·011	0·009
1	3	0·168	0·174	0·170
1	4	0·285	0·240	0·240
*1	5	—	0·062	0·077
2	2	0·208	0·206	(0·207)
2	3	0·020	0·017	0·018
2	4	0·165	0·170	0·169
2	5	0·262	0·194	0·194
*2	6	—	0·045	0·037

[a]Miller (1970b); [b]Secrest and Johnson (1966). *Classically forbidden; ( ) values obtained by difference.

more sophisticated analysis. The optimum approach in this case is by means of a uniform approximation to the integral representation of S, which is most readily justified from the semi-classical wavefunction point of view.

The above remarks refer to the theoretical justification for the transition probabilities derived in previous sections. It is also illuminating to consider the relative advantages of the various numerical approximations introduced in Section 9.2. Tables 9.1–9.3 contain a selection of results from extensive

† The terms "isolated trajectory", "stationary phase" and "primitive semi-classical" are synonymous in this context.

studies of the harmonic oscillator excitation problem in the model defined by Eqns (7.34)–(7.42). The first point, clearly demonstrated by Table 9.1, is that the primitive semi-classical forms (Eqns (9.56) and 9.68)) has no advantage over the uniform Airy approximation (Eqn (9.65)), because the two expressions are calculated from the same four parameters P_a, P_b, Δ_a and Δ_b. It is also seen from Table 9.1 that the uniform Airy approximation, based in this case on a

TABLE 9.2. Harmonic oscillator excitation probabilities $m = 2/3$, $\alpha = 3/10$, $E = 3$

n_1	n_2	Semi-classical[a]	Uniform Airy[a]	Bessel[c]	Numerical[a]	Exact[b]
*1	0	—	0·012	0·013	0·025	0·022
1	1	1·633	1·457	0·973	0·973	(0·977)
*1	2	—	0·013	0·013	0·0016, 0·0004	(0·001)

[a]Miller (1970b); [b]Secrest and Johnson (1966); [c]equation (9.74), with $N = 0{\cdot}23$ derived from Fig. 4 of Miller (1970b).
*classically forbidden; ( ) values obtained by difference.

TABLE 9.3. Harmonic oscillator excitation probabilities $m = 2/3$, $\alpha = 3/10$, $E = 4$

n_1	n_2	Uniform Airy[a] (complex trajectories)	Exact[b]	Numerical[c]
*0	1	$1{\cdot}08 \times 10^{-1}$	$1{\cdot}08 \times 10^{-1}$	—
*0	2	$1{\cdot}20 \times 10^{-3}$	$1{\cdot}22 \times 10^{-3}$	—
*1	2	$4{\cdot}41 \times 10^{-2}$	$4{\cdot}18 \times 10^{-2}$	$5{\cdot}3 \times 10^{-2}$, $4{\cdot}3 \times 10^{-2}$
*1	3	$1{\cdot}51 \times 10^{-5}$	$1{\cdot}46 \times 10^{-5}$	$2{\cdot}5 \times 10^{-4}$, $1{\cdot}8 \times 10^{-6}$
*2	3	$1{\cdot}48 \times 10^{-3}$	$1{\cdot}33 \times 10^{-3}$	$1{\cdot}7 \times 10^{-3}$, $4{\cdot}6 \times 10^{-4}$

[a]Stine and Marcus (1972); [b]Secrest and Johnson (1966); [c]Wong and Marcus (1971). The second column gives $P_{n_2 n_1}$.

functional fit to $n_2(n_1, w_1)$ and $w_2(n_1, w_1)$ over the classical range (Miller, 1970b), gives tolerable agreement with exact results for the classically forbidden events. Table 9.2 however illustrates the expected weakness of the uniform Airy approximation in a case of weak interaction. Here the crude Bessel approximation (Eqn (9.74)) is markedly superior, and the uniform Bessel formula (Eqn (9.73)) may be expected to provide a further improvement. The direct approach, by numerical integration of (9.43), is also shown to give good agreement with exact results in this case, although the asymmetry of $P_{n_1 n_2}$ for the 1–2 transition may be noted. Finally Table 9.3 indicates the high level of reliability of the uniform Airy approximation based on complex trajectories for the classically forbidden events. The same level of accuracy is obtained for transition probabilities as low as 10^{-11} (Stine and Marcus,

1972). The overall conclusion is that the uniform Airy approximation based on exact trajectories gives a high level of accuracy but that the uniform Bessel formula (Eqn (9.73)) may be superior in cases of weak interaction (near elastic collisions). Kreek and Marcus (1974) find that an extension of the uniform Bessel method becomes superior to Airy based approximations in problems with more than one degree of freedom.

One serious problem in extending this type of calculation to systems with many internal degrees of freedom is however the double ended nature of the boundary conditions, which entail a search for the trajectories which focus on the required final quantum number n_2. This search may be avoided by performing a direct numerical evaluation of the appropriate multidimensional integral analogous to that in (9.43), but reliable results may be expected only for the classically allowed transition probabilities.

A final observation concerns the possibility of obtaining "partially averaged" (Doll and Miller, 1973) transition probabilities in situations where it is intended to treat only one of several degrees of freedom by quantum mechanics, and where the individual transition probabilities are adequately regarded as a sum of stationary phase contributions. In this case, by generalization of (9.56),

$$P_{\mathbf{n}_1\mathbf{n}_2} = \sum_{\nu} P^{(\nu)}_{\mathbf{n}_1\mathbf{n}_2} = \sum_{\nu} \left| \frac{\partial(\bar{w}_a, \bar{w}_b \ldots \bar{w}_m)}{\partial(n_{2a}, n_{2b} \ldots n_{2m})} \right|, \tag{9.121}$$

after neglect of the interference terms; a similar expression containing only a single term is also obtained via the generalization (9.69), if the transition $\mathbf{n}_1 \rightarrow \mathbf{n}_2$ is forbidden by classical mechanics. It follows that the mean transition probability, obtained by integrating over the final quantum numbers $(n_{2b}, n_{2c} \ldots n_{2m})$ is made up of contributions

$$\bar{P}_{\mathbf{n}_1 \rightarrow n_{2a}} = C \int\int \cdots \int \left| \frac{\partial(\bar{w}_{1a}, \bar{w}_{1b} \ldots \bar{w}_{1m})}{\partial(n_{2a}, n_{2b} \ldots n_{2m})} \right| \mathrm{d}n_{2b}\, \mathrm{d}n_{2c} \ldots \mathrm{d}n_{2m}, \tag{9.122}$$

which may be transformed, by a change of integration variables from $(n_{2b}, n_{2c} \ldots n_{2m})$ to $(\bar{w}_{1b}, \bar{w}_{1c} \ldots \bar{w}_{1m})$ to yield

$$\begin{aligned}
\bar{P}_{\mathbf{n}_1 \rightarrow n_{2a}} &= C \int\int \cdots \int \left| \frac{\partial(\bar{w}_{1a}, \bar{w}_{1b} \ldots \bar{w}_{1m})}{\partial(n_{2a}, n_{2d} \ldots n_{2m})} \right| \left| \frac{\partial(n_{2b} \ldots n_{2m}}{\partial(\bar{w}_{1b} \ldots \bar{w}_{1m})} \right| \mathrm{d}\bar{w}_{1b} \ldots \mathrm{d}\bar{w}_{1m} \\
&= C \int\int \cdots \int \left| \frac{\partial(\bar{w}_{1a}, \bar{w}_{1b} \ldots \bar{w}_{1m})}{\partial(n_{2a}, \bar{w}_{1b} \ldots \bar{w}_{1m})} \right| \mathrm{d}\bar{w}_{1b} \ldots \mathrm{d}\bar{w}_{1m} \\
&= C \int\int \cdots \int \left| \frac{\partial \bar{w}_{1a}}{\partial n_{2a}} \right| \mathrm{d}\bar{w}_{1b} \ldots \mathrm{d}\bar{w}_{1m},
\end{aligned} \tag{9.123}$$

where the final partial derivative is evaluated at constant values of the initial phases, $(\bar{w}_{1b} \ldots \bar{w}_{1m})$, of the remaining degrees of freedom. The advantage of this form over that given by (9.122) is that it is necessary to apply double ended boundary conditions only to motion in the coordinate of particular interest, and to perform a simple phase average over the remaining degrees of freedom.

CHAPTER 10

Reactive Scattering

Reactive or rearrangement processes differ from inelastic events by the appearance of new molecular species with different masses and different coordinate systems. Of these it is the change in coordinate system which presents the most severe theoretical challenge. In view of the present limited development of the theory, attention is restricted below to two aspects of the problem.

The first concerns the familiar case of a reaction accompanied by a smooth change in the overall electronic state, and hence governed by a single electronic energy surface. Here the introduction of a "natural" curvilinear coordinate system (Marcus, 1966), adapted to the nature of the surface, has been shown to give tractable equations of motion in one physical dimension (Wu and Levine, 1971), and an extension of this approach to three dimensions has also been suggested (Wyatt, 1972; Harms and Wyatt, 1973). The alternative for a correct quantum-mechanical description is to solve the equations of motion by a numerical finite difference method (Mortensen and Pitzer, 1962; Truhlar and Kupperman, 1972); this is insensitive to the choice of coordinate system but being purely numerical it may tend to obscure the physical origin of the results. At the opposite extreme lie the classical Monte Carlo calculations described by Bunker (1970), and Karplus (1970). These are purely classical, and a detailed description lies outside the scope of this book. Again being purely numerical these calculations require no particular choice of coordinate system, and now have the advantage that physical motion along a classical trajectory is readily visualized. Furthermore such classical results are open in principle to semi-classical corrections, by the methods of the previous chapter.

The second type of system, discussed in Section 10.3 below, involves a significant probability of transition from one adiabatic surface to another. Since the relevant transition region may be quite sharply localized in space, the complications of a change in coordinate system are less severe, and certain significant conclusions may be deduced from the conservation laws.

10.1 General Theory

As mentioned above, the major complication in moving from an inelastic to a reactive (or rearrangement) process arises from a coordinate change between the reactant and product configurations. The simplest system†

$$A + BC \rightarrow AB + C \tag{10.1}$$

is illustrated in Fig. 10.1, the conditions for the rearrangement being that the

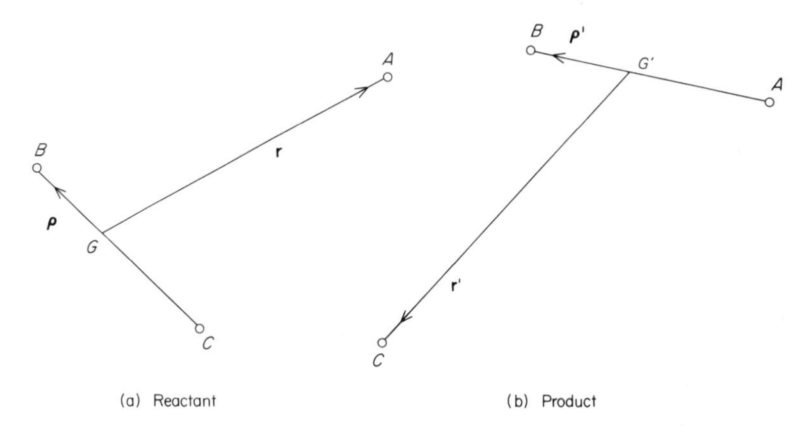

FIG. 10.1. (a) The initial and (b) the final coordinate systems.

potential function is open to dissociation as both $r \rightarrow \infty$ and $r' \rightarrow \infty$, in the sense that

$$V \overset{r\rightarrow\infty}{\sim} V_{\infty} + V_{BC}(\boldsymbol{\rho}), \tag{10.2a}$$

$$\overset{r'\rightarrow\infty}{\sim} V'_{\infty} + V_{AB}(\boldsymbol{\rho}'), \tag{10.2b}$$

and that the available energy exceeds both $V_{\infty} + E_0$ and $V'_{\infty} + E'_0$, where E_0 and E'_0 are the reactant and product zero-point internal energies respectively. This change from

$$\begin{aligned} \mathbf{r} &= \mathbf{r}_a - \mathbf{r}_g = \mathbf{r}_a - (m_b\mathbf{r}_b + m_c\mathbf{r}_c)/(m_b + m_c) \\ \boldsymbol{\rho} &= \mathbf{r}_b - \mathbf{r}_c \end{aligned} \tag{10.3}$$

† The more general case would also include the possibility of a second rearrangement channel corresponding to the alternative products $AC + B$, and also the possibility of a simultaneous electronic transition to yield such products as $A^* + BC$, $A + BC^*$, $AB + C^*$ etc.

to

$$\mathbf{r}' = \mathbf{r}_c - \mathbf{r}'_g = \mathbf{r}_c - (m_a\mathbf{r}_a + m_b\mathbf{r}_b)/(m_a + m_b)$$
$$\boldsymbol{\rho}' = \mathbf{r}_b - \mathbf{r}_a \tag{10.4}$$

results in the following alternative forms for the Hamiltonian;

$$H = -\frac{\hbar^2}{2m}\nabla_r^2 - \frac{\hbar^2}{2\mu}\nabla_\rho^2 + V(\mathbf{r}, \boldsymbol{\rho}) \tag{10.5a}$$

or

$$H = -\frac{\hbar^2}{2m'}\nabla_{r'}^2 - \frac{\hbar^2}{2\mu'}\nabla_{\rho'}^2 + V'(\mathbf{r}', \boldsymbol{\rho}'), \tag{10.5b}$$

where

$$m = m_a(m_b + m_c)/M, \qquad m' = (m_a + m_b)m_c/M$$
$$\mu = m_b m_c/(m_b + m_c), \qquad \mu' = m_a m_b/(m_a + m_b)$$
$$M = m_a + m_b + m_c. \tag{10.6}$$

It should be stressed that (10.5a) and (10.5b) are strict alternatives, and that any eigenfunction of one is necessarily an eigenfunction of the other. As a matter of convenience however (10.5a) is preferable in the reaction region, because it is, by virtue of (10.2a) separable in $\mathbf{r}$ and $\boldsymbol{\rho}$ as $r \to \infty$, while (10.5b) is separable in $\mathbf{r}'$ and $\boldsymbol{\rho}'$ in the asymptotic products zone, $r' \to \infty$.

These two asymptotic forms may therefore be used to set the boundary conditions on the problem. In the case of scattering from the reactant internal state $\phi_0(\boldsymbol{\rho})$ we must allow for the possibility of inelastic transitions to other reactant states $\phi_n(\boldsymbol{\rho})$ and also for rearrangement processes leading to product states $\phi'_m(\boldsymbol{\rho}')$. This implies as usual, an incident plane wave in the entrance channel and outgoing waves elsewhere, so that

$$\Psi \overset{r\to\infty}{\sim} \phi_0(\boldsymbol{\rho})\, e^{ik_0z} + \sum_n \phi_n(\boldsymbol{\rho}) f_{0n}(\theta, \phi)\, e^{ik_nr}/r$$
$$\overset{r'\to\infty}{\sim} \sum_m \phi'_m(\boldsymbol{\rho}) f'_{0m}(\theta', \phi')\, e^{ik'_mr'}/r' \tag{10.7}$$

where

$$k_n^2 = 2m(E - E_n - V_\infty)/\hbar^2,$$
$$k_m'^2 = 2m'(E - E'_m - V'_\infty)/\hbar^2. \tag{10.8}$$

The terms in the first line of (10.7) are readily verified to be asymptotic eigenfunctions of H given by (10.5a), while those in the second are eigenfunctions of (10.5b) as $r' \to \infty$.

Given the solution (10.7), the scattering amplitudes $f_{0n}(\theta, \phi)$ are converted in the usual way into elastic ($n = 0$) and inelastic cross-sections,

$$\frac{d\sigma_{0n}}{d\Omega} = I_{0n}(\theta, \phi) = \left(\frac{k_n}{k_0}\right) |f_{0n}(\theta, \phi)|^2$$

$$\sigma_{0n} = \int_0^{2\pi} \int_0^{\pi} I_{0n}(\theta, \phi) \sin\theta \, d\theta \, d\phi, \tag{10.9}$$

while $f'_{0m}(\theta', \phi')$ yield the corresponding reactive quantities

$$I'_{0m}(\theta', \phi') = \left(\frac{mk'_m}{m'k_0}\right) |f'_{0m}(\theta', \phi')|^2$$

$$\sigma'_{0m} = \int_0^{2\pi} \int_0^{\pi} I'_{0m}(\theta', \phi') \sin\theta' \, d\theta' \, d\phi' \tag{10.10}$$

The above expressions for the Hamiltonian and for the boundary conditions define the scattering problem, but they do nothing to indicate a practical means for its solution, particularly as in many molecular contexts the possibility of reactive scattering cannot be considered as a perturbation to the elastic and inelastic events. Although it is immaterial from a formal viewpoint which coordinate representation is employed, because any function in $(\mathbf{r}, \boldsymbol{\rho})$ may be converted to $(\mathbf{r}', \boldsymbol{\rho}')$ by the transformation

$$\begin{pmatrix} \mathbf{r}' \\ \boldsymbol{\rho}' \end{pmatrix} = \begin{pmatrix} -\dfrac{m_a}{m_a + m_b}, & -\dfrac{m_b M}{(m_a + m_b)(m_b + m_c)} \\ -1, & \dfrac{m_c}{m_b + m_c} \end{pmatrix} \begin{pmatrix} \mathbf{r} \\ \boldsymbol{\rho} \end{pmatrix} \tag{10.11}$$

implied by (10.3) and (10.4), the normal "inelastic" approach by expansion in reactant states must lead to heavy coupling between highly excited channels if it is to provide an adequate description of the products, while a mixed expansion in reactant and product states raises problems of orthogonality and over completeness in the basis set. One practical approach is by direct numerical solution of the partial differential equation by a finite difference method as first employed by Mortensen and Pitzer (1962) and more recently by Diestler and McKoy (1968) and Truhlar and Kupperman (1972). Present applications have been limited to collinear systems containing only two coordinates, the magnitudes of $\mathbf{r}$ and $\boldsymbol{\rho}$ (or $\mathbf{r}'$ and $\boldsymbol{\rho}'$) in Fig. 10.1, but the general approach is capable of extension to any system. The main disadvantage, common to any purely numerical method, is that it tends to obscure the physical origin of the results. These difficulties may be overcome for the collinear $A + BC \rightarrow AB + C$ system by a transformation due to

Marcus (1966) to a system of "natural collision coordinates" (s, x) designed to pass smoothly from $(r, \boldsymbol{\rho})$ to $(r', \boldsymbol{\rho}')$ as the reaction proceeds.

10.2 Natural Collision Coordinates

The origin of these natural collision coordinates lies in the familiar contour representation of the potential function, plotted, for later convenience, in Fig. 10.2 as a function of r and the scaled vibrational coordinate

$$\tilde{\rho} = (\mu/m)^{\frac{1}{2}}\rho. \tag{10.12}$$

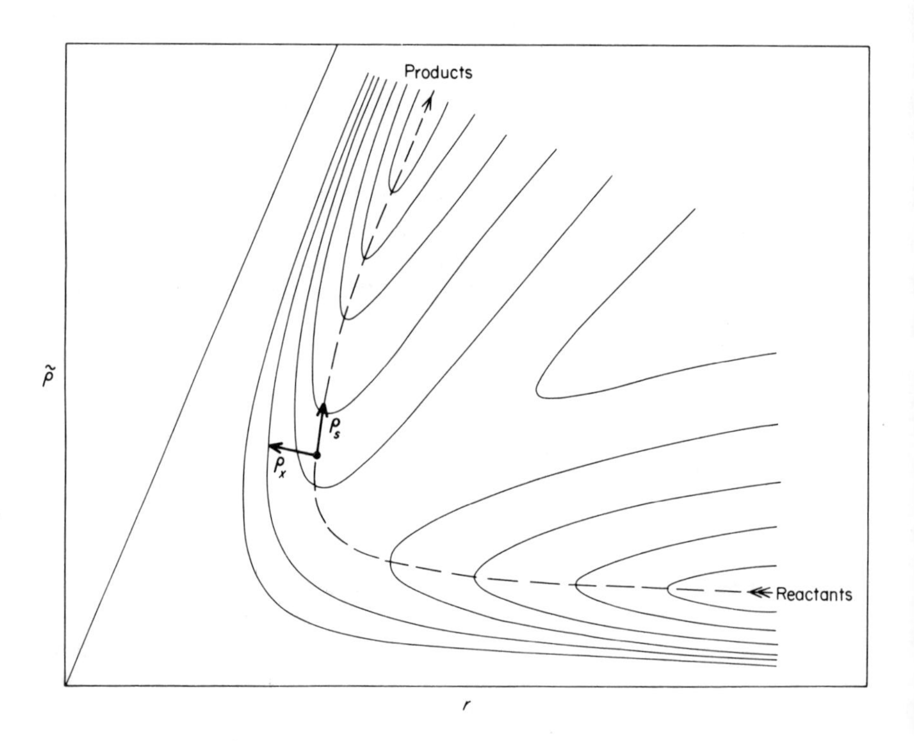

FIG. 10.2. Contours of the electronic energy surface, showing the minimum energy reaction path. The natural collision coordinates of Marcus (1966) are referred to the moving dihedral $(\boldsymbol{\rho}_s, \boldsymbol{\rho}_x)$.

The main points of interest are first the angle of skew

$$\chi = \tan^{-1}(Mm_b/m_a m_c)^{\frac{1}{2}}, \tag{10.13}$$

between the reactant $(r_a \to \infty)$ and product $(r_c \to \infty)$ asymptotes which are given according to (10.3) and (10.11) by

$$\tilde{\rho} \doteq 0 \quad \text{and} \quad \tilde{\rho} = (Mm_b/m_a m_c)^{\frac{1}{2}} r \tag{10.14}$$

respectively, and secondly the existence of a natural minimum energy path from the reactants to the products regions.

The purpose of the scaling in (10.11) is to reduce the kinetic energy, written classically

$$T = \tfrac{1}{2}m\dot{r}^2 + \tfrac{1}{2}\mu\dot{\rho}^2 = \tfrac{1}{2}m(\dot{r}^2 + \dot{\tilde{\rho}}) = \frac{1}{2m}(p_r^2 + p_{\tilde{\rho}}^2), \tag{10.15}$$

to a form invariant to any orthogonal transformation on $(r, \tilde{\rho})$. Furthermore a similar scaling on the products side, designed to accommodate the change in translational reduced mass yields

$$\tilde{\mathbf{r}}' = (m'/m)^{\frac{1}{2}} r, \qquad \tilde{\rho}' = (\mu'/m)\rho', \tag{10.16}$$

which are orthogonal transforms of $(r, \tilde{\rho})$,

$$\begin{pmatrix} \tilde{r}' \\ \tilde{\rho}' \end{pmatrix} = \begin{pmatrix} -\cos\chi, & -\sin\chi \\ -\sin\chi, & \cos\chi \end{pmatrix} \begin{pmatrix} r \\ \tilde{\rho} \end{pmatrix}. \tag{10.17}$$

χ is again the angle of skew defined by (10.14), as may be verified by substitution in (10.11). Hence T may also be written

$$T = \frac{1}{2m}(\dot{\tilde{r}}'^2 + \dot{\tilde{\rho}}'^2) = \frac{1}{2m}(p_{\tilde{r}'}^2 + p_{\tilde{\rho}'}^2). \tag{10.18}$$

The purpose of the natural collision coordinates, based by Marcus (1966) on the moving dihedral $(\boldsymbol{\rho}_s, \boldsymbol{\rho}_x)$ in Fig. 10.2 is to follow a smooth rotation along the reaction path such that (s, x) coincide initially with $(r, \tilde{\rho})$ and finally with $(\tilde{r}', \tilde{\rho}')$. The coordinates of a point P in Fig. 10.3 in this system are precisely defined by the distance x along a perpendicular to the path† (with x positive on the convex side) and the distance, s, measured along the path from a fixed point 0 to the foot of this perpendicular A. By this definition the cartesian coordinates $(r, \tilde{\rho})$ at P, abbreviated to $\mathbf{r}$, may be written

$$\mathbf{r} = \mathbf{a} + x\boldsymbol{\rho}_x, \tag{10.19}$$

and the dihedral $(\boldsymbol{\rho}_s, \boldsymbol{\rho}_x)$ is tied to the plane curve C by the Frenet formula (Corben and Stehle, 1960),

$$\frac{\mathrm{d}\boldsymbol{\rho}_x}{\mathrm{d}s} = \kappa\boldsymbol{\rho}_s \tag{10.20}$$

† The possibility of ambiguity in the choice of perpendicular for points on the concave side of the path with x greater than the radius of curvature, σ, may be discounted in practice because the dynamical equations, (see (10.26)) are protected by a singularity at $x = -\sigma$.

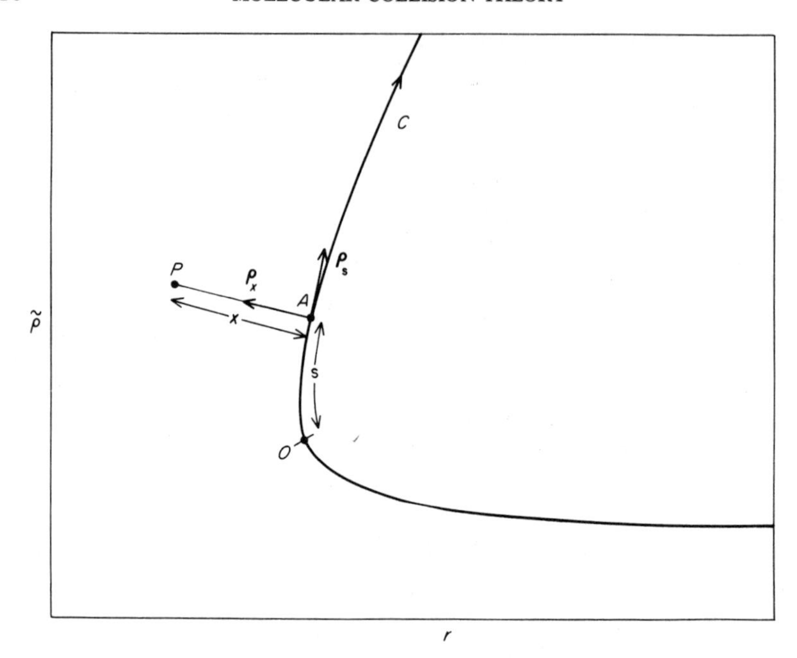

FIG. 10.3. The natural coordinate system. s and x denote the curvilinear coordinates of the point P.

linking the change in $\boldsymbol{\rho}_x$ to the local curvature $\kappa(s)$ of C, and the requirement that as A moves along the path

$$\frac{\mathrm{d}\mathbf{a}}{\mathrm{d}s} = \boldsymbol{\rho}_s. \tag{10.21}$$

The next step is to determine the momenta (P_s, P_x) conjugate to (s, x), for which purpose we use a classical generator of the form defined in Appendix F.3,

$$F_3 = -\mathbf{p}\,.\,\mathbf{r} = -\mathbf{p}\,.\,(\mathbf{a} + x\boldsymbol{\rho}_x), \tag{10.22}$$

so that in the light of (10.20) and (10.21)

$$P_s = -\frac{\partial F_3}{\partial s} = \mathbf{p}\,.\left(\frac{\partial \mathbf{a}}{\partial s} + x\,\frac{\partial \boldsymbol{\rho}_x}{\partial s}\right) = (1 + \kappa x)\,\mathbf{p}\,.\,\boldsymbol{\rho}_s = (1 + \kappa x)\,\mathbf{p}_s$$

$$P_x = -\frac{\partial F_3}{\partial x} = \mathbf{p}\,.\,\boldsymbol{\rho}_x = p_x. \tag{10.23}$$

Here we distinguish between the new conjugate momenta (P_s, P_x) and the components (p_s, p_x) of the old linear momentum $\mathbf{p}$ along $(\boldsymbol{\rho}_s, \boldsymbol{\rho}_x)$. It remains to note that (p_s, p_x) may be obtained by an orthogonal transformation of $(p_r, p_{\tilde{\rho}})$ in (10.15); hence the kinetic energy now becomes

$$T = \frac{1}{2m}(p_s^2 + p_x^2) = \frac{1}{2m}\left(\frac{1}{\eta^2}P_s^2 + P_x^2\right), \tag{10.24}$$

where

$$\eta = 1 + \kappa(s)x \tag{10.25}$$

giving for the classical Hamiltonian

$$H = \frac{1}{2m}\left(\frac{1}{\eta^2}P_s^2 + P_x^2\right) + V(s, x). \tag{10.26}$$

Note that H reduces to the normal rectilinear form in the asymptotic regions because $\kappa \to 0$ as $s \to \pm\infty$.

The physical significance of the curvature term is most clearly demonstrated in classical mechanics by the Hamiltonian equation

$$\dot{P}_x = -\frac{\partial H}{\partial x} = \frac{\kappa P_s^2}{m(1 + \kappa x)^3} - \frac{\partial V}{\partial x}, \tag{10.27}$$

for the effective force, $F_x = \dot{P}_x$, normal to the path. The curvature term may be seen as a local centrifugal force, with value $m\dot{s}^2/\sigma$ at $x = 0$, where σ is the radius of curvature; it acts outwards to force the motion off the designated reaction path (along which $\partial V/\partial x = 0$). This leads Marcus (1966) to distinguish between the reaction path, $x = 0$, and the dynamically displaced reaction coordinate, $x(s)$, along which $\dot{P}_x$ given by (10.27) vanishes. The motion of a typical representative particle projected initially along the path is shown in Fig. 10.4; it is swept initially outside the path to be left, as the curvature diminishes, oscillating more or less violently around the final asymptotic path according to the available tangential (or translational) kinetic energy $(P_s^2/2m)$ over the curvature region. This classical "bobsled" mechanism for the excitation of product vibrational energy has been amply verified in practice (see Polyani and Wong, 1969; Mok and Polyani, 1969, 1970).

Here we have assumed classical behaviour with positive translational energy. We may also note that while the quantum mechanical results at positive kinetic energies are broadly similar (see McCullough and Wyatt, 1971), the possibility of quantum mechanical tunnelling at negative translational energies introduces a novel effect. In this case the centrifugal term in (10.27) change sign and tends to force the motion to the concave side of the path (Marcus, 1966b).

The full quantum description is based on the operator form

$$H = -\frac{\hbar^2}{2m}\left[\frac{1}{\eta}\frac{\partial}{\partial s}\left(\frac{1}{\eta}\frac{\partial}{\partial s}\right) + \frac{1}{\eta}\frac{\partial}{\partial x}\left(\eta\frac{\partial}{\partial x}\right)\right] + V(s, x) \tag{10.28}$$

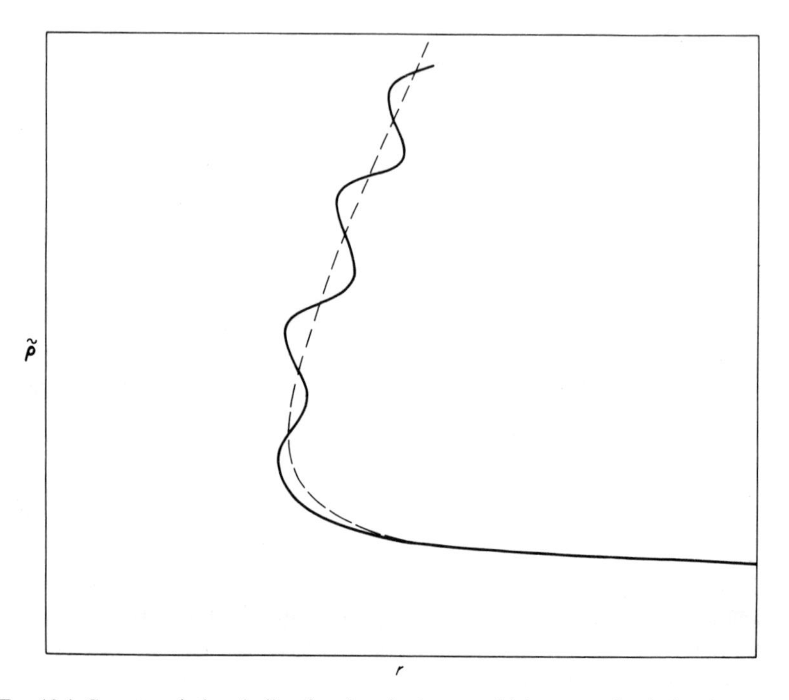

FIG. 10.4. Curvature induced vibrational excitation on a high energy classical trajectory.

derived from (10.26), with the wavefunction subject to the boundary conditions

$$\Psi(s, x) \overset{s \to -\infty}{\sim} k_0^{-\frac{1}{2}}\, \phi_0(x)\, \mathrm{e}^{ik_0 s} + \sum_n k_n^{-\frac{1}{2}}\, \phi_n(x)\, S_{0n}\, \mathrm{e}^{-ik_n s}$$

$$\overset{s \to \infty}{\sim} \sum_m k_m'^{-\frac{1}{2}}\, \phi_m(x)\, S_{0m}\, \mathrm{e}^{ik_m' s} \tag{10.29}$$

where

$$\begin{aligned} k_n^2 &= 2m[E - E_n - V(-\infty, 0)]/\hbar^2 \\ k_m'^2 &= 2m[E - E_m' - V(\infty, 0)]/\hbar^2. \end{aligned} \tag{10.30}$$

Note that the form (10.28) avoids ambiguity in the Hamiltonian, and that, with the translational coordinate open at both ends, the incident and transmitted (reactively scattered) waves have positive momenta $k_0\hbar$ and $k'_m\hbar$ as $s \to -\infty$ and $s \to \infty$ respectively, while the reflected (elastically and inelastically scattered) waves move in the negative s direction as $s \to -\infty$. The values of the S and S' matrix elements determine the elastic, P^R_{00}, inelastic, P^R_{0n}, and reactive, P^T_{0m}, transition probabilities.

$$\begin{aligned} P^R_{0n} &= |S_{0n}|^2 \\ P^T_{0m} &= |S'_{0m}|^2. \end{aligned} \tag{10.31}$$

Wu and Levine (1971) have outlined a solution to the scattering equations defined by (10.28) and (10.29) by means of an expansion

$$\Psi(s, x) = \eta^4 \sum_n \psi_n(s)\, \chi_n(s, x), \tag{10.32}$$

based on the local adiabatic vibrational eigenfunctions, $\chi_n(s, x)$ given by

$$\left[-\frac{\hbar^2}{2m}\frac{d^2}{dx^2} + V(s, x) - V(s, 0)\right]\chi_n(s; x) = \varepsilon_n(s)\,\chi_n(s, x). \tag{10.33}$$

This ensures that

$$\begin{aligned} \chi_n(s; x) &\overset{s \to -\infty}{\sim} \phi_n(x), \qquad \varepsilon_n(s) \overset{s \to -\infty}{\sim} E_n \\ &\overset{s \to \infty}{\sim} \phi'_n(x), \qquad \overset{s \to \infty}{\sim} E'_n, \end{aligned} \tag{10.34}$$

and leads to the following equations for the translational components $\psi_n(s)$:

$$\left\{\frac{d^2}{ds^2} + \frac{2m}{\hbar^2}[E - \varepsilon_n(s) - V(s, 0)]\right\}\psi_n(s) = \sum_m \left\{X_{nm}(s)\frac{d}{ds} + Y_{nm}(s)\right\}\psi_m(s) \tag{10.35}$$

with the matrices $\mathbf{X}$ and $\mathbf{Y}$ given by

$$\begin{aligned} \mathbf{X} &= -\mathbf{R}^{-1}\mathbf{P} \\ \mathbf{Y} &= -\mathbf{R}^{-1}\mathbf{Q} \end{aligned}$$

where

$$\begin{aligned} R_{nm} &= \langle \chi_n | \eta^2 | \chi_m \rangle, \\ P_{nm} &= \left\langle \chi_n \left| 2\eta^2 \frac{\partial}{\partial s} + 7\eta\left(\frac{\partial \eta}{\partial s}\right) \right| \chi_m \right\rangle, \\ Q_{nm} &= \frac{2m}{\hbar^2}\langle \chi_n | (\eta^4 - 1)(E - \varepsilon_m(s) - V(s, 0)] | \chi_m \rangle \\ &\quad + \left\langle \chi_n \left| 8\left(\frac{\partial \eta}{\partial s}\right)^2 + 4\eta\frac{\partial^2 \eta}{\partial s^2} + 16\kappa^2\eta^2 \right| \chi_m \right\rangle \end{aligned}$$

$$+\left\langle \chi_n \left| \eta^2 \frac{\partial^2}{\partial s^2} + 7\eta\left(\frac{\partial \eta}{\partial s}\right)\frac{\partial}{\partial s} + 9\eta^3\kappa\frac{\partial}{\partial x}\right| \chi_m \right\rangle. \qquad (10.36)$$

Two types of coupling term may now be distinguished, those due to non-zero curvature ($\eta \neq 1$) and those due to the parametric dependence of $\chi_m(s; x)$ on the translational coordinate s. Although both increase in importance, by virtue of the term $X_{nm}(s)(\mathrm{d}/\mathrm{d}s)$ in (10.35), as the translational velocity increases, the former are termed the "dynamic" and the latter the "static" coupling terms. Neglect of both $X_{nm}(s)$ and $Y_{nm}(s)$ for $n \neq m$ yields the "vibrationally adiabatic approximation" according to which the motion is reduced to one dimension with a different mean potential $V(s, 0) + \varepsilon_n(s)$ in each channel; the elastic and reactive scattering probabilities, P^R_{nn} and P^T_{nn}, are then given by the reflection and transmission coefficients respectively.

A measure of the validity of this approximation may be obtained by a distorted wave solution of (10.35), made subject, by a change in the form of the Green's function to accommodate the boundary conditions (10.36) rather than to those employed in Section 7.2. This yields in the case of scattering from the nth reactant channel

$$\psi_n(s) = \psi_n^{(+)}(s)$$

$$\psi_m(s) \overset{s\to\infty}{\sim} -\frac{1}{2k'_m}\psi_m^{(+)}(s)\int_{-\infty}^{\infty} \psi_m^{(-)}(s)U_{mn}(s)\psi_n^{(+)}(s)\ \mathrm{d}s, \qquad (10.37)$$

where

$$U_{mn}(s) = X_{mn}(s)\frac{\mathrm{d}}{\mathrm{d}s} + Y_{mn}(s) \qquad (10.38)$$

and $\psi_m^{(+)}(s)$ and $\psi_n^{(\pm)}(s)$ are the adiabatic waves in channels m and n respectively subject to the boundary conditions

$$\psi_m^{(\pm)}(s) \overset{s\to\infty}{\sim} \exp(\pm ik'_m s)$$

$$\psi_n^{(\pm)}(s) \overset{s\to-\infty}{\sim} \exp(\pm ik_n s). \qquad (10.39)$$

The corresponding transition amplitude is therefore given by

$$T_{nm} = \frac{1}{2(k_n k'_m)^{\frac{1}{2}}}\int_{-\infty}^{\infty} \psi_m^{(-)}(s)\, U_{mn}(s)\, \psi_n^{(+)}(s)\, \mathrm{d}s. \qquad (10.40)$$

The major factors governing the coupling strength may now be demonstrated with the help of certain simplifying approximations. We shall assume, for the sake of illustration, (a) that the system is symmetric; (b) that the translational velocities are constant over the coupling region, so that

$$\psi_m^{(\pm)}(s) = \mathrm{e}^{\pm ik_m s}; \qquad (10.41)$$

(c) that the vibrations perpendicular to the path are harmonic, with variable frequency $\omega(s)$; (d) that $\eta \simeq 1$, but $(\partial\eta/\mathrm{d}s) = (\partial\kappa/\mathrm{d}s)x$; and (e) that the dominant part of $U_{mn}(s)$ is the term $X_{mn}(s)\,\mathrm{d}/\mathrm{d}s$. The coupling then depends on the forms of the curvature $\kappa(s)$, and frequency, $\omega(s)$, functions; $\kappa(s)$ may be assumed to increase from zero at $s = \pm\infty$ to a maximum value, κ_0 at $s = 0$, while $\omega(s)$ may either increase or decrease from its asymptotic value, ω_∞, as $|s| \to 0$. Convenient parametrizations to represent this behaviour are given by

$$\kappa(s) = \kappa_0/(1 + b^2 s^2) \tag{10.40}$$

$$\omega(s) = \omega_\infty \exp\left[c/(1 + d^2 s^2)\right] \tag{10.41}$$

so that the parameters b^{-1} and d^{-1} govern the ranges of the dynamic and static coupling forces respectively. It may also be noted that (10.40) implies the assumption of a path with the total angle of skew given by

$$\chi = \int_{-\infty}^{\infty} \kappa(s)\,\mathrm{d}s = \pi\kappa_0/b, \tag{10.42}$$

and that c in (10.41) may be written

$$c = \ln(\omega_0/\omega_\infty). \tag{10.43}$$

The matrix elements required in (10.36), have been evaluated for this model by Wu and Levine (1971), with the result that

$$\begin{aligned} X_{mn}(s) = P_{nm}(s) &= \left\langle \chi_m \left| 2\frac{\partial}{\mathrm{d}s} + 7\frac{\partial\eta}{\partial s} \right| \chi_n \right\rangle \\ &= -\frac{2cd^2}{(1 + d^2s^2)^2}\,u_{nm} - \frac{14\kappa_0 x_0 b^2}{(1 + b^2s^2)^2}\,v_{nm} \end{aligned} \tag{10.44}$$

where x_0 is the zero-point vibrational amplitude (assumed constant over the dynamic coupling region)

$$x_0 = (\hbar/\mu\omega)^{\frac{1}{2}}$$

and

$$\begin{aligned} u_{nm} &= [n(n-1)/2]^{\frac{1}{2}}\delta_{n,n-2} - [(n+1)(n+2)/2]^{\frac{1}{2}}\delta_{n,n+2} \\ v_{nm} &= [n/2]^{\frac{1}{2}}\delta_{n,n-1} + [(n+1)/2]^{\frac{1}{2}}\delta_{n,n+1}. \end{aligned} \tag{10.45}$$

Note that the static and dynamic coupling terms, governed by u_{nm} and v_{nm}, are subject to selection rules $\Delta n = \pm 2$ and $\Delta n = \pm 1$ respectively. Hence the transition amplitude (10.40) may be reduced by means of the standard integral (Dwight, 1961),

$$\int_0^\infty \frac{x \sin px}{(q^2 + x^2)^2} \mathrm{d}x = \frac{\pi p \,\mathrm{e}^{-pq}}{4q} \tag{10.46}$$

to the form

$$|T_{nm}| = \frac{\pi}{2}\left(\frac{kn}{km}\right)^{\frac{1}{2}} \{u_{nm} \ln(\omega_0/\omega_\infty)\, y_{nm} \exp(-y_{nm})$$
$$+ 7v_{nm}\kappa_0 x_0 z_{nm} \exp(-z_{nm})\}, \tag{10.47}$$

where

$$y_{nm} = \left|\frac{k_n - k_m}{d}\right| = \left|\frac{E_m - E_n}{\hbar\bar{v}d}\right| = \frac{\omega_{nm}}{\bar{v}d}$$
$$z_{nm} = \left|\frac{k_n - k_m}{b}\right| = \frac{\omega_{nm}}{\bar{v}b} \tag{10.48}$$

$\bar{v}$ being the arithmetic mean of the velocities in the two channels. These factors y_{nm} and z_{nm} may be recognized as the Massey adiabatic parameters (see Section 6.1), for the static and dynamic couplings respectively; each appears as a ratio of the interaction time, $\bar{v}d$ or $\bar{v}b$, to the time period, ω_{nm}^{-1}, of the internal motion. One condition on the vibrationally adiabatic, zero coupling, model is therefore that these ratios y_{nm} and z_{nm} should be large. The magnitude of the static coupling term is also seen to depend on the numerical factor u_{nm} and the relative values of $\omega(s)$ at $s = 0$ and $s \to \infty$, while the second (dynamic coupling) term is governed by v_{nm} and the ratio of the vibrational amplitude x_0 to the maximum radius of curvature, κ_0^{-1}, along the path.

We turn now to a short discussion of a close coupled (nine channel) numerical solution of Eqns (10.34) by Wu and Levine (1971), based on a potential surface for the H_3 system due to Porter and Karplus (1964) modified by a harmonic approximation to the local oscillations about the path. The finite difference methods of Truhlar and Kupperman (1972) applied to the exact surface lead to similar results. The significant features of the surface are first the shape of the reaction path, along which the maximum curvature is achieved at two points symmetrically disposed about the saddle point, $s = 0$, and secondly a drastic change in the adiabatic vibrational energy separation from an asymptotic value of 0·53 eV, to 0·27 eV at $s = 0$. The latter suggests strong vibrational coupling, both from the static terms (10.36) because of the large frequency change and from the dynamic terms because a low frequency over the high curvature region implies a large vibrational amplitude, $x_0 \simeq 0{\cdot}15$ Å. This is confirmed by the sharp increase in P^T_{01} shown in Fig. 10.5, as the first product excitation channel opens.

The variations in the 0–0 transmission and reflection probabilities, P^T_{00} and P^R_{00}, after the expected rise in P^T_{00} and fall in P^R_{00} on crossing the reaction threshold are also of some interest. While the general oscillatory behaviour at high energies may possibly be understood, as suggested by Truhlar and Kupperman (1972), in terms of semi-classical interference discussed in the previous chapter, the strong resonance geature at $E = 0{\cdot}94$ eV is more simply explained by the shapes of the adiabatic energy curves $V(s, 0) + \varepsilon_n(s)$ depicted in Fig. 10.6, the central dips for $n > 0$ being due to the vibrational frequency change discussed above. Clearly as confirmed by Wu *et al.* (1973), this feature may be seen as a Feshbach resonance arising from interaction between the zero-point channel continuum and bound (or pseudo-bound) translational states associated with vibrational levels $n = 1$ and $n = 2$ respectively. A final point of interest concerns the detailed threshold behaviour of P^T_{00} which is seen to be inconsistent with tunnelling through the simple adiabatic barrier $V(x, 0) + \varepsilon_0(s)$ because the energy threshold, 0·53 eV, does not yield 50% transmission probability as predicted by (C.52). The reason may be traced to the diagonal coupling term $Y_{00}(s)$ in (10.34) which provide an additional dynamic (energy dependent) contribution to the barrier, equivalent to the energy change resulting from dynamic displacement of the classical path. The quantum mechanical interpretation is however more

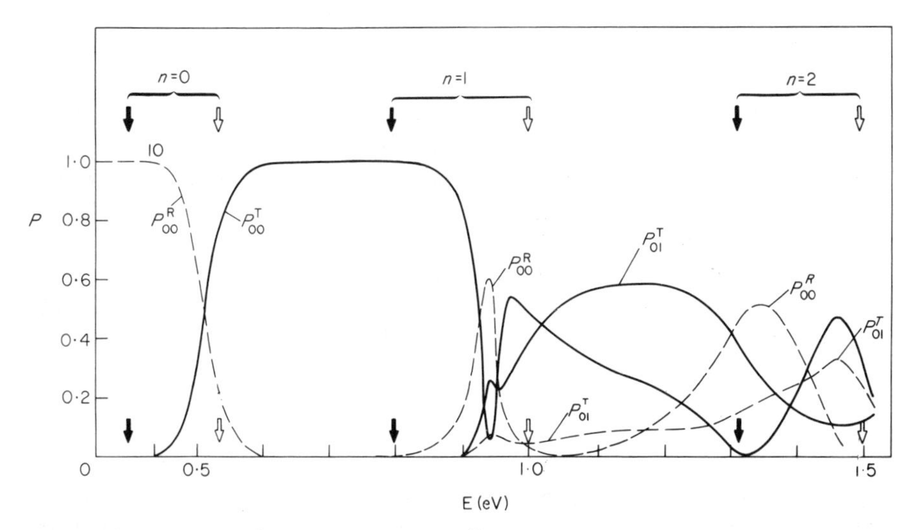

FIG. 10.5. Reflection, P^R_{0n}, and transmission, P^T_{0n}, probabilities for the collinear H_3 system reported by Wu and Levine (1971). Closed and open arrows indicate the strict and adiabatic thresholds for the levels in question respectively; compare Fig. 10.6.

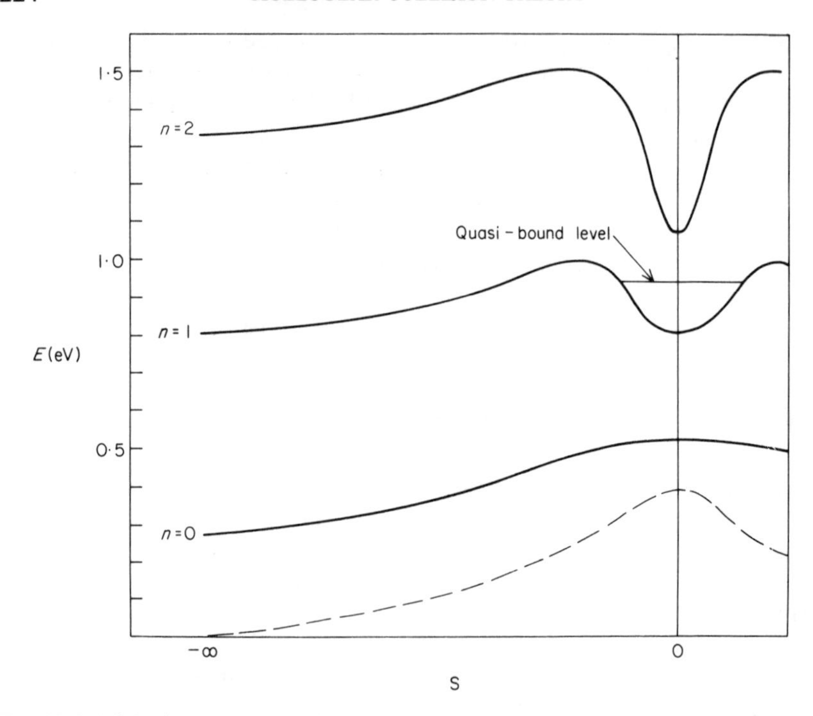

FIG. 10.6. Adiabatic energy curves for the collinear H_3 system. The dashed line indicates the energy variation along the reaction path.

complicated. Wu and Levine (1971) report that the curvature terms in (10.34) alone cause an apparent increase in the barrier height, but that the inclusion of both curvature and frequency variation terms results in an increase in the transmission coefficient, consistent with a reduction in the barrier. Detailed investigation (Wu *et al.*, 1973) shows that the dynamic correction term $Y_{00}(s)$ in (10.35) oscillates with s, and that such oscillations may lead to weak potential resonances of a tunnelling type (Child, 1967, Connor, 1970), supported by a central dip in the dynamically corrected barrier.

10.3 Multi-surface Processes

It is assumed in the previous section that motion of the representative particle is governed by a single electronic energy surface, the implication being that the electronic wavefunction is only weakly dependent on nuclear position. Important molecular systems may however give rise to potential surface intersections and hence to the possibility of rapid changes in the electronic state; notable examples are rearrangements of the type

$$M + X_2 \rightarrow MX + X \tag{10.49}$$

which are among the best characterized experimentally (see Herschbach, 1966), the crossing being in this case between a covalent, $M + X_2$, and ionic, $M^+ + X_2^-$, surface as illustrated in Fig. 10.7. A large interaction at R would of course lead to a single significant adiabatic surface but there is evidence from the observation of ionized by-products (Helbing and Rothe (1969), Baede *et al.* (1969)), that a single surface mechanism is inadequate for a complete description of such systems. The theory of such multi-surface processes is far from complete, but certain points may be illustrated by application to the systems described by (10.49).

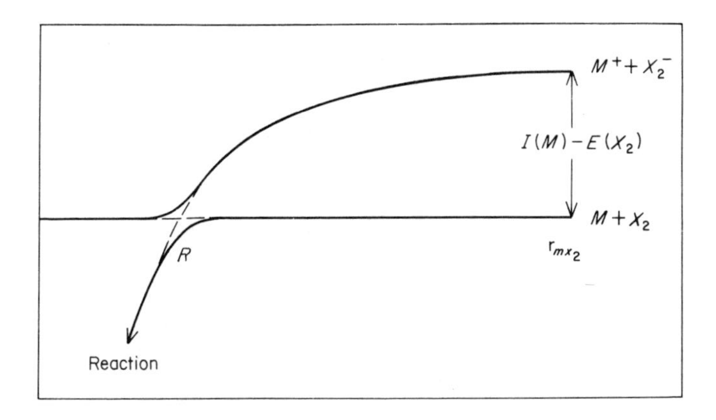

FIG. 10.7. Electronic energy curves for the harpoon model.

Here the form of analysis depends on the treatment of the crossing zone and the assumed fate of the negative ion, X_2^-. At the simplest level lies the "harpoon" (Magee, 1940), or spectator stripping (Herschbach, 1966) model, according to which the electron transfer at R, $K \rightarrow Br_2$, occurs with unit probability, and the molecule X_2 simply drifts apart on ionization leaving the velocity of the neutral (spectator) fragment, X, undisturbed, as illustrated in the centre of mass frame in Fig. 10.8. It is also convenient for the present illustrative discussion to ignore the effects of reactant internal energy, and to assume zero intermolecular potential in the covalent state and electrostatic forces between the ions. This leads to a reaction cross-section

$$\sigma = \pi R^2; \tag{10.50}$$

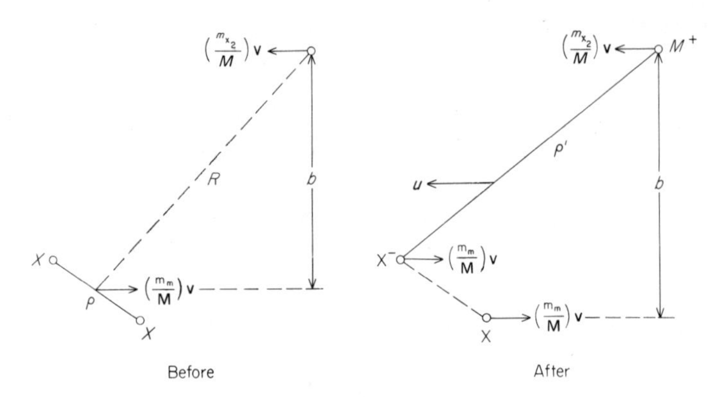

FIG. 10.8. Kinematics of the spectator stripping model.

with R dependent only on the ionization potential–electron affinity difference $I(M) - E(X_2)$;

$$R = e^2/[I(M) - E(X_2)]. \tag{10.51}$$

The strength of this model is that the final product motion is entirely governed by conservation laws. Thus the centre of mass velocity of the product molecule, $\mathbf{u}$, given by

$$m_{MX}\mathbf{u} = [m_M(m_{X_2} - m_X)/M]\mathbf{v} \tag{10.52}$$

necessarily lies in the same (forward) direction as the initial motion of M. Secondly, with primes taken to refer to product quantities, the value of the product relative velocity

$$\mathbf{v}' = \mathbf{u} - (m_M/M)\mathbf{v} = (m_M/m_{MX})\mathbf{v} \tag{10.53}$$

and the energy conservation condition

$$\frac{1}{2}\left(\frac{m_M m_{X_2}}{M}\right)v^2 + Q = \frac{1}{2}\left(\frac{m_{MX}\, m_X}{M}\right)v'^2 + E'_{\text{int}}, \tag{10.54}$$

(where Q is the zero-point exothermicity of the reaction) require that the product internal energy takes the form

$$E'_{\text{int}} = Q + \frac{m_M}{2M}\left(\frac{m_{X_2}m_{MX} - m_X m_M}{m_{MX}}\right)v^2. \tag{10.55}$$

Finally, for a given MX potential function, the distribution of this internal energy between vibrational and rotational motion is set by conservation of the internal angular momentum $\mathbf{J}'$; the mean squared magnitude of J', averaged over all orientations (with X^- taken as the nearer atom to M) being readily shown, in the notation of Fig. 10.8 to be

$$\langle J'^2 \rangle = \left(\frac{m_M m_X}{m_M + m_X} \right)^2 v^2 \left[b^2 + \frac{1}{6}\rho^2 - \frac{1}{2}\frac{b^2 \rho}{R} \right]. \tag{10.56}$$

This means, since b lies in the range $0 < b < R$, that the rotational contribution to the internal energy may be extremely high on the normal molecular scale. These conclusions are clearly open to relatively simple modification in detail by inclusion of the reactant internal motion and van der Waal's

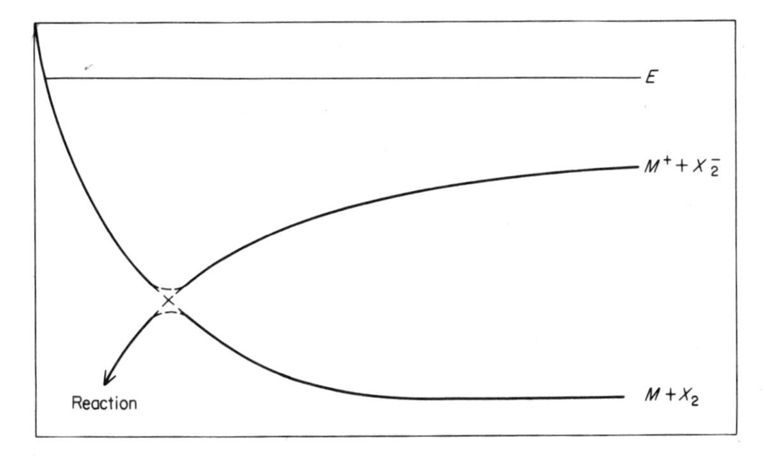

FIG. 10.9. Electronic energy curves for the charge exchange mechanism.

attraction in the covalent state, but even in its present form the model is remarkably successful in explaining certain broad thermal energy features of these reactions (Gillen, Rulis and Bernstein (1971)) namely the high reaction cross-section and low product translation energy in the strong forward peak in the differential cross-section. Predictions about the product rotational to vibrational energy ratio are not at present open to test.

At higher energies one must take explicit account of the multi-surface characteristics of the system in order to account for the formation of ionized by-products, $M^+ + X_2^-$. The obvious mechanism involves partial direct transmission through the crossing region, followed by an adiabatic switch to the ionic state on the outward trajectory governed by the potential curves depicted in Fig. 10.9. This is readily treated by the curve crossing theory outlined in Section 8.5, on the assumptions that the covalent-ionic interaction term $V_{12}(r)$ is independent of the internal coordinates, and that the potential function governing the X_2 vibration suffers no change on ionization. Each initial covalent internal (rotational, vibrational) state is then coupled to a single ionic level, and the probabilities of reaction and ionization arising from a given trajectory may be written

$$P_{\text{reaction}} = P_{12}$$
$$P_{\text{ionization}} = P_{12}(1 - P_{12}) \tag{10.57}$$

where P_{12} denotes the single passage transition probability, given in the Landau–Zener approximation (Eqn (8.113)) by

$$P_{12}(b) = 1 - \exp\{-2\pi[V_{12}(R)]^2/\hbar v_b|F_1 - F_2|\}, \tag{10.58}$$

v_b being the radial velocity component at R, and F_i the slopes of the diagonal terms $V_{ii}(r)$. In a case of zero intermolecular potential in the covalent state $P_{12}(b)$ is also conveniently expressed by means of the formula,

$$P_{12} = 1 - \exp\{-[s(1 - x)]^{-\frac{1}{2}}\} \tag{10.59}$$

in terms of the dimensionless parameters

$$s = E/U,$$
$$x = b^2/R^2, \tag{10.60}$$

where E is the translational energy at the crossing point and,

$$U = 2\pi^2 m[V_{12}(R)]^2/\hbar^2(F_1 - F_2)^2. \tag{10.61}$$

This gives for the reaction and ionization cross-sections,

$$\sigma_{\text{reaction}} = 2\pi \int_0^R P_{12}(b)b\,\mathrm{d}b = \pi R^2 \int_0^1 [1 - \exp\{-[s(1 - x)]^{-\frac{1}{2}}\}]\,\mathrm{d}x$$

$$\sigma_{\text{ionization}} = 2\pi \int_0^R P_{12}(b)\,[1 - P_{12}/b)]b\,\mathrm{d}b$$

$$= \pi R^2 \int_0^1 \exp\{-[s(1 - x)]^{-\frac{1}{2}}\}\,[1 - \exp\{-[s(1 - x)]^{-\frac{1}{2}}\}]\,\mathrm{d}x \tag{10.62}$$

since $P_{12}(b) = 0$ for $b > R$.† These formulae apply at energies above the ionization threshold. At lower energies the ionization channel is closed, and as shown by Child (1969) and Nikitin (1968),

$$\sigma_{\text{reaction}} = 2\pi R^2 \int_0^1 \left[\frac{1 - \exp\{-[s(1 - x)]^{-\frac{1}{2}}\}}{2 - \exp\{-[s(1 - x)]^{-\frac{1}{2}}\}}\right] \mathrm{d}x. \tag{10.63}$$

† The inclusion of dispersion forces in the model, would lead to the introduction of $(y - x)$ in place of $(1 - x)$ in Eqns (10.59) and (10.62), where $y = [E - V_{11}(R)]/E$, and to upper integration limits $Ry^{\frac{1}{2}}$ and y in Eqns (10.62).

Thus the reaction cross-section is predicted by both (10.62) and (10.63) to decrease with increasing energy, at a rate governed by the parameter U defined by (10.61), with a discontinuity at the ionization threshold, while, after an abrupt rise at threshold, $\sigma_{\text{ionization}}$ may pass through a maximum at $E \simeq 5{\cdot}5\, U$ or decrease monotonically to zero according to the value of the

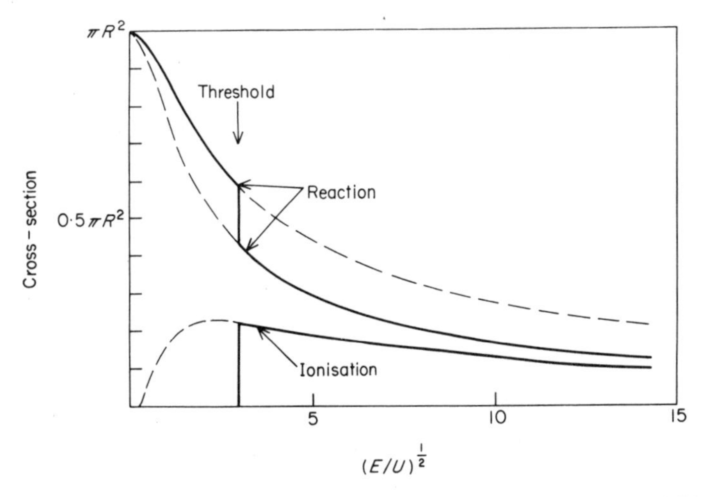

FIG. 10.10. Reaction and ionization cross-sections derived from Eqns (10.62) and (10.63).

threshold ratio $[I(M) - E(X_2)]/U$. This typical behaviour is illustrated in Fig. 10.10.

These relatively simple formulae are unfortunately again limited in practical application, first by the validity of the Landau–Zener formula, and more seriously by neglect of orientation dependence in the interaction term V_{12} (Herschbach, 1966), and by the assumption of the same vibrational states for both X_2 and X_2^-. As a measure of the limitations of the latter we may consider the more normal situation in which there is strong Franck–Condon overlap between say the $v = 0$ state of X_2 and many vibrational and possibly continuum states of X_2^-. The resulting network of potential curves—one for each interacting vibronic state—illustrated in Fig. 10.11, severely complicates the picture (Bauer *et al.*, 1969). The corresponding equations of motion, written below for simplicity in the time dependent classical trajectory form, must allow for coupling between many channels. Thus the expansions coefficients $c_{1n}(t)$ and $c_{2m}(t)$ in the total wavefunction

$$\Psi(\mathbf{q}, \boldsymbol{\rho}, t) = \xi_1(\mathbf{q}) \sum_n c_{1n}(t)\phi_{1n}(\boldsymbol{\rho}) + \xi_2(\mathbf{q}) \sum_m c_{2m}(t)\phi_{2m}(\boldsymbol{\rho}) \tag{10.64}$$

in which **q** and **ρ** denote electronic and internal nuclear coordinates respectively, must satisfy,

$$\left[i\hbar\frac{\mathrm{d}}{\mathrm{d}t} - E_{1n} - V_{11}(t)\right]c_{1n} = V_{12}(t)\sum_{m'}\Omega_{nm'}c_{2m'}(t)$$
$$\left[i\hbar\frac{\mathrm{d}}{\mathrm{d}t} - E_{2m} - V_{22}(t)\right]c_{2m} = V_{21}(t)\sum_{n'}\Omega_{mn'}c_{2n'}(t), \tag{10.65}$$

where E_{1n} and E_{2m} denote the relevant vibrational energies, and Ω_{nm} are the elements of the (orthogonal) vibrational overlap matrix. At sufficiently low energies and weak interaction strengths, the transition zones, with widths determined by (8.110) and (8.111), may be taken to be sufficiently narrow

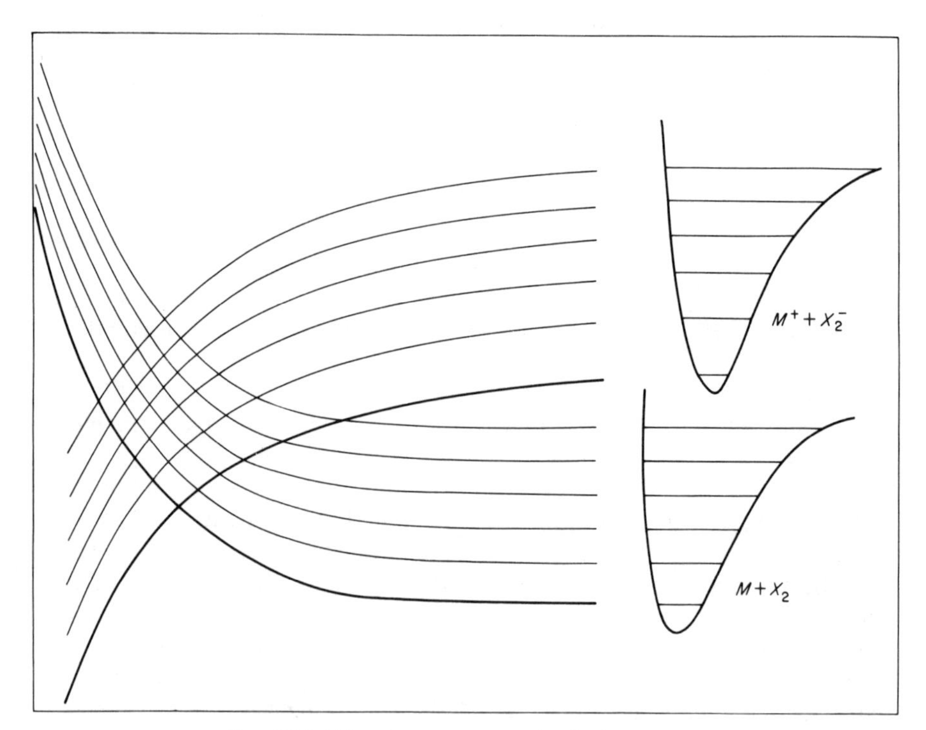

FIG. 10.11. Vibronic energy curves for the harpoon model.

that each crossing point may be treated separately (Osherov, 1965; Demkov, 1966; Nikitin, 1966). Hence the total reaction and ionization probabilities for a given trajectory may be written

$$P_{\text{reaction}}(b) = 1 - \prod_{n=1}^{N}(1 - P_{12}^{(m)}(b))$$

$$P_{\text{ionization}}(b) = P_{\text{reaction}}(b)(1 - P_{\text{reaction}}(b)), \tag{10.66}$$

where $P_{12}^{(m)}(b)$ denotes the adiabatic switching probability at the mth crossing point.

At higher energies however one must recognize an increase in the dynamic width parameter, set in the Landau–Zener approximation by

$$\Delta x_d = [\hbar v/|F_1 - F_2|)^{\frac{1}{2}}, \tag{10.67}$$

or more generally, according to the uncertainty arguments of Section 8.5 by the condition

$$|\Delta x_d[V_1(R \pm \Delta x_d) - V_2(R \pm \Delta x_d)]/\hbar v| \simeq 1. \tag{10.68}$$

Thus the transition zones around adjacent points begin to overlap and no general solution of (10.65) is possible.

Coupled with the increase in dynamic width, on the other hand, we may recognize a decrease in the interaction time, t_0, since according to (10.67)

$$t_0 = \frac{\Delta x_d}{v} = \left[\frac{\hbar}{v|F_1 - F_2|}\right]^{\frac{1}{2}}. \tag{10.69}$$

Thus as the collision velocity increases further, a point is eventually reached at which t_0 becomes small compared with the vibrational time period. In this limit we may expect some form of Franck–Condon transition probability distribution over the vibrational levels of the product ionic state. This expectation may be confirmed in the following way.

Imagine the preparation of a non-stationary vibrational state in the ionic channel,

$$\psi_{2n}(\boldsymbol{\rho}) = \sum_m \Omega_{nm}\phi_m(\boldsymbol{\rho}) \tag{10.70}$$

with mean energy

$$\bar{E}_{2n} = \sum_m \Omega_{nm} E_{2m} \Omega_{mn}, \tag{10.71}$$

designed to overlap exactly with the nth covalent vibrational state $\phi_{1n}(\boldsymbol{\rho})$. Such a state will tend to decay at a rate which may be determined by expanding the total wavefunction for the isolated ionic channel in the form

$$\Psi_2(\mathbf{q}, \boldsymbol{\rho}, t) = \xi_2(\mathbf{q}) \sum_n a_{2n}(t)\, \psi_{2n}(\boldsymbol{\rho}), \tag{10.72}$$

with the result, according to normal time dependent theory that the coefficients $a_{2n}(t)$ must satisfy

$$\left[i\hbar\frac{\mathrm{d}}{\mathrm{d}t} - \bar{E}_{2n}\right] a_{2n} = \sum_{n' \neq n} X_{nn'} a_{2n'}(t) \tag{10.73}$$

where

$$X_{nn'} = \sum_{m} \Omega_{nm}(E_{2m} - \bar{E}_{2n})\Omega_{mn} \tag{10.74}$$

The rate of spontaneous decay therefore depends on the magnitude of the mixing parameters $X_{nn'}$.

Furthermore the introduction of such states in (10.65), by means of the transformation

$$a_{2n}(t) = \sum_{m'} \Omega_{mn'} c_{2m'}(t), \tag{10.75}$$

leads to the equations

$$\begin{aligned} \left[i\hbar \frac{\mathrm{d}}{\mathrm{d}t} - W_{1n}(t) \right] c_{1n}(t) &= V_{12}(t)\, a_{2n}(t) \\ \left[i\hbar \frac{\mathrm{d}}{\mathrm{d}t} - \overline{W}_{2n}(t) \right] a_{2n}(t) &= V_{21}(t)\, c_{1n}(t) + \sum_{n' \neq n} X_{nn'} a_{2n'}(t) \end{aligned} \tag{10.76}$$

where

$$\begin{aligned} W_{1n}(t) &= E_{1n} + V_{11}(t) \\ \overline{W}_{2n}(t) &= \bar{E}_{2n} + V_{22}(t). \end{aligned} \tag{10.77}$$

These are identical, apart from the spontaneous decay terms in the second line of (10.76), with those for a single curve crossing problem. Hence we wish to demonstrate that these leakage terms have negligible importance for sufficiently short interaction times. This may be achieved by means of the substitutions,

$$\begin{aligned} c_{1n}(t) &= b_{1n}(t) \exp\left[-\frac{i}{\hbar} \int_{t_x}^{t} W_{1n}(t)\, \mathrm{d}t \right] \\ a_{2n}(t) &= b_{2n}(t) \exp\left[-\frac{i}{\hbar} \int_{t_x}^{t} \overline{W}_{2n}(t)\, \mathrm{d}t \right], \end{aligned} \tag{10.78}$$

where t_x is the crossing time at which $W_{1n}(t) = \overline{W}_{2n}(t)$, and the introduction of a dimensionless independent variable

$$\tau = (t - t_x)/t_0 \tag{10.79}$$

scaled by the interaction time t_0. Consequently

$$\begin{aligned} i\frac{\mathrm{d}b_{1n}}{\mathrm{d}\tau} &= \frac{t_0 V_{12}(t)}{\hbar} \exp\,[i\alpha_n(\tau)] b_{2n}(\tau) \\ i\frac{\mathrm{d}b_{2n}}{\mathrm{d}\tau} &= \frac{t_0 V_{21}(t)}{\hbar} \exp\,[-i\alpha_n(\tau)] b_{1n}(\tau) + \sum_{n' \neq n} \frac{t_0 X_{nn'}}{\hbar} \exp\,[i\beta_{nn'}(\tau)]\, b_{2n'}(t), \end{aligned} \tag{10.80}$$

where

$$\alpha_n(\tau) = \frac{t_0}{\hbar}\int_0^\tau [W_{1n}(\tau) - \overline{W}_{2n}(\tau)]\, d\tau$$
$$\beta_{nn'}(\tau) = \frac{t_0}{\hbar}(\overline{E}_{2n} - E_{2n'})\,\tau. \tag{10.81}$$

This form also underlines the mathematical significance of the interaction time, t_0, since it appears by the definitions (10.68) and (10.69) as the time at which

$$|d\alpha_n/d\tau| = 1. \tag{10.82}$$

Thus the onset of rapid oscillations in (10.80), which damp out any significant contribution to the transition probability, may be taken to start at this point. The immediate conclusion is that the condition for negligible spontaneous decay of the prepared state during the interaction time may be stated in the form†

$$\frac{t_0 X_{nn'}}{\hbar} = \sum_m \Omega_{nm}\left[\frac{t_0(E_{2m} - \overline{E}_{2n})}{\hbar}\right]\Omega_{mn'} \ll 1. \tag{10.83}$$

This will clearly be satisfied for interaction times small compared with the internal time period $\hbar/(E_{2m} - \overline{E}_{2n})$.

Provided the condition (10.83) is satisfied, the overall solution again takes a simple form, equivalent to that for a single curve crossing between potentials appropriate to the initial covalent vibronic energy E_{1n}, and the appropriate non-stationary ionic state ψ_{2n} with energy $\overline{E}_{2m}$. The resulting transition amplitude, given in the Landau–Zener approximation by

$$|T_{12}^{(n)}| = |b_{2n}(\infty)| = \exp[-\pi[V_{12}(t_x)]^2/\hbar v|F_1 - F_2|], \tag{10.84}$$

is readily converted, by inverting (10.77) and (10.78), into a set of Franck–Condon transition amplitudes

$$|T_{nm}| = |c_{2m}(\infty)| = |\Omega_{mn} T_{12}^{(n)}|, \tag{10.85}$$

† It may be verified that the coupling terms $X_{nn'}$ are readily evaluated from knowledge of the nuclear hamiltonians H_1 and H_2 for the initial and final channels respectively, because by the definition (10.74), if $n \neq n'$,

$$\begin{aligned} X_{nn'} &= \sum_m \langle n|m\rangle E_m \langle m|n'\rangle \\ &= \sum_m \langle n|m\rangle\langle m|H_2|n'\rangle \\ &= \langle n|H_2|n'\rangle = \langle n|H_1 + V_2 - V_1|n'\rangle \\ &= \langle n|V_2 - V_1|n'\rangle. \end{aligned}$$

Thus the coupling terms appear as matrix elements of the potential function difference, evaluated over the wavefunctions of the entrance channel.

for the stationary vibrational levels of the ionic state. Finally, by virtue of the closure relation

$$\sum_m \Omega_{nm}\Omega_{mn} = 1, \tag{10.86}$$

the total probability of scattering into the ionic channel is given by

$$P_{12} = \sum_m |T_{nm}|^2 = |T_{12}^{(n)}|^2. \tag{10.87}$$

Thus the single curve-crossing approximation employed in Eqns (10.57)–(10.63) above is justified, even in the presence of a substantial difference between the potential functions of X_2 and X_2^-, provided first that the velocity is sufficiently high to satisfy (10.83) and secondly that the electron affinity employed in (10.51) is given the correct, near vertical value

$$E(X_2) = E(X) + D(X_2^-) - D(X_2) + E_{1n} - \bar{E}_{2n}. \tag{10.88}$$

We may also anticipate, on physical grounds that orientation dependence of the interaction potential, which is ignored above, may be accommodated in the theory by an average over fixed orientations, when the rotational period becomes large compared with the interaction time.

APPENDIX A

Continuum Wavefunctions

The continuum solutions of the equation

$$\left[\frac{d^2}{dr^2} + k^2 - U_l(r)\right]\psi_{kl}(r) = 0, \tag{A.1}$$

where it is assumed that

$$\lim_{r\to\infty} rU_l(r) = 0 \tag{A.2}$$

$$\lim_{r\to 0} r^2 U_l(r) = l(l+1), \tag{A.3}$$

correspond with positive energy states, $k^2 > 0$. One pair of independent solutions, $\psi_{kl}^{(\pm)}(r)$, may be taken in view of (A.2) to have asymptotic forms†

$$\psi_{kl}^{(\pm)}(r) \overset{r\to\infty}{\sim} e^{\pm ikr}; \tag{A.4}$$

they are related in general by the equations (see Alfaro and Regge, 1965),

$$\psi_{kl}^{(-)}(r) = [\psi_{k^*l}^{(+)}(r)]^* = \psi_{-kl}^{(+)}(r) \tag{A.5}$$

so that for real values of k

$$\psi_{kl}^{(-)}(r) = [\psi_{kl}^{(+)}(r)]^*. \tag{A.6}$$

These are termed the asymptotically incoming and outgoing solutions of (A.1) by virtue of their properties under the radial momentum operator

$$\hat{p}_r\psi_{kl}^{(\pm)}(r) = -i\hbar\frac{d}{dr}\psi_{kl}^{(\pm)} \overset{r\to\infty}{\sim} \pm k\hbar\,\psi_{kl}^{(\pm)}(r). \tag{A.7}$$

This directional property makes them particularly useful in collision problems.

† See Eqn (3.36) and Mott and Massey, 1965 p. 22 for problems which may arise if the convergence of $U_l(r)$ is slower than that implied by (A.2).

These travelling solutions suffer however from the disadvantage that they diverge at the origin, since it is readily verified, subject to (A.3), that the solutions of (A.1) must behave as r^{l+1} or r^{-l} at the origin. Hence there can be only one solution which vanishes at the origin, and this solution must be real, apart from a multiplying constant, for real values of k; otherwise its complex conjugate would be a second independent solution. These conditions exclude $\psi_{kl}^{(\pm)}(r)$ individually, but the combination

$$\psi_{kl}^{(0)}(r) = C[\psi_{kl}^{(+)}(r)\psi_{kl}^{(-)}(0) - \psi_{kl}^{(-)}(r)\,\psi_{kl}^{(+)}(0)] \tag{A.8}$$

clearly satisfies the above requirements. This solution may also be written

$$\psi_{kl}^{(0)}(r) = C[(-i)^l f_l(k)\,\psi_{kl}^{(+)}(r) - (i)^l f_l(-k)\,\psi_{kl}^{(-)}(r)], \tag{A.9}$$

where the Jost functions $f_l(\pm k)$ are given by

$$f_l(\pm k) = (\pm i)^l \psi_{kl}^{(\mp)}(0), \tag{A.10}$$

the factors $(\pm i)^l$ being inserted for conventional reasons. The constant C in (A.8) and A.9) is most conveniently chosen so that

$$\psi_{kl}^{(0)}(r) \overset{r\to\infty}{\sim} \sin(kr + \delta_l). \tag{A.11}$$

The function defined in this way, or equivalently by the equation

$$\psi_{kl}^{(0)}(r) = \frac{1}{2i}[e^{i\delta_l}\,\psi_{kl}^{(+)}(r) - e^{-i\delta_l}\,\psi_{kl}^{(-)}(r)] \tag{A.12}$$

is termed the bounded solution of (A.1). Equations (A.4), (A.9) and (A.12) show that δ_l, which is closely related to the phase shift of Section 3.1 may be determined from the behaviour of $\psi_{kl}^{(\pm)}(r)$ at the origin,

$$\psi_{kl}^{(\pm)}(0) = (\mp i)^l f_l(\pm k) = A_{kl}\,e^{\pm i\delta_l} \tag{A.13}$$

The questions of normalization and orthogonality between continuum wavefunctions (which remain finite at infinity) raise some points of difficulty. The normalization given in (A.4) and (A.9), by comparison in the asymptotic region with a function of known properties, is actually the most convenient for many purposes. In contemplating an expansion in terms of continuum states however, it is necessary to examine the properties of integrals of the form

$$I_l(k, k') = \int_0^\infty \psi_{kl}^{(0)}(r)\,\psi_{k'l}^{(0)}(r)\,dr. \tag{A.14}$$

The significance of such integrals may be seen from the fact that the $\psi_{kl}^{(0)}(r)$ satisfy (A.1); thus

$$\psi_{kl}^{(0)}\left[\frac{d^2}{dr^2} + k'^2 - U_l(r)\right]\psi_{k'l}^{(0)} = \psi_{k'l}^{(0)}\left[\frac{d^2}{dr^2} + k^2 - U_l(r)\right]\psi_{kl}^{(0)}, \tag{A.15}$$

and hence

$$(k^2 - k'^2)\int_0^\infty \psi_{kl}^{(0)}(r)\,\psi_{k'l}^{(0)}(r)\,\mathrm{d}r = \int_0^\infty \left[\psi_{kl}^{(0)}\frac{\mathrm{d}^2\psi_{k'l}^{(0)}}{\mathrm{d}r^2} - \psi_{k'l}^{(0)}\frac{\mathrm{d}^2\psi_{kl}^{(0)}}{\mathrm{d}r^2}\right]\mathrm{d}r$$

$$= \left[\psi_{kl}^{(0)}\frac{\mathrm{d}\psi_{k'l}^{(0)}}{\mathrm{d}r} - \psi_{k'l}^{(0)}\frac{\mathrm{d}\psi_{kl}^{(0)}}{\mathrm{d}r}\right]_0^\infty, \tag{A.16}$$

after integrating once by parts. This means, since the $\psi_{kl}^{(0)}(r)$ vanish at the origin and behave asymptotically as (A.11), that

$$\int_0^\infty \psi_{kl}^{(0)}(r)\,\psi_{k'l}^{(0)}(r)\,\mathrm{d}r = \lim_{R\to\infty}\frac{1}{2}\left\{\frac{\sin\left[(k-k')R+\delta_l-\delta_l'\right]}{k-k'} - \frac{\sin\left[(k+k')R+\delta_l+\delta_l'\right]}{k+k'}\right\} \tag{A.17}$$

$$= \frac{1}{2}\int_0^\infty \{\cos\left[(k-k')r+\delta_l-\delta_l'\right] - \cos\left[(k+k')r+\delta_l+\delta_l'\right]\}\,\mathrm{d}r.$$

The first term on the right-hand side of (A.17) oscillates rapidly with k', at a fixed value of k, and diverges at the point $k = k'$. The second term oscillates rapidly for all positive values of k'. It is shown by Dirac (1958) that the oscillatory regions make no contribution to any integral over k', and hence

$$\int_0^\infty F(k')\,I_l(k,k')\,\mathrm{d}k' = AF(k), \tag{A.18}$$

This behaviour is represented by writing

$$I_l(k,k') = \int_0^\infty \psi_{kl}^{(0)}(r)\,\psi_{k'l}^{(0)}(r)\,\mathrm{d}r = A\delta(k-k'), \tag{A.19}$$

where the delta function has the property that

$$\left.\begin{aligned}\int_a^b F(k')\delta(k-k')\,\mathrm{d}k' &= F(k)\text{ if } a < k < b\\ &= 0,\text{ otherwise}\end{aligned}\right\} \tag{A.20}$$

The numerical value of A in (A.18) and (A.19) follows from (A.17) and the identity (Dirac, 1958),

$$\int_0^\infty \cos px\,\mathrm{d}x = \tfrac{1}{2}\,\mathrm{Re}\left\{\int_{-\infty}^\infty \mathrm{e}^{ipx}\,\mathrm{d}x\right\} = \pi\delta(p). \tag{A.21}$$

This means that

$$\int_0^\infty \psi_{kl}^{(0)}(r)\,\psi_{k'l}^{(0)}(r)\,\mathrm{d}r = \frac{\pi}{2}\,\delta(k-k'). \tag{A.22}$$

The bounded solutions defined by (A.11) are therefore said to be normalized on the k scale to $\pi/2$ times a delta function. It sometimes happens that we wish to perform an integration over energy, $E' = k'^2\hbar^2/2m$, rather than over k' as in (A.18). Such an integral is however readily transformed

$$\int_0^\infty F(E')\,I(k,k')\,\mathrm{d}E' = \frac{\hbar^2}{m}\int_0^\infty F(E')k'\,I(k,k')\,\mathrm{d}k' = \frac{\hbar^2}{m}AkF(E). \tag{A.23}$$

Written in another way, with $A = \pi/2$, $k\hbar = mv$, this implies that

$$\int_0^\infty \psi_{kl}^{(0)}(r)\,\psi_{k'l}^{(0)}(r)\,\mathrm{d}r = \frac{\pi\hbar v}{2}\,\delta(E - E'); \tag{A.24}$$

hence the $\psi_{kl}^{(0)}(r)$ may also be said to be normalized to $\pi\hbar v/2$ times a delta function of energy.

The practical significance of (A.22) and (A.24) is that if a function which is zero at the origin and finite elsewhere, is expanded in the form

$$Q(r) = \int_0^\infty q(k)\,\psi_{kl}^{(0)}(r)\,\mathrm{d}k, \tag{A.25}$$

then the expansion function $q(k)$ is given by

$$q(k) = \frac{2}{\pi}\int_0^\infty Q(r)\,\psi_{kl}^{(0)}(r)\,\mathrm{d}r. \tag{A.26}$$

APPENDIX B

Green's Functions

The Green's function upon which the Born, Distorted Wave and other perturbation methods are based may be regarded as a tool for solution of the inhomogeneous equation

$$(k^2 - h)\psi = X, \tag{B.1}$$

where h denotes a reduced Hamiltonian, $h = 2mH/\hbar^2$. It is related to the Green's operator G, such that

$$\psi = GX, \tag{B.2}$$

which may be loosely written

$$G = (k^2 - h)^{-1}, \tag{B.3}$$

although this prescription is in fact ambiguous, as might be expected from the presence of arbitrary constants in the general solution of (B.1). It will be shown in Section B.3 that a more precise definition due to Lippman and Schwinger, 1950, yields an operator such that the solutions given by (B.2) also satisfy prescribed boundary conditions.

Our immediate concern is with the Green's function, $\mathcal{G}(\mathbf{r}, \mathbf{r}')$, or coordinate representative of G, such that

$$\psi(\mathbf{r}) = \int \mathcal{G}(\mathbf{r}, \mathbf{r}')\, X(\mathbf{r}'). \tag{B.4}$$

This must, by comparison with (B.1), satisfy

$$[k^2 - h(\mathbf{r})]\, \mathcal{G}(\mathbf{r}, \mathbf{r}') = \delta(\mathbf{r} - \mathbf{r}'). \tag{B.5}$$

We restrict attention initially to the one dimensional problem

$$\left[\frac{d^2}{dr^2} + k^2 - U(r)\right]\psi(r) = X(r), \tag{B.6}$$

with $U(r)$ subject to (A.2) and (A.3), and $X(r)$ assumed to satisfy

$$\lim_{r\to\infty} X(r) = 0$$
$$\lim_{r\to 0} rX(r) = \text{const.} \tag{B.7}$$

Different forms for $\mathscr{G}(r, r')$, are found to be convenient according to whether k^2 lies in the discrete ($k^2 < 0$) or continuous ($k^2 > 0$) spectrum of $h(r)$. In the former case there is only one physically significant function, $\mathscr{G}^{(0)}(r, r')$, because the boundary conditions require that $\psi(r) \to 0$ as $r \to 0$ and $r \to \infty$. The latter admits however of three types, $\mathscr{G}^{(+)}(r, r')$, $\mathscr{G}^{(-)}(r, r')$ and $\mathscr{G}^{(0)}(r, r')$ chosen to generate solutions of (B.6) which are finite at the origin and behave at infinity as outgoing, incoming and standing waves respectively.

B.1 Negative Energy Solutions ($k^2 < 0$)

Since the negative energy solutions of (B.6) must vanish at the origin and remain finite elsewhere it is logical to seek an expansion for $\psi(r)$ in terms of a complete set, satisfying the same boundary conditions. This set comprises the bound eigenfunctions of h, $\psi_n(r)$, with energies, k_n^2, and normalized to unity,

$$\int_0^\infty \psi_n^*(r)\,\psi_m(r)\,\mathrm{d}r = \delta_{nm}, \tag{B.8}$$

and the bounded continuum functions $\psi_{k'}^{(0)}(r)$ defined by (A.8)–(A.12); we therefore set

$$\psi(r) = \sum_n a_n\psi_n(r) + \int_0^\infty \alpha(k')\,\psi_{k'}^{(0)}(r)\,\mathrm{d}k'. \tag{B.9}$$

This yields on substitution in (B.6) and the use of (A.26) and (B.8),

$$a_n = \int_0^\infty \frac{\psi_n^*(r')\,X(r')\,\mathrm{d}r'}{k^2 - k_n^2}$$
$$\alpha(k') = \frac{2}{\pi}\int_0^\infty \frac{\psi_{k'}^{(0)}(r')\,X(r')\,\mathrm{d}r'}{k^2 - k'^2}. \tag{B.10}$$

In other words $\psi(r)$ takes the form (B.4)

$$\psi(r) = \int_0^\infty \mathscr{G}^{(0)}(r, r')\,X(r')\,\mathrm{d}r', \tag{B.11}$$

with the Green's function $\mathscr{G}^{(0)}(r, r')$ given by,

$$\mathscr{G}^{(0)}(r, r') = \sum_n \frac{\psi_n(r)\,\psi_n^*(r')}{k^2 - k_n^2} + \frac{2}{\pi}\int_0^\infty \frac{\psi_{k'}^{(0)}(r)\,\psi_{k'}^{(0)}(r')\,\mathrm{d}k'}{k^2 - k'^2}. \tag{B.12}$$

This is termed the bilinear form of $\mathscr{G}^{(0)}(r, r')$.

In view of the singularity in (B.12) Eqn (B.6) has no solution subject to the stated boundary conditions when k^2 coincides with an eigenvalue k_n^2 unless

$$\int_0^\infty \psi_n^*(r')\,X(r')\,\mathrm{d}r' = 0.$$

There can of course be no similar singularities in the continuum part of $\mathscr{G}^{(0)}(r, r')$ when $k^2 < 0$.

B.2 Positive Energy Solutions ($k^2 > 0$)

The boundary conditions in the positive energy case may be taken to require that $\psi(r)$ vanishes at the origin and behaves as an outgoing, incoming or standing wave at infinity. These may be generated from analogous bilinear forms by a careful treatment of the singularity which now appears in the continuum part of (B.12); details are discussed in Section B.3.

It is however convenient first to obtain alternative forms by seeking a solution of (B.6) in terms of a pair of independent solutions $\psi_a(r)$, $\psi_b(r)$ of the homogeneous counterpart of (B.6), (i.e. with $X(r) = 0$);

$$\psi(r) = c_a(r)\psi_a(r) + c_b(r)\psi_b(r). \tag{B.13}$$

Substitution of this form in (B.6) cannot however alone determine the two coefficients $c_a(r)$ and $c_b(r)$. One further condition, conveniently taken to be

$$\psi_a \frac{\mathrm{d}c_a}{\mathrm{d}r} + \psi_b \frac{\mathrm{d}c_b}{\mathrm{d}r} = 0, \tag{B.14}$$

is also required. Different choices of $\psi_a(r)$ and $\psi_b(r)$ then lead to forms for $\psi(r)$ in (B.13), which satisfy different boundary conditions.

In every case, substitution of (B.13) in (B.6) leads after elimination of terms implied by (B.14), to the equation

$$\frac{\mathrm{d}\psi_a}{\mathrm{d}r}\frac{\mathrm{d}c_a}{\mathrm{d}r} + \frac{\mathrm{d}\psi_b}{\mathrm{d}r}\frac{\mathrm{d}c_b}{\mathrm{d}r} = X. \tag{B.15}$$

Hence on combining (B.14) and (B.15),

$$\frac{\mathrm{d}c_a}{\mathrm{d}r} = -\psi_b X \bigg/ \left[\psi_a \frac{\mathrm{d}\psi_b}{\mathrm{d}r} - \psi_b \frac{\mathrm{d}\psi_a}{\mathrm{d}r}\right], \tag{B.16}$$

where the Wronskian in the denominator

$$W(\psi_a, \psi_b) = \psi_a \frac{d\psi_b}{dr} - \psi_b \frac{d\psi_a}{dr}, \tag{B.17}$$

is in fact constant because ψ_a and ψ_b satisfy the same homogeneous equation. This means that

$$c_a(r) = -\frac{1}{W(\psi_a, \psi_b)} \int_\beta^r \psi_b(r')\, X(r')\, dr', \tag{B.18}$$

and by a similar argument

$$c_b(r) = \frac{1}{W(\psi_a, \psi_b)} \int_\alpha^r \psi_a(r')\, X(r')\, dr', \tag{B.19}$$

with overall result that

$$\psi(r) = \frac{1}{W(\psi_a, \psi_b)} \left[\psi_b(r) \int_\alpha^r \psi_a(r')\, X(r')\, dr' - \psi_a(r) \int_\beta^r \psi_b(r')\, X(r')\, dr' \right]. \tag{B.20}$$

This is the general solution of (B.6), containing as arbitrary constants the two integration limits α and β.

We come now to the boundary conditions. The behaviour at the origin may be satisfied by setting $\alpha = 0$ and taking $\psi_a(r)$ as the bounded solution defined by (A.8) and (A.11);

$$\psi_a(r) = \psi_k^{(0)}(r) \overset{r \to \infty}{\sim} \sin(kr + \delta). \tag{B.21}$$

The choice $\beta = \infty$ then ensures that the second term in (B.20) vanishes at infinity, and hence that the behaviour of $\psi(r)$ is governed by that of $\psi_b(r)$. This gives a solution of the desired form,

$$\psi(r) = \int_0^\infty \mathscr{G}(r, r')\, X(r')\, dr', \tag{B.22}$$

where

$$\begin{aligned} \mathscr{G}(r, r') &= \frac{1}{W(\psi_a, \psi_b)} \psi_a(r)\, \psi_b(r') \qquad r < r' \\ &= \frac{1}{W(\psi_a, \psi_b)} \psi_b(r)\, \psi_a(r') \qquad r > r', \end{aligned} \tag{B.23}$$

with the understanding that $\psi_a(r)$ is given by (B. 21).

The choice of $\psi_b(r)$ as an outgoing wave,

$$\psi_b(r) = \psi_k^{(+)}(r) \overset{r \to \infty}{\sim} e^{ikr + i\delta} \tag{B.24}$$

now yields the outgoing Green's function,

$$\mathscr{G}^{(+)}(r,r') = -k^{-1}\psi_k^{(0)}(r)\,\psi_k^{(+)}(r') \qquad r < r'$$
$$= -k^{-1}\psi_k^{(+)}(r)\psi_k^{(0)}(r') \qquad r > r', \tag{B.25}$$

since, by (B.17), (B.21) and (B.24)

$$W(\psi_k^{(0)}, \psi_k^{(+)}) = \sin(kr+\delta)\frac{\mathrm{d}}{\mathrm{d}r}[\mathrm{e}^{ikr+i\delta}] - \mathrm{e}^{ikr+i\delta}\frac{\mathrm{d}}{\mathrm{d}r}[\sin(kr+\delta] = -k. \tag{B.26}$$

The incoming Green's function is the complex conjugate of this:

$$\mathscr{G}^{(-)}(r,r') = -k^{-1}\psi_k^{(0)}(r)\,\psi_k^{(-)}(r') \qquad r < r'$$
$$= -k^{-1}\psi_k^{(-)}(r)\,\psi_k^{(0)}(r') \qquad r > r', \tag{B.27}$$

where

$$\psi_k^{(-)}(r) \overset{r\to\infty}{\sim} \mathrm{e}^{-ikr-i\delta}, \tag{B.28}$$

and the standing wave Green's function takes the form

$$\mathscr{G}^{(0)}(r,r') = -k^{-1}\psi_k^{(0)}(r)\,\psi_k^{(1)}(r') \qquad r < r'$$
$$= -k^{-1}\psi_k^{(1)}(r)\,\psi_k^{(0)}(r') \qquad r > r' \tag{B.29}$$

where

$$\psi_k^{(1)}(r) \overset{r\to\infty}{\sim} \cos(kr+\delta) \tag{B.30}$$

These results are readily generalized to cover the special multidimensional problem

$$[k^2 - h(r,\boldsymbol{\rho})]\Psi(r,\boldsymbol{\rho}) = X(r,\boldsymbol{\rho}), \tag{B.31}$$

for which $h(r,\boldsymbol{\rho})$ is separable in r and $\boldsymbol{\rho}$,

$$h(r,\boldsymbol{\rho}) = -\frac{\mathrm{d}^2}{\mathrm{d}r^2} + U(r) + h_{\text{int}}(\boldsymbol{\rho}), \tag{B.32}$$

and the spectrum of $h_{\text{int}}(\boldsymbol{\rho})$ is discrete,

$$h_{\text{int}}(\boldsymbol{\rho})\,\phi_m(\boldsymbol{\rho}) = \varepsilon_m\phi_m(\boldsymbol{\rho}). \tag{B.33}$$

The expansions

$$\Psi(r,\boldsymbol{\rho}) = \sum_m \psi_m(r)\,\phi_m(\boldsymbol{\rho})$$
$$X(r,\boldsymbol{\rho}) = \sum_m x_m(r)\,\phi_m(\boldsymbol{\rho}) \tag{B.34}$$

in which

$$x_m(r) = \int \phi_m^*(\boldsymbol{\rho}')\, X(r, \boldsymbol{\rho}')\, \mathrm{d}\boldsymbol{\rho}' \tag{B.35}$$

then reduce (B.32) to the familiar one dimensional form

$$\left[\frac{\mathrm{d}^2}{\mathrm{d}r^2} + k^2 - \varepsilon_m - U(r)\right]\psi_m(r) = x_m(r) \tag{B.36}$$

the solution of which depends on the sign of $k^2 - \varepsilon_m$. For $k^2 < \varepsilon_m$ the mth channel is closed, and the only physical solution is that given by (B.12), with k^2 replaced by $k^2 - \varepsilon_m$; this contribution to $\Psi(r, \boldsymbol{\rho})$, which vanishes as $r \to \infty$, may be denoted

$$\psi_m(r) = \int_0^\infty \mathscr{G}_m^{(0)}(r, r')\, x_m(r')\, \mathrm{d}r'. \tag{B.37}$$

For $k^2 > \varepsilon_m$, on the other hand, we may again have incoming, outgoing, or standing wave solutions. Thus in the outgoing case

$$\psi_m(r) = \int_0^\infty \mathscr{G}_m^{(+)}(r, r')\, x_m(r')\, \mathrm{d}r', \tag{B.38}$$

with $\mathscr{G}_m^{(+)}(r, r')$ defined by (B.25) evaluated at energy $k^2 - \varepsilon_m$ rather than k^2. This means, on substitution in (B.34) that

$$\psi(r, \boldsymbol{\rho}) = \int\int G^{(+)}(r, \boldsymbol{\rho}, r', \boldsymbol{\rho}')X(r', \boldsymbol{\rho}')\, \mathrm{d}r'\, \mathrm{d}\boldsymbol{\rho}', \tag{B.39}$$

where

$$G^{(+)}(r, \boldsymbol{\rho}, r', \boldsymbol{\rho}') = \sum_{\substack{\text{closed}\\ \text{channels}}} \mathscr{G}_m^{(0)}(r, r')\, \phi_m(\boldsymbol{\rho})\, \phi_m^*(\boldsymbol{\rho}') + \sum_{\substack{\text{open}\\ \text{channels}}} \mathscr{G}_m^{(+)}(r, r')\, \phi_m(\boldsymbol{\rho})\, \phi_m^*(\boldsymbol{\rho}'). \tag{B.40}$$

Defined in this way $G^{(+)}(r, \boldsymbol{\rho}, r', \boldsymbol{\rho}')$ is the outgoing multidimensional Green's function because, as indicated above, contributions to $\Psi(r, \boldsymbol{\rho})$ generated by the first term in (B.40) disappear at infinity. Similar arguments may be used to express the analogous functions $G^{(-)}(r, \boldsymbol{\rho}, r', \boldsymbol{\rho}')$ and $G^{(0)}(r, \boldsymbol{\rho}, r', \boldsymbol{\rho}')$ in terms of $\mathscr{G}_m^{(-)}(r, r')$ and $\mathscr{G}_m^{(0)}(r, r')$ of (B.25) and (B.29) respectively.

B.3 Lippman–Schwinger Form

The following discussion is intended to elucidate the relation between the representatives $\mathscr{G}^{(\pm)}(r, r')$ and $\mathscr{G}^{(0)}(r, r')$ and the corresponding Lippman–Schwinger operators $G^{(\pm)}$ and $G^{(0)}$. This involves resolving the ambiguities in equations (B.3) and (B.12). The essential points may be demonstrated by assuming h in (B.1) to refer to free motion in one dimension.

The first step is to define the inverse operator $(k^2 - h)^{-1}$ in terms of its effect on the bounded eigenfunctions of h; in the present free motion model,

$$\psi_{k'}^{(0)} = \sin k'r. \tag{B.41}$$

Thus

$$(k^2 - h)^{-1}\,\psi_{k'}^{(0)} = (k^2 - k'^2)^{-1}\,\psi_{k'}^{(0)}. \tag{B.42}$$

The function $X(r)$ is now expanded in terms of the $\psi_{k'}^{(0)}(r)$,

$$X(r) = \int_0^\infty x(k') \sin k'r \,\mathrm{d}k' \tag{B.43}$$

so that by (A.26)

$$x(k') = \frac{2}{\pi}\int_0^\infty X(r') \sin k'r' \,\mathrm{d}r', \tag{B.44}$$

and hence

$$\begin{aligned} X(r) &= \frac{2}{\pi}\int_0^\infty X(r')\left[\int_0^\infty \sin k'r' \sin k'r \,\mathrm{d}k'\right]\mathrm{d}r' \\ &= \frac{1}{\pi}\int_0^\infty X(r')\left[\int_{-\infty}^\infty \sin k'r' \sin k'r \,\mathrm{d}k'\right]\mathrm{d}r', \end{aligned} \tag{B.45}$$

the integration over k' having been extended to the range $-\infty < k' < \infty$ for later convenience. This means in the loose notation of (B.3) that

$$\begin{aligned} \psi(r) &= [k^2 - h(r)]^{-1}\,X(r) \\ &= \int_0^\infty \mathscr{G}(r, r')X(r')\,\mathrm{d}r', \end{aligned} \tag{B.46}$$

where

$$\mathscr{G}(r, r') = \frac{1}{\pi}\int_{-\infty}^\infty \frac{\sin k'r' \sin k'r}{k^2 - k'^2}\,\mathrm{d}k'. \tag{B.47}$$

This is of course simply the bilinear form (B.12) for the present free motion model, but it has as it stands no precise meaning because the poles of the integrand lie on the real axis (see Phillips, 1957 p. 122); it is necessary also to specify how the integration contour is to be taken, different contours being found to lead to different types of Green's function.

We shall first see how the prescription

$$G^{(+)} = \lim_{\varepsilon\to 0}\,[k^2 - h + i\varepsilon]^{-1}, \tag{B.48}$$

with ε real and positive, leads to the outgoing Green's function $\mathscr{G}^{(+)}(r, r')$ defined by (B.25), on the understanding that the limit is taken after evaluation of the integral in (B.47). Equation (B.47) therefore now becomes

$$\mathscr{G}^{(+)}(r, r') = \lim_{\varepsilon \to 0} \frac{1}{\pi} \int_{-\infty}^{\infty} \frac{\sin k'r' \sin k'r \, dk'}{\tilde{k}^2 - k'^2}, \tag{B.49}$$

where

$$\tilde{k}^2 = k^2 + i\varepsilon, \tag{B.50}$$

so that the poles at $k' = \pm\tilde{k}$ now lie in the first and third quadrants of Fig. B.1(a). The introduction of the term $+i\varepsilon$ in (B.48) is therefore formally equivalent to performing the integration in (B.47) along the contour shown in Fig. B.1(b).

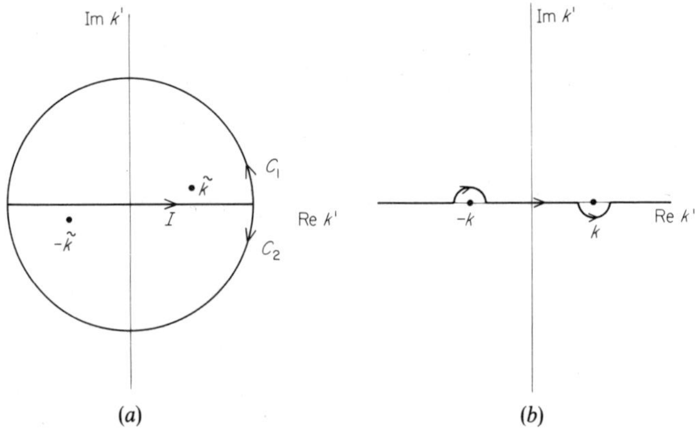

FIG. B.1. (a) Poles, and integration contours for the evaluation of (B.49); and (b) the equivalent contour.

The value of this integral now depends on the sign of $r - r'$. If $r < r'$, (B.49) is conveniently rewritten,

$$\mathscr{G}^{(+)}(r, r') = \lim_{\varepsilon \to 0} \frac{1}{4\pi} \int_{-\infty}^{\infty} \left\{ \frac{e^{ik'(r'+r)} - e^{ik'(r'-r)}}{(k' + \tilde{k})(k' - \tilde{k})} + \frac{e^{-ik'(r'+r)} - e^{-ik'(r'-r)}}{(k' + \tilde{k})(k' - \tilde{k})} \right\} dk', \tag{B.51}$$

so that the first and second terms in the integrand tend to zero on the large semi-circles C_1 and C_2 respectively. This means, by applying the Cauchy residue theorem (see Phillips, 1958) to the closed contours $I + C_1$ (counter-clockwise) and $I + C_2$ (clockwise), that the first and second terms depend on the residues at $k' = \tilde{k}$ and $k' = -\tilde{k}$ respectively. On taking the limit $\varepsilon \to 0$, so that $\tilde{k} \to k$, the result becomes

$$\mathscr{G}^{(+)}(r, r') = \frac{i}{2}\left[\frac{e^{ik(r'+r)} - e^{ik(r'-r)}}{2k} - \frac{e^{ik(r'+r)} - e^{ik(r'-r)}}{(-2k)}\right]$$

$$= -k^{-1} \sin kr \, e^{ikr'} \quad \text{for} \quad r < r'. \tag{B.52}$$

On the other hand when $r > r'$ the alternative rearrangement

$$\mathscr{G}^{(+)}(r, r') = \lim_{\varepsilon \to 0} \frac{1}{4\pi} \int_{-\infty}^{\infty} \left\{ \frac{e^{ik'(r+r')} - e^{ik'(r-r')}}{(k' + \tilde{k})(k' - \tilde{k})} + \frac{e^{-ik'(r+r')} - e^{-ik'(r-r')}}{(k' - \tilde{k})(k' + \tilde{k})} \right\} dk' \tag{B.53}$$

leads, by a similar argument, to

$$\mathscr{G}^{(+)}(r, r') = -k^{-1} e^{ikr} \sin kr' \quad \text{for} \quad r > r'. \tag{B.54}$$

Equations (B.53) and (B.54) may be seen to be identical for the present model with the previous result (B.25). It is readily verified that the definition

$$G^{(-)} = \lim_{\varepsilon \to 0} [k^2 - h - i\varepsilon]^{-1} \tag{B.55}$$

leads to the equivalent form for $\mathscr{G}^{(-)}(r, r')$ to that given by (B.27).

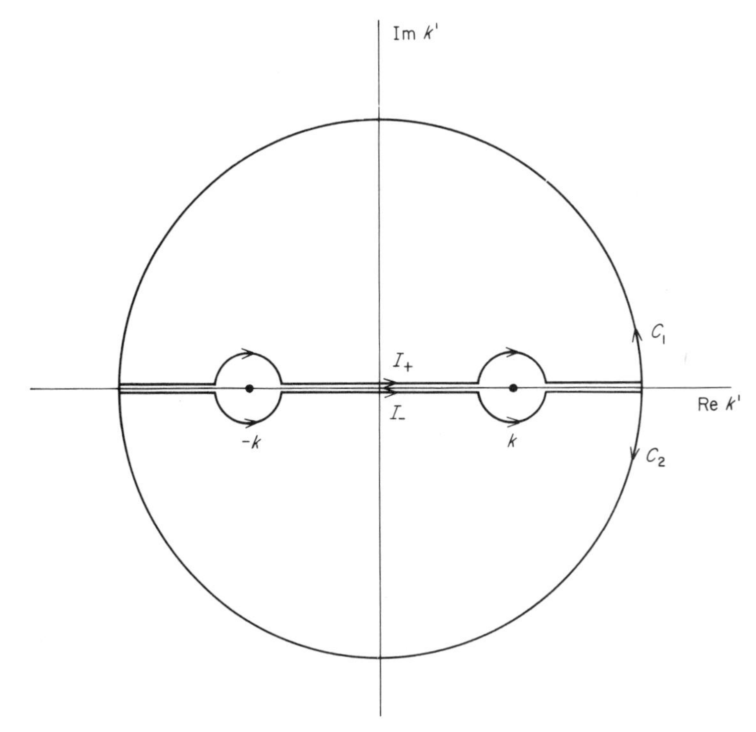

FIG. B.2. Integration contours for the Cauchy Principal value.

Finally the standing wave operator $G^{(0)}$ (written G^P by Newton, 1966) may be obtained by specifying the Cauchy Principal value

$$G^{(0)} = P(k^2 - h)^{-1}, \tag{B.56}$$

so that

$$\mathscr{G}^{(0)}(r, r') = \frac{P}{\pi}\int_{-\infty}^{\infty} \frac{\sin k'r' \sin k'r}{(k^2 - k'^2)}\, dk', \tag{B.57}$$

where P implies the mean value of the integral taken over the two contours $I_\pm$ shown in Fig. B.2.

It is again natural to combine the contours $I_\pm$ with C_1 or with C_2 when considering the first or second terms respectively in (B.51) and (B.53). Both poles now however lie within $I_- + C_1$ and $I_+ + C_2$ but outside $I_+ + C_1$ and $I_- + C_2$. Hence, as is readily verified,

$$\begin{aligned} \mathscr{G}^{(0)}(r, r') &= -k^{-1} \sin kr \cos kr' \quad \text{for} \quad r < r' \\ &= -k^{-1} \cos kr \sin kr' \quad \text{for} \quad r > r', \end{aligned} \tag{B.58}$$

a result which corresponds exactly with (B.29).

Extension of the above argument from the present free motion model to the general case rests on two properties of the bounded solution $\psi_{k'}^{(0)}$. First, in order to justify extension of the integration range in (B.45),

$$\psi_{-k}^{(0)}(r) = -\psi_k^{(0)}(r), \tag{B.59}$$

as may be verified with the help of (A.5) and (A.9), and secondly that the component terms $\psi_{k'}^{(\pm)}(r)$ of $\psi_{k'}^{(0)}(r)$, which take the place of $e^{\pm ik'r}$, must vanish on the large semi-circles C_1 and C_2 respectively.

APPENDIX C

Semi-Classical Connection Formulae

C.1 The JWKB Approximation

The JWKB semi-classical approximation to the solution of the one dimensional equation

$$\left[\frac{d^2}{dr^2} + \frac{p^2(r)}{\hbar^2}\right]\psi(r) = 0, \tag{C.1}$$

where

$$p^2(r) = 2m(E - V(r)), \tag{C.2}$$

is taken in the form

$$\psi(r) = e^{iS(r)/\hbar}; \tag{C.3}$$

hence on substituting in (3.68),

$$-(dS/dr)^2 + i\hbar(d^2S/dr^2) + p^2(r) = 0. \tag{C.4}$$

This means, if the term in $\hbar$ is neglected, as a first approximation, that

$$S(r) \simeq \pm \int p(r)\, dr, \tag{C.5}$$

the condition on this equation being that

$$\left.\begin{aligned} &|\hbar(d^2S/dr^2)| \ll |dS/dr|^2 \\ \text{or} \quad & \\ &\left|\frac{d}{dr}\left(\frac{\hbar}{p(r)}\right)\right| \ll 1 \end{aligned}\right\}. \tag{C.6}$$

This implies that the generalized de Broglie wavelength

$$\lambda(r) = h/p(r), \tag{C.7}$$

varies sufficiently slowly that

$$|\mathrm{d}\lambda(r)/\mathrm{d}r| \ll 2\pi. \tag{C.8}$$

Equations (C.6) and (C.8) determine the validity of the semi-classical approximation.

Higher corrections to (C.5) are derived from the expansion

$$S(r) = S_0(r) + \hbar S_1(r) + \hbar^2 S_2(r) + \ldots, \tag{C.9}$$

by equating coefficients of powers of $\hbar$ in (C.1). The first correction for example is obtained from the terms in $\hbar$

$$\left.\begin{aligned} &2(\mathrm{d}S_0/\mathrm{d}r)/(\mathrm{d}S_1/\mathrm{d}r) - i(\mathrm{d}^2S_0/\mathrm{d}r^2) = 0, \\ &S_1(r) = \frac{i}{2}\ln\left(\frac{\mathrm{d}S_0}{\mathrm{d}r}\right) = \frac{i}{2}\ln p(r) \end{aligned}\right\}. \tag{C.10}$$

It follows from (C.3), (C.5) and (C.10) that

$$\psi(r) \simeq C[p(r)]^{-\frac{1}{2}} \exp\left[\pm\frac{i}{\hbar}\int p(r)\,\mathrm{d}r\right], \tag{C.11}$$

where the factor $[p(r)]^{-\frac{1}{2}}$ may be interpreted in physical terms as the correction required to maintain a constant particle flux, $v|\psi(r)|^2$. Higher corrections are given by Landau and Lifshitz (1965).

It is seen from this analysis that the motion may be semi-classical, in the sense of (C.8), on either side of a classical turning point, but that it must break down at the point itself because $\lambda(r) \to \infty$ as $p(r) \to 0$. For another view of the origin of this breakdown, we may turn to the uncertainty principle according to which no classical turning point, at which both p and r are known exactly, can exist. The point must be replaced by a turning region Δr over which the momentum uncertainty arises from the possibility of incoming or outgoing motion. Suppose for simplicity that the potential $V(r)$ is linear in r;

$$V(r) = E - F(r - a), \tag{C.12}$$

so that at a given displacement Δr from the turning point a,

$$p(r) = \pm(2mF)^{\frac{1}{2}}(\Delta r)^{\frac{1}{2}}, \tag{C.13}$$

or

$$\Delta p = (2mF)^{\frac{1}{2}}(\Delta r)^{\frac{1}{2}}. \tag{C.14}$$

Δp and Δr defined in this way may be accommodated within the uncertainty principle provided that

$$\Delta p \Delta r = (2mF)^{\frac{1}{2}}(\Delta r)^{\frac{3}{2}} \gg \hbar, \tag{C.15}$$

in other words provided that

$$\Delta r = (r - a) \gg (\hbar^2/2mF)^{\frac{1}{3}}. \tag{C.16}$$

It is readily verified that (C.6) implies the same restriction.

The problem therefore is that according to (C.11), $\psi(r)$ may be approximated outside any non-semi-classical region in the form

$$\psi(r) = [k(r)]^{-\frac{1}{2}}\left\{C'_\alpha \exp\left[i\int_\alpha^r k(r)\,\mathrm{d}r\right] + C''_\alpha \exp\left[-i\int_\alpha^r k(r)\,\mathrm{d}r\right]\right\}, \tag{C.17}$$

where $k(r) = p(r)/\hbar$, if r is classically accessible, or in the form

$$\psi(r) = |k(r)|^{-\frac{1}{2}}\left\{X_\beta \exp\left[-\int_\beta^r |k(r)|\,\mathrm{d}r\right] + Y_\beta \exp\left[\int_\beta^r |k(r)|\,\mathrm{d}r\right]\right\} \tag{C.18}$$

if r lies in a classically inaccessible region, but that the coefficients C'_α, C''_α, X_β and Y_β cannot be determined by semi-classical methods. Furthermore it should be noted that their values must depend on the chosen phase reference points α and β. The notation in (C.17) with single and double primes used to denote outgoing (final) and incoming (initial) motion respectively is suggested by spectroscopic analogy. The purpose of the JWKB connection formulae is to establish relations between such coefficients, in order that a solution may be chosen to match any given boundary conditions.

It is assumed in what follows that $V(r)$ in (C.2) may be approximated over the appropriate forbidden region by a simple model form.

There are two possible routes to the connection formula. The first, so called "phase integral method" (see Heading, 1962) is entirely based on properties of asymptotic functions such as (C.17) and (C.18). As such it illuminates, as shown in Section C.2 below, an important limitation of any asymptotic method. It is however incapable of determining all the phases in the connection matrix if $V(r)$ is quadratic in r. The phase integral method is therefore applied to the case of an isolated turning point in C.2.

The second method employs established asymptotic forms of exact solutions for the model problem in hand. It is more complicated in detail but capable of extension by the association method of Miller and Good (1953) to other than model problems. The quadratic potential barrier and potential well problems are treated in this way in Sections C.3–C.4. An important review of the same general method has been given by Berry and Mount (1972).

C.2 Isolated Turning Points

C.2.1. *Left-hand turning point*

In seeking connection formulae to bridge the non semi-classical region

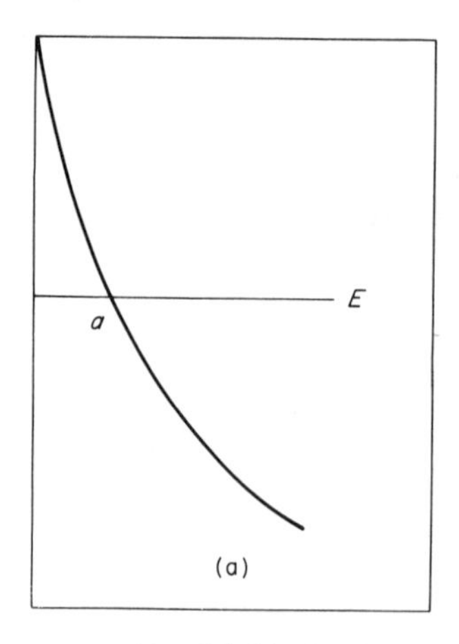

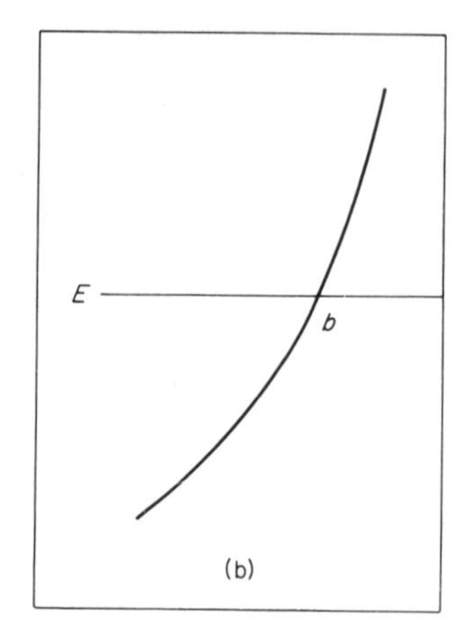

FIG. C.1. (a) Right-hand and (b) left-hand turning points.

around a left-hand turning point, a in Fig. C.1(a), we make the model approximation

$$V(r) = E - F(r - a), \tag{C.19}$$

which together with the substitution

$$z = (2mF/\hbar^2)^{\frac{1}{3}}(r - a), \tag{C.20}$$

reduces (C.1) to the form

$$\frac{d^2\psi}{dz^2} + z\psi = 0. \tag{C.21}$$

The phase integral route to the connection formulae between the coefficients C_a', C_a'', X_a and Y_a in (C.17) and (C.18) is discussed in detail by Heading (1962). It is based on following the variation in the general asymptotic solution of (C.21),

$$\psi(z) \overset{|z| \gg 1}{\sim} z^{-\frac{1}{4}}[c_+ \exp(\tfrac{2}{3} iz^{\frac{3}{2}}) + c_- \exp(-\tfrac{2}{3} iz^{\frac{3}{2}})], \tag{C.22}$$

in circumscribing a large circle, $|z| \gg 1$, around the turning point, $z = 0$. The connection between (C.22) on one hand and (C.17) and (C.18) on the other rests on the identities

$$\left.\begin{aligned} \int_a^r k(r)\,dr &= (2mF/\hbar^2)^{\frac{1}{2}} \int_a^r (r - a)^{\frac{1}{2}}\,dr = \tfrac{2}{3} z^{\frac{3}{2}}, \quad r > a \\ \int_a^r |k(r)|\,dr &= -(2mF/\hbar^2)^{\frac{1}{2}} \int_r^a (a - r)^{\frac{1}{2}}\,dr = -\tfrac{2}{3}|z^{\frac{3}{2}}|, \quad r < a \end{aligned}\right\} \tag{C.23}$$

There are two important features of the method.

The first arises from the variation with arg z of the relative magnitudes of the terms in (C.22), which implies that there are three directions in the z plane, termed the anti-Stokes lines, in which one term dominates the other. These are the directions arg $z = \pi/3, 5\pi/3$ for which

$$\pm \tfrac{2}{3} i z^{\frac{3}{2}} = \mp \tfrac{2}{3}|z|^{\frac{3}{2}}, \qquad \text{(C.24a)}$$

so that the second term in (C.22) is dominant, and the direction arg $z = \pi$ for which

$$\pm \tfrac{2}{3}\, i z^{\frac{3}{2}} = \pm \tfrac{2}{3}|z|^{\frac{3}{2}}, \qquad \text{(C.24b)}$$

with the dominance transferred to the first term. This raises a question concerned with the validity of the asymptotic approximation, since according to Stokes (see Heading, 1962), the approximation of the dominant solution of (C.21) by its first term, $\exp(\tfrac{2}{3}|z|^{\frac{3}{2}})$, introduces an error proportional to the first term in the sub-dominant solution. This means that the coefficient of the sub-dominant in (C.22) is not uniquely defined in the anti-Stokes directions. Since however this behaviour is associated only with these directions we can say that the coefficient of the sub-dominant term acquires an increment proportional to that of the dominant term in crossing an anti-Stokes line. With the proportionality (or Stokes) constants denoted α, β, and γ for the directions arg $z = \pi/3, \pi$, and $5\pi/3$ respectively, this leads to the pattern of coefficients shown in Fig. C.2(a), the upper and lower coefficients being associated with the first and second terms in (C.22) respectively.

The second important point fixes the proportionality constants α, β and γ. It is that $\psi(z)$ is single valued, hence since at arg $z = 2\pi$

$$\left.\begin{aligned} z^{-\frac{1}{4}} &= |z|^{-\frac{1}{4}} e^{-i\pi/4}, \\ \tfrac{2}{3} z^{\frac{3}{2}} &= -\tfrac{2}{3}|z|^{\frac{3}{2}}, \end{aligned}\right\} \qquad \text{(C.25)}$$

it follows from (C.22) that

$$\left.\begin{aligned} c_+ + \alpha c_- + \gamma[c_- + \beta(c_+ + \alpha c_-)] &= i c_- \\ c_- + \beta(c_+ + \alpha c_-) &= i c_+ \end{aligned}\right\} \qquad \text{(C.26)}$$

Thus on equating coefficients of c_+ and c_-

$$\alpha = \beta = \gamma = i. \qquad \text{(C.27)}$$

The corresponding values of the coefficients are given in Fig. C.2(b). Clearly c_+ and c_- here may be identified with C_a' and C_a'' in (C.17) and $c_+ + ic_-$ with X_a in (C.18). There is however an ambiguity in the coefficient Y_a of the sub-dominant term which varies between c_- for $\pi/3 < \arg z < \pi - \varepsilon$ and ic_+ for $\pi + \varepsilon < c < 5\pi/3$. This ambiguity, associated with the requirement for a connection formula involving coefficients on an anti-Stokes lines, nullifies

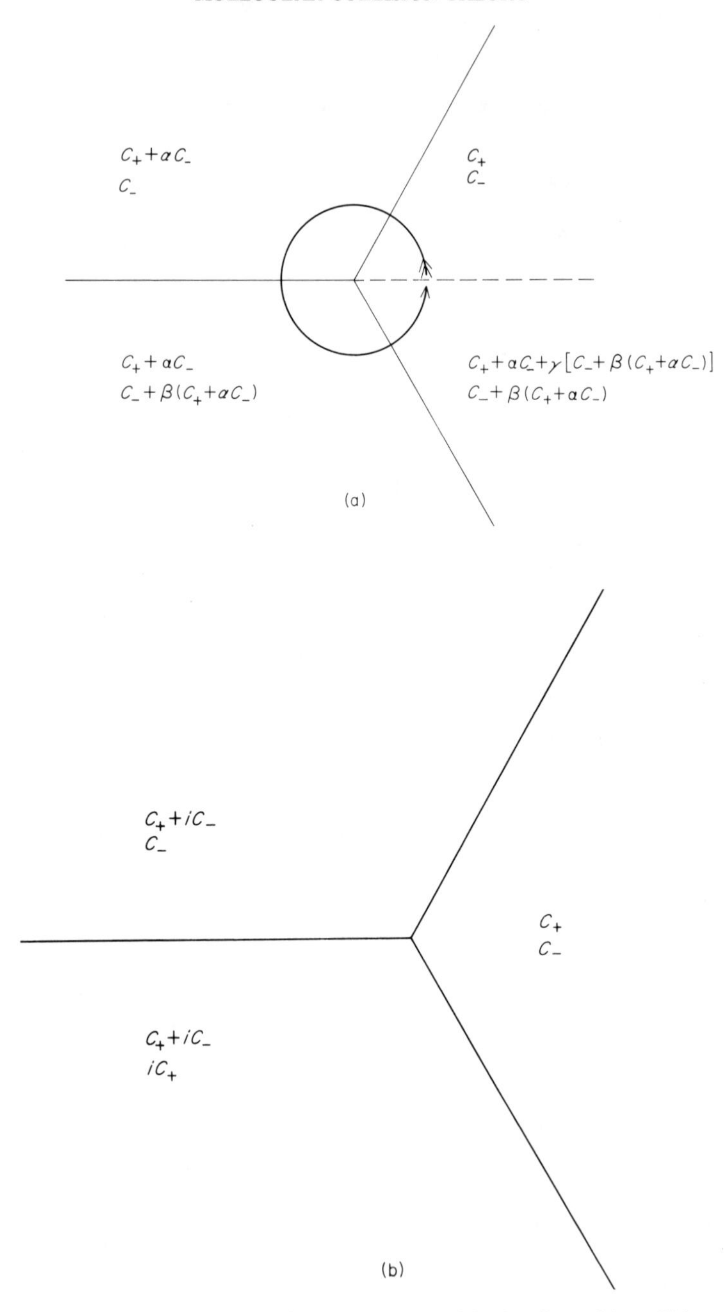

FIG. C.2. (a) Coefficient changes associated with the anti-Stokes lines; (b) coefficient values consistent with the single valued nature of $\psi(z)$.

any attempt to establish a rigorous general connection formula (see Heading, 1962 p. 79).

One can however choose particular values, $c_- = ic_+$, in order to remove this ambiguity. The coefficient $(c_+ + ic_-)$ of the dominant term then vanishes. The standard choice

$$c_\pm = \frac{\pi^{-\frac{1}{2}}}{2} \exp(\mp i\pi/4)$$

leads to a solution such that

$$\begin{aligned}\psi_1 &\overset{z\gg 1}{\sim} \pi^{-\frac{1}{2}} z^{-\frac{1}{4}} \sin\left(\tfrac{2}{3}z^{\frac{3}{2}} + \pi/4\right)\\ &\overset{z\ll -1}{\sim} \frac{\pi^{-\frac{1}{2}}}{2} |z|^{-\frac{1}{4}} \exp\left(-\tfrac{2}{3}|z|^{\frac{3}{2}}\right).\end{aligned} \tag{C.28}$$

This behaviour, with its exponential decrease into the non-classical region ($z < 0$ or $r < a$ in Fig. C.1(a)) is characteristic of the first Airy function $\mathrm{Ai}(-z)$ (Abramowitz and Stegun, 1965). A second solution may be defined by taking the mean $\frac{1}{2}(c_- + ic_+)$, of the sub-dominant coefficients on the negative real axis in Fig. C.2(b). Hence on setting

$$c_\pm = \frac{\pi^{-\frac{1}{2}}}{2} \exp(\pm i\pi/4)$$

we obtain a solution which varies like the second Airy function, $\mathrm{Bi}(-z)$.

$$\begin{aligned}\psi_2(z) &\overset{z\gg 1}{\sim} \pi^{-\frac{1}{2}} z^{-\frac{1}{4}} \cos\left(\tfrac{2}{3}z^{\frac{3}{2}} + \pi/4\right)\\ &\overset{z\ll -1}{\sim} \pi^{-\frac{1}{2}} |z|^{-\frac{1}{4}} \exp\left(\tfrac{2}{3}|z|^{\frac{3}{2}}\right).\end{aligned} \tag{C.29}$$

Written in terms of r the two solutions in (C.28) and (C.29) imply

$$\left.\begin{aligned}\psi_1(r) &\overset{r\gg a}{\sim} C[k(r)]^{-\frac{1}{2}} \sin\left(\int_a^r k(r)\,\mathrm{d}r + \pi/4\right)\\ &\overset{r\ll a}{\sim} \tfrac{1}{2}C|k(r)|^{-\frac{1}{2}} \exp\left(\int_a^r |k(r)|\,\mathrm{d}r\right)\end{aligned}\right\} \tag{C.30}$$

$$\left.\begin{aligned}\psi_2(r) &\overset{r\gg a}{\sim} C[k(r)]^{-\frac{1}{2}} \cos\left[\int_a^r k(r)\,\mathrm{d}r + \pi/4\right)\\ &\overset{r\ll a}{\sim} C|k(r)|^{-\frac{1}{2}} \exp\left(-\int_a^r |k(r)|\,\mathrm{d}r\right)\end{aligned}\right\} \tag{C.31}$$

Finally the choice of the average sub-dominant coefficient, due to Jeffreys, 1923, implies a weak connection between the coefficients in (C.17) and (C.18) of the form

$$\begin{pmatrix} X_a \\ Y_a \end{pmatrix} \Leftarrow \begin{pmatrix} \frac{1}{2}\mathrm{e}^{i\pi/4}, & \frac{1}{2}\mathrm{e}^{-i\pi/4} \\ \mathrm{e}^{-i\pi/4}, & \mathrm{e}^{i\pi/4} \end{pmatrix} \begin{pmatrix} C'_a \\ C''_a \end{pmatrix}, \tag{C.32}$$

the connection here being unique only in the direction of the arrow.

C.2.2. Right-hand turning point

The application of similar arguments around a right hand turning point, $r = b$ in Fig. C.1(b), leads to solutions of (C.1) with the asymptotic forms

$$\left.\begin{aligned} \psi_1(r) &\overset{r \ll b}{\sim} C[k(r)]^{-\frac{1}{2}} \sin\left(\int_r^b k(r)\,\mathrm{d}r + \pi/4\right) \\ &\overset{r \gg b}{\sim} \tfrac{1}{2}C|k(r)|^{-\frac{1}{2}} \exp\left(-\int_b^r |k(r)|\,\mathrm{d}r\right) \end{aligned}\right\} \tag{C.33}$$

$$\left.\begin{aligned} \psi_2(r) &\overset{r \ll b}{\sim} C[k(r)]^{-\frac{1}{2}} \cos\left(\int_r^b k(r)\,\mathrm{d}r + \pi/4\right) \\ &\overset{r \gg b}{\sim} C|k(r)|^{-\frac{1}{2}} \exp\left(\int_b^r |k(r)|\,\mathrm{d}r\right) \end{aligned}\right\}. \tag{C.34}$$

The corresponding weak connection formula, analogous to (C.32) is

$$\begin{pmatrix} X_b \\ Y_b \end{pmatrix} \Leftarrow \begin{pmatrix} \frac{1}{2}\mathrm{e}^{-i\pi/4} & \frac{1}{2}\mathrm{e}^{i\pi/4} \\ \mathrm{e}^{i\pi/4} & \mathrm{e}^{-i\pi/4} \end{pmatrix} \begin{pmatrix} C'_b \\ C''_b \end{pmatrix}. \tag{C.35}$$

C.3 Potential Barrier

The most convenient approximation to a potential barrier is the quadratic form

$$V(r) = V_{\mathrm{max}} - \tfrac{1}{2}\kappa x^2, \tag{C.36}$$

where $x = r - r_{\mathrm{max}}$. The analysis below follows the arguments of Connor

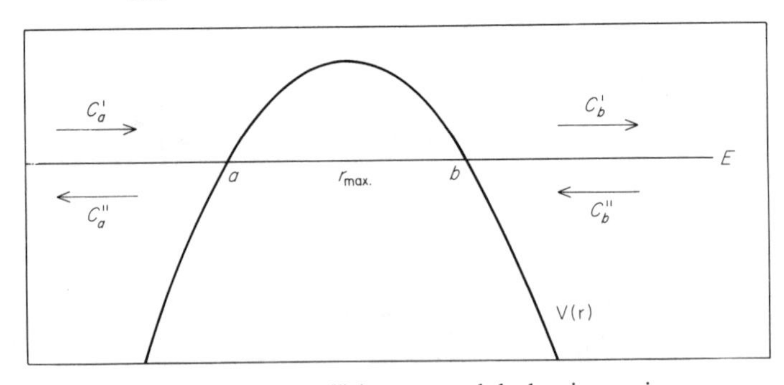

FIG. C. 3. Amplitude coefficients around the barrier maximum.

(1968). Dickinson (1970) has applied similar methods to the problem of an asymmetric model barrier. The first step is to perform the substitutions

$$\left.\begin{aligned} x &= (\hbar^2/4m\kappa)^{\frac{1}{4}}\, e^{i\pi/4} z, \\ \varepsilon &= (E - V_{max})/\hbar\omega^* \\ \omega^* &= (\kappa/m)^{\frac{1}{2}} \end{aligned}\right\} \qquad \text{(C.37)}$$

in terms of which Eqn (C.1) reduces to the standard form of the Weber equation (Whittaker and Watson, 1962),

$$\left(\frac{d^2}{dz^2} + i\varepsilon - \tfrac{1}{4}z^2\right)\psi(z) = 0, \qquad \text{(C.38)}$$

the general solution of which may be written in terms of parabolic cylinder functions $D_n(z)$,

$$\psi(z) = PD_{i\varepsilon-\frac{1}{2}}(z) + QD_{-i\varepsilon-\frac{1}{2}}(-iz). \qquad \text{(C.39)}$$

Here, the argument $-iz$ in the second term is chosen in conjunction with the factor $e^{i\pi/4}$ in (C.37) in order to make use of the following established asymptotic forms† (Whittaker and Watson, 1962) for real values of x,

$$D_n(z) \sim z^n \exp(-\tfrac{1}{4}z^2), \qquad |\arg z| < 3\pi/4 \qquad \text{(C.40a)}$$

$$\sim z^n \exp(-\tfrac{1}{4}z^2) - \frac{(2\pi)^{\frac{1}{2}}}{\Gamma(-n)} e^{n\pi i} z^{-n-1} \exp(\tfrac{1}{4}z^2),$$

$$\pi/4 < \arg z < 5\pi/4 \qquad \text{(C.40b)}$$

$$\sim z^n \exp(-\tfrac{1}{4}z^2) - \frac{(2\pi)^{\frac{1}{2}}}{\Gamma(-n)} e^{-n\pi i} z^{-n-1} \exp(\tfrac{1}{4}z^2),$$

$$-5\pi/4 < \arg z < -\pi/4. \qquad \text{(C.40c)}$$

The next step is to use these asymptotic expansions in order to determine the asymptotic behaviour of the general solution $\psi(z)$ in (C.39) for real

† The general forms of (C.40) are readily understood in terms of the Stokes phenomenon discussed in Section C.2. The anti-Stokes lines for (C.38) coincide with the real and imaginary axes, and $D_n(z)$ is defined as the (unique) solution which is sub-dominant on the positive real axis. $D_n(z)$ therefore becomes dominant when $\arg z = \pm\pi/2$; this accounts for the appearance of the second terms in (C.40b) and (C.40c). While being accurate over the specified ranges of arg z, to the extent that sub-dominant terms are neglected (C.40 a, b and c) are more safely employed for the purposes of obtaining connection formulae over the non-overlapping ranges $-\pi/2 < \arg z < \pi/2$, $\pi/2 < \arg z < \pi$ and $-\pi < \arg z < -\pi/2$ respectively. Furthermore the mean of (C.40b) and (C.40c), which represents a Jeffreys (1923) solution,

$$D_n(z) \sim \sin(n + \tfrac{1}{2})\pi |z|^n \exp(-\tfrac{1}{4}z^2) + \frac{(2\pi)^{\frac{1}{2}}}{\Gamma(-n)} |z|^{-n-1} \exp(\tfrac{1}{4}z^2) \qquad \text{(C.40d)}$$

is best employed on the negative real axis.

positive and negative values of x; in other words, according to (C.37), in the directions $\arg z = -\pi/4$ and $\arg z = 3\pi/4$ respectively. With P and Q replaced by

$$\left.\begin{aligned} A_+ &= P - i(2\pi)^{\frac{1}{2}}[\Gamma(i\varepsilon + \tfrac{1}{2})]^{-1} \exp\left[-\frac{\pi\varepsilon}{2} + \frac{i\pi}{4}\right] Q, \\ B_+ &= \exp\left(-\frac{\pi\varepsilon}{2} + \frac{i\pi}{4}\right) Q \end{aligned}\right\}, \tag{C.41}$$

the result may be written, after use of the identity (Abramowitz and Stegun, 1965)

$$\Gamma(i\varepsilon + \tfrac{1}{2})\,\Gamma(-i\varepsilon + \tfrac{1}{2}) = 2\pi[\mathrm{e}^{\pi\varepsilon} + \mathrm{e}^{-\pi\varepsilon}]^{-1}, \tag{C.42}$$

in the form

$$\left.\begin{aligned} \psi(z) &\overset{x\to\infty}{\sim} A_+ z^{i\varepsilon-\frac{1}{2}} \exp(-\tfrac{1}{4}z^2) + B_+ z^{-i\varepsilon-\frac{1}{2}} \exp(\tfrac{1}{4}z^2) \\ &\overset{x\to-\infty}{\sim} A_- z^{i\varepsilon-\frac{1}{2}} \exp(-\tfrac{1}{4}z^2) + B_- z^{-i\varepsilon-\frac{1}{2}} \exp(\tfrac{1}{4}z^2) \end{aligned}\right\}, \tag{C.43}$$

where

$$\begin{pmatrix} A_- \\ B_- \end{pmatrix} = \begin{pmatrix} 1, & i(2\pi)^{\frac{1}{2}}[\Gamma(\frac{1}{2}+i\varepsilon)]^{-1} \\ i(2\pi)^{\frac{1}{2}}[\Gamma(\frac{1}{2}-i\varepsilon)]^{-1}\exp(-\pi\varepsilon), & -\exp(-2\pi\varepsilon) \end{pmatrix} \begin{pmatrix} A_+ \\ B_+ \end{pmatrix}. \tag{C.44}$$

Finally Eqn (C.43) must be compared with the general JWKB solutions, given by (C.17) to the left and right of the barrier,

$$\left.\begin{aligned} \psi(r) &\overset{r\ll a}{\sim} [k(r)]^{-\frac{1}{2}} \left\{ C_a' \exp\left[i\int_a^r k(r)\,\mathrm{d}r \right] + C_a'' \exp\left[-i\int_a^r k(r)\,\mathrm{d}r \right] \right\} \\ &\overset{r\gg b}{\sim} [k(r)]^{-\frac{1}{2}} \left\{ C_b' \exp\left[i\int_b^r k(r)\,\mathrm{d}r \right] + C_b'' \exp\left[-i\int_b^r k(r)\,\mathrm{d}r \right] \right\} \end{aligned}\right\} \tag{C.45}$$

so that the above connection between $A_\pm$ and $B_\pm$ may be used to relate the left-hand coefficients C_a' and C_a'' to those on the right, C_b' and C_b''.

Suppose first for simplicity that $E < V_{\max}$ (hence $\varepsilon < 0$) so that the turning points a and b are real. The substitutions

$$\left.\begin{aligned} y &= (m\kappa/\hbar^2)^{\frac{1}{4}} x = 2^{-\frac{1}{2}}\,\mathrm{e}^{i\pi/4} z \\ y_0 &= (2|\varepsilon|)^{\frac{1}{2}} \end{aligned}\right\} \tag{C.46}$$

with $y = \mathrm{e}^{i\pi}|y|$ for negative values, then reduce the integrals in the exponents in (C.45) to the following forms for large $|x|$,

$$\pm i \int_b^r (k(r)\,\mathrm{d}r = \pm i \int_{y_0}^{y} (y^2 - 2|\varepsilon|)^{\frac{1}{2}}\,\mathrm{d}y,$$

$$\overset{y\to\infty}{\sim} \pm \frac{i}{2}(y^2 + \varepsilon + 2\varepsilon \ln 2y - \varepsilon \ln 2|\varepsilon|)$$

$$\sim \pm(-\tfrac{1}{4}z^2 + i\varepsilon \ln z - \frac{\pi\varepsilon}{4} + \frac{i\varepsilon}{2} - \frac{i\varepsilon}{2}\ln|\varepsilon|),$$

$$\pm i \int_a^r k(r)\,\mathrm{d}r = \pm i \int_{-y_0}^{y} (y^2 - 2|\varepsilon|)^{\frac{1}{2}}\,\mathrm{d}y. \tag{C.47a}$$

$$\overset{y\to-\infty}{\sim} \mp \frac{i}{2}(|y|^2 + \varepsilon + 2\varepsilon \ln 2|y| - \varepsilon \ln 2|\varepsilon|)$$

$$\sim \mp\left(-\tfrac{1}{4}z^2 + i\varepsilon \ln z + \frac{3\pi\varepsilon}{4} + \frac{i\varepsilon}{2} - \frac{i\varepsilon}{2}\ln|\varepsilon|\right). \tag{C.47b}$$

The same equations, with ε in place in $|\varepsilon|$, are obtained for $\varepsilon > 0$ if the turning points a and b are replaced by the barrier maximum, $r_{\max}$ in Fig. C.3 ($y = 0$ (C.46)). It follows for all values of ε, on combining (C.45) and (C.47) and comparing coefficients of $z^{\pm i\varepsilon - \frac{1}{2}} \exp(\mp\frac{1}{4}z^2)$ that

$$\begin{pmatrix} A_+ \\ B_+ \end{pmatrix} = \begin{pmatrix} \exp\left(-\frac{\pi\varepsilon}{4} + \frac{i\varepsilon}{2} - \frac{i\varepsilon}{2}\ln|\varepsilon|\right), & 0 \\ 0, & \exp\left(\frac{\pi\varepsilon}{4} - \frac{i\varepsilon}{2} + \frac{i\varepsilon}{2}\ln|\varepsilon|\right) \end{pmatrix} \begin{pmatrix} C'_b \\ C''_b \end{pmatrix}, \tag{C.48a}$$

and

$$\begin{pmatrix} A_- \\ B_- \end{pmatrix} = \begin{pmatrix} 0, & i\exp\left(\frac{3\pi\varepsilon}{4} + \frac{i\varepsilon}{2} - \frac{i\varepsilon}{2}\ln|\varepsilon|\right) \\ i\exp\left(-\frac{3\pi\varepsilon}{4} - \frac{i\varepsilon}{2} + \frac{i\varepsilon}{2}\ln|\varepsilon|\right), & 0 \end{pmatrix} \begin{pmatrix} C'_a \\ C''_a \end{pmatrix}, \tag{C.48b}$$

the factor i in (C.48b) being due to the term $|k(r)|^{-\frac{1}{2}}$ in (C.45). Hence in the light of (C.42) and (C.44)

$$\begin{pmatrix} C'_a \\ C''_a \end{pmatrix} = \begin{pmatrix} (1 + \mathrm{e}^{-2\pi\varepsilon})^{\frac{1}{2}}\,\mathrm{e}^{i\phi(\varepsilon)}, & i\,\mathrm{e}^{-\pi\varepsilon} \\ -i\,\mathrm{e}^{-\pi\varepsilon}, & (1 + \mathrm{e}^{-2\pi\varepsilon})^{\frac{1}{2}}\,\mathrm{e}^{-i\phi(\varepsilon)} \end{pmatrix} \begin{pmatrix} C'_b \\ C''_b \end{pmatrix}, \tag{C.49}$$

where

$$\phi(\varepsilon) = \arg\Gamma(i\varepsilon + \tfrac{1}{2}) - \varepsilon \ln|\varepsilon| + \varepsilon. \tag{C.50}$$

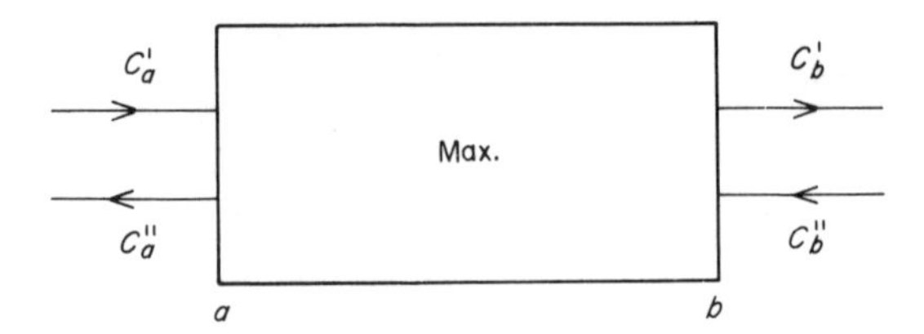

FIG. C.4. Connection formula diagram for a maximum. The amplitude coefficients are related by Eqn (C.49).

This is the required connection formula, a diagrammatic form of which is shown in Fig. C.4. Table C.4 gives selected values of the phase factor $\phi(\varepsilon)$.

A more useful form of (C.49), for some purposes is obtained by expressing the outgoing amplitudes, C_a'', C_b' in Fig. C.4, in terms of their incoming counterparts

$$\begin{pmatrix} C_b' \\ C_a'' \end{pmatrix} = e^{-i\phi(\varepsilon)} \begin{pmatrix} (1 + e^{-2\pi\varepsilon})^{\frac{1}{2}}, & -i(1 + e^{2\pi\varepsilon})^{-\frac{1}{2}} \\ -i(1 + e^{2\pi\varepsilon})^{-\frac{1}{2}}, & (1 + e^{-2\pi\varepsilon})^{-\frac{1}{2}} \end{pmatrix} \begin{pmatrix} C_a' \\ C_b'' \end{pmatrix} \qquad \text{(C.51)}$$

Thus, with $C_b'' = 0$ for example, the barrier reflection and transmission probabilities become

$$\begin{aligned} R &= |C_a''/C_a'|^2 = (1 + e^{-2\pi\varepsilon})^{-1} \\ T &= |C_b'/C_a'|^2 = (1 + e^{2\pi\varepsilon})^{-1}, \end{aligned} \qquad \text{(C.52)}$$

with

$$R + T = 1. \qquad \text{(C.53)}$$

Finally the parameter ε, defined as a reduced energy in (C.37), may also be identified with the integrated modulus of the classical action,

$$\pi\varepsilon = \int_a^b |k(r)|\, dr. \qquad \text{(C.54)}$$

TABLE C.1. Values of $\phi(\varepsilon)$

ε	$\phi(\varepsilon)$	ε	$\phi(\varepsilon)$
0	0	0·6	0·080
0·1	0·137	0·7	0·068
0·2	0·150	0·8	0·058
0·3	0·135	0·9	0·051
0·4	0·115	1·0	0·045
0·5	0·096		

By (C.50) $\phi(\varepsilon) = -\phi(-\phi)$

$\phi(\varepsilon) \simeq 1/24\varepsilon + 7/2880\varepsilon^3$ for $\varepsilon > 1·0$.

Furthermore it may be verified by the methods of Miller and Good (1953) that with ε defined by (C.54), (C.49) remains valid for a non-quadratic barrier with the substitution of factors $\pm ie^{\pm i\theta}$ in place of $\pm i$ in the off-diagonal terms, where

$$\theta = \int_{r_+}^{r_{\max}} k(r)\,\mathrm{d}r + \int_{r_-}^{r_{\max}} k(r)\,\mathrm{d}r, \tag{C.55}$$

and $r_\pm$ are the (possibly complex) roots of $k(r)$.

C.4 Potential Well

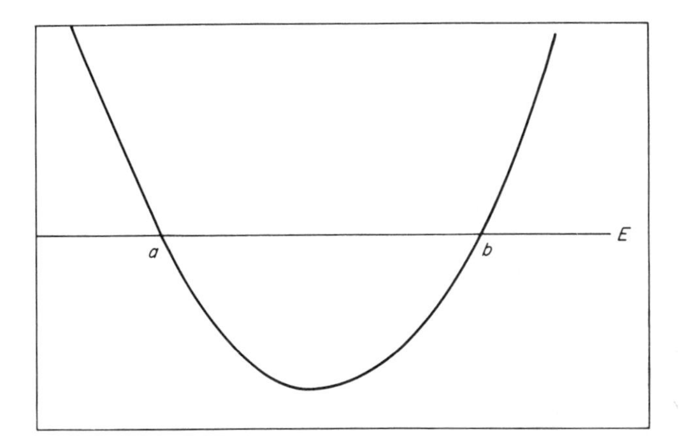

FIG. C.5. The potential well.

The analysis for a quadratic potential well

$$V(r) = V_{\min} + \tfrac{1}{2}kx^2 \tag{C.56}$$

follows the lines of the discussion in Section C.3, except that in this case, since the solutions are compared on the anti-Stokes lines (see Section C.2 and the footnote to p 257) with one solution of (C.1) dominating the other, the final connection formulae are less secure. The final result, obtained by Connor (1969) with the help of (C.49d) is that if

$$\begin{aligned} \psi(r) &\overset{r \ll a}{\sim} |k(r)|^{-\frac{1}{2}} \left\{ X_a \exp\left[-\int_a^r |k(r)|\,\mathrm{d}r \right] + Y_a \exp\left[\int_a^r |k(r)|\,\mathrm{d}r \right] \right\}, \\ &\overset{r \gg b}{\sim} |k(r)|^{-\frac{1}{2}} \left\{ X_b \exp\left[-\int_b^r |k(r)|\,\mathrm{d}r \right] + Y_b \exp\left[\int_b^r |k(r)|\,\mathrm{d}r \right] \right\}, \end{aligned} \tag{C.57}$$

then

$$\begin{pmatrix} X_a \\ Y_a \end{pmatrix} = \begin{pmatrix} \cos \pi\tilde{\varepsilon}\, e^{-\chi(\tilde{\varepsilon})}, & \sin \pi\tilde{\varepsilon} \\ -\sin \pi\tilde{\varepsilon}, & \cos \pi\tilde{\varepsilon}\, e^{\chi(\tilde{\varepsilon})} \end{pmatrix} \begin{pmatrix} X_b \\ Y_b \end{pmatrix}, \qquad (C.58)$$

where

$$\begin{aligned} \tilde{\varepsilon} &= (E - V_{\text{min}})/\hbar\omega \\ \omega &= (k/m)^{\frac{1}{2}} \\ \chi(\tilde{\varepsilon}) &= -\tilde{\varepsilon} \ln \tilde{\varepsilon} + \tilde{\varepsilon} + \ln \Gamma(\tfrac{1}{2} + \tilde{\varepsilon}) - \tfrac{1}{2} \ln(\pi/2), \qquad \tilde{\varepsilon} > 0 \\ &= -\tilde{\varepsilon} \ln |\tilde{\varepsilon}| + \tilde{\varepsilon} - \ln \Gamma(\tfrac{1}{2} - \tilde{\varepsilon}) - \ln(\cos \pi\tilde{\varepsilon}) \\ &\qquad + \tfrac{1}{2} \ln(2\pi), \qquad \tilde{\varepsilon} < 0. \end{aligned} \qquad (C.59)$$

APPENDIX D

Curve Crossing in the Momentum Representation

Analysis of the curve crossing problem in the momentum representation allows an investigation of the constant interaction, linear crossing model employed by Landau (1932) and Zener (1932), without the restrictive assumption of a constant velocity inherent in their semi-classical treatments.

With the origins of energy and displacement taken at the crossing-point, and the assumptions

$$\left.\begin{aligned} V_i(x) &= E - F_i x \qquad i = 1, 2 \\ V_{12}(x) &= A = \text{const.} \end{aligned}\right\}, \tag{D.1}$$

the equations for the momentum representatives $u_i(k)$ take the form

$$\left.\begin{aligned} \left[k^2 - \varepsilon - if_1 \frac{\mathrm{d}}{\mathrm{d}k}\right] u_1(k) &= -\alpha u_2(k) \\ \left[k^2 - \varepsilon - if_2 \frac{\mathrm{d}}{\mathrm{d}k}\right] u_2(k) &= -\alpha u_1(k) \end{aligned}\right\}, \tag{D.2}$$

where

$$\begin{aligned} \varepsilon &= 2mF/\hbar^2, \\ f_i &= 2mF_i/\hbar^2, \\ \alpha &= 2mA/\hbar^2. \end{aligned} \tag{D.3}$$

The cases of $f_1 f_2 > 0$ and $f_1 f_2 < 0$ require different treatment. It is assumed throughout that $f_1 > 0$ and $f_1 > f_2$. The case $f_1 f_2 > 0$ is first discussed in detail; corresponding results for $f_1 f_2 < 0$ are then quoted from the literature.

D.1 Forces with the Same Sign: $f_1 f_2 > 0$

Since in the absence of coupling ($\alpha = 0$),

$$u_i(k) = C \exp\left[\frac{i}{f_i}(\varepsilon k - k^3/3)\right], \tag{D.4}$$

solutions to (D.2) are sought in the form

$$u_i(k) = (2/f_i)\, A_i(k) \exp\left[\frac{i}{f_i}(\varepsilon k - k^3/3)\right], \tag{D.5}$$

the factor $(2/f_i)^{\frac{1}{2}}$ being inserted for later convenience. The coefficient $A_i(k)$ is therefore identified as the probability amplitude for channel i, and the problem is to relate the final (positive momentum) amplitudes $A_i(+\infty)$ to their initial counterparts $A_i(-\infty)$.

Subsequent analysis is simplified by the introduction of the dimensionless variable t and parameters a^2 and b^2 defined by

$$\left.\begin{aligned} t &= (2\alpha/f)k, \\ a^2 &= f(f_1 - f_2)/8\alpha^3 \\ b^2 &= \varepsilon(f_1 - f_2)/2\alpha f \end{aligned}\right\} \tag{D.6}$$

where

$$f = (f_1 f_2)^{\frac{1}{2}}, \tag{D.7}$$

in terms of which the $A_i(t)$ may be verified to satisfy the coupled equations

$$\left.\begin{aligned} \frac{dA_1}{dt} &= -\frac{i}{2} A_2(t) \exp\left[-i(a^2t^3/3 - b^2t)\right] \\ \frac{dA_2}{dt} &= -\frac{i}{2} A_1(t) \exp\left[i(a^2t^3/3 - b^2t)\right] \end{aligned}\right\}. \tag{D.8}$$

On eliminating $A_2(t)$,

$$\frac{d^2A_1}{dt^2} + i(a^2t^2 - b^2)\frac{dA_1}{dt} + \tfrac{1}{4}A_1 = 0, \tag{D.9}$$

so that with the further substitution

$$A_1(t) = B_1(t) \exp\left[-\frac{i}{2}\left(\frac{a^2t^3}{3} - b^2t\right)\right], \tag{D.10}$$

$B_1(t)$ is given by

$$\frac{d^2B_1}{dt^2} + \left[\tfrac{1}{4} - ia^2t + \tfrac{1}{4}(a^2t^2 - b^2)^2\right] B_1 = 0. \tag{D.11}$$

The solution of (D.11) depends on the positions of the roots of the function

$$K(t) = \tfrac{1}{4} - ia^2t + \tfrac{1}{4}(a^2t^2 - b^2)^2. \tag{D.12}$$

Values of a and b for which all four roots lie in the same region of space, to the extent that the asymptotic condition (see (C.6))

$$\left|\frac{\mathrm{d}}{\mathrm{d}t}[K(t)]^{-\frac{1}{4}}\right| \ll 1, \tag{D.13}$$

cannot be applied between them, allow of no simple general solution. Two cases of practical importance, illustrated in Fig. D.1, give rise however to an

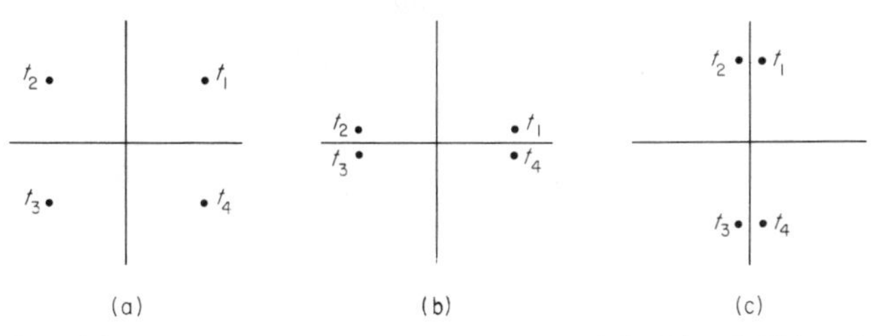

FIG. D.1. Disposition of the roots of $K(t)$. (a) in general; (b) for $b^2 \gg 1$, $b^3/a \gg 1$, and (c) for $b^2 \ll -1$, $|b^3/a| \gg 1$.

adequate separation between the roots in pairs. The necessary condition is that (D.13) should apply at $t = 0$. These correspond to

$$b^2 \gg 1, \qquad b^3/a \gg 1 \tag{D.14a}$$

and

$$b^2 \ll -1, \qquad |b^3/a| \gg 1. \tag{D.14b}$$

(D.14a) and (D.14b) correspond according to (D.8) to energies well above ($\varepsilon > 0$) and well below ($\varepsilon < 0$) the crossing point respectively.

The technique in these cases is to approximate the solution of (D.11) in the localized regions around these root pairs in terms of parabolic cylinder functions and to use the asymptotic approximation

$$B_1(t) = |K(t)|^{-\frac{1}{4}} \exp\left(\pm i \int K^{\frac{1}{2}}(t)\,\mathrm{d}t\right), \tag{D.15}$$

(the validity of which depends on (D.13)), away from these points.

The two cases covered by (D.14a) and (D.14b) again require different treatment.

D.1.1. High energy limit: $b^2 \gg 1$, $b^3/a \gg 1$

Under present conditions the inequality

$$|(a^2t^2 - b^2)| \gg |1 - 4ia^2t|, \tag{D.16}$$

valid for $t \to \pm\infty$ and $t = 0$, allows a simple reduction of the exponent in (D.15).

$$\int K^{\frac{1}{2}}(t)\,dt \simeq \tfrac{1}{2}(a^2t^3/3 - b^2t) + \delta \ln\left|\frac{t - b/a}{t + b/a}\right| - \frac{i}{2}\ln(a^2t^2 - b^2), \quad \text{(D.17)}$$

where

$$\delta = \frac{1}{8ab} \quad \text{(D.18)}$$

$A_1(t)$ defined by (D.10) and (D.15) is therefore conveniently represented in these regions by the forms

$$A_1(t) \overset{t\to\pm\infty}{\sim} A_1(\pm\infty)\left(\frac{t - b/a}{t + b/a}\right)^{i\delta} + \frac{A_2(\pm\infty)}{2(a^2t^2 - b^2)}\left(\frac{t - b/a}{t + b/a}\right)^{-i\delta} \exp\left[-i\left(\frac{a^2t^3}{3} - b^2t\right)\right] \quad \text{(D.19)}$$

$$A_1(t) \overset{t\to 0}{\sim} X_1\left(\frac{b/a - t}{b/a + t}\right)^{i\delta} + \frac{X_2}{2(a^2t^2 - b^2)}\left(\frac{b/a - t}{b/a + t}\right)^{-i\delta} \exp\left[-i\left(\frac{a^2t^3}{3} - b^2t\right)\right] \quad \text{(D.20)}$$

It is therefore readily verified, using (D.8) that

$$A_2(t) \overset{t\to\pm\infty}{\sim} A_2(\pm\infty). \quad \text{(D.21)}$$

The problem now is to connect the coefficients $A_i(-\infty)$, X_i and $A_i(+\infty)$ around the points $t = \pm b/a$ where the above approximations break down.

Consider in detail the connection between $A_i(-\infty)$ and X_i around the point $t = -b/a$. The first step is to apply the approximations

$$\left.\begin{aligned} at - b &\simeq -2b \\ at + b &\simeq a\tau \end{aligned}\right\}, \quad \text{(D.22)}$$

in the function $K(t)$, with the result, on neglecting terms of order a/b^3 compared with unity, that (D.11) takes the form

$$\frac{d^2B_1}{dz^2} + [\tfrac{1}{4} + iab + a^2b^2\tau^2]\,B_1 = 0. \quad \text{(D.23)}$$

This is reduced by the substitution

$$z = (2ab)^{\frac{1}{2}} \exp(-i\pi/4)\tau = (4\delta)^{-\frac{1}{2}} \exp(-i\pi/4)\tau, \quad \text{(D.24)}$$

to the standard form, (Whittaker and Watson (1962))

$$\frac{d^2 B_1}{dz^2} + [-\tfrac{1}{2} + i\delta - \tfrac{1}{4}z^2] B_1 = 0, \tag{D.25}$$

the general solution of which is given by the parabolic cylinder functions

$$B_1(z) = PD_{-1+i\delta}(z) + QD_{-i\delta}(-iz). \tag{D.26}$$

It remains to determine the coefficients P and Q in terms of $A_1(-\infty)$ and $A_2(-\infty)$ by extending this solution into the asymptotic region $\tau \to -\infty$, and hence to relate $A_i(-\infty)$ with X_i in (D.20) by means of the form of (D.26) as $\tau \to +\infty$.

As $\tau \to -\infty$ for example (with $\arg \tau = \pi$) $\arg z = 3\pi/4$ and, according to the asymptotic properties of $D_n(z)$ given by (C.40), $B_1(z)$ is given by

$$B_1(z) \sim \left[\frac{(2\pi)^{\frac{1}{2}}}{\Gamma(1 - i\delta)} e^{-\pi\delta} P + Q e^{-\pi\delta/2}\right] z^{-i\delta} \exp(\tfrac{1}{4}z^2) + Pz^{-1+i\delta} \exp(-\tfrac{1}{4}z^2). \tag{D.27}$$

Similarly as $\tau \to +\infty$, $\arg z = -\pi/4$ and

$$B_1(z) \sim Q e^{-\pi\delta/2} z^{-i\delta} \exp(\tfrac{1}{4}z^2) + \left[P - \frac{i(2\pi)^{\frac{1}{2}}}{\Gamma(i\delta)} e^{-\pi\delta/2} Q\right] z^{-1+i\delta} \exp(-\tfrac{1}{4}z^2). \tag{D.28}$$

This means, after use of the identity (Abramowitz and Stegun, 1965)

$$\frac{2\pi i\, e^{-\pi\delta}}{\Gamma(i\delta)\Gamma(1 - i\delta)} = e^{-2\pi\delta} - 1, \tag{D.29}$$

that if

$$\left.\begin{aligned} B_1(z) &\overset{\tau\to-\infty}{\sim} F_1 z^{-i\delta} \exp(\tfrac{1}{4}z^2) + F_2 z^{-1+i\delta} \exp(-\tfrac{1}{4}z^2) \\ \text{and} \quad B_1(z) &\overset{\tau\to+\infty}{\sim} G_1 z^{-i\delta} \exp(\tfrac{1}{4}z^2) + G_2 z^{-1+i\delta} \exp(-\tfrac{1}{4}z^2), \end{aligned}\right\} \tag{D.30}$$

then

$$\begin{pmatrix} G_1 \\ G_2 \end{pmatrix} = \begin{pmatrix} 1, & -(2\pi)^{\frac{1}{2}}[\Gamma(1 - i\delta)]^{-1} \exp(-\pi\delta) \\ -i(2\pi)^{\frac{1}{2}}[\Gamma(i\delta)]^{-1}, & \exp(-2\pi\delta) \end{pmatrix} \begin{pmatrix} F_1 \\ F_2 \end{pmatrix}. \tag{D.31}$$

We may verify however, on introducing the substitutions (D.22) and (D.24) in (D.19) and (D.20), that

$$B_1 \overset{\tau\to-\infty}{\sim} A_1(-\infty) \exp\left(\frac{-3\pi\delta}{4} + i\theta\right) z^{-i\delta} \exp(\tfrac{1}{4}z^2) - A_2(-\infty)\delta^{\frac{1}{2}} \left(\frac{3\pi\delta}{4} - i\theta - \frac{i\pi}{4}\right) z^{-1+i\delta} \exp(-\tfrac{1}{4}z^2), \tag{D.32}$$

and

$$B_1 \overset{\tau\to\infty}{\sim} X_1 \exp\left(\frac{\pi\delta}{4} + i\theta\right) z^{-i\delta} \exp\left(\tfrac{1}{4}z^2\right),$$

$$- X_2 \delta^{\frac{1}{2}} \exp\left(\frac{-\pi\delta}{4} - i\theta - \frac{i\pi}{4}\right) z^{-1+i\delta} \exp\left(\tfrac{1}{4}z^2\right),$$

where

$$\theta = \tfrac{1}{3}(b^3/a) + \delta \ln (2b/a) - (\delta/2) \ln 4\delta. \tag{D.33}$$

Hence by comparison with (D.30)

$$\begin{pmatrix} F_1 \\ F_2 \end{pmatrix} = \begin{pmatrix} \exp\left(\dfrac{-3\pi\delta}{4} + i\theta\right), & 0 \\ 0, & -\delta^{\frac{1}{2}} \exp\left(\dfrac{3\pi\delta}{4} - i\theta - \dfrac{i\pi}{4}\right) \end{pmatrix} \begin{pmatrix} A_1(-\infty) \\ A_2(-\infty) \end{pmatrix}, \tag{D.34}$$

and

$$\begin{pmatrix} G_1 \\ G_2 \end{pmatrix} = \begin{pmatrix} \exp\left(\dfrac{\pi\delta}{4} + i\theta\right), & 0 \\ 0, & -\delta^{\frac{1}{2}} \exp\left(\dfrac{-\pi\delta}{4} - i\theta - \dfrac{i\pi}{4}\right) \end{pmatrix} \begin{pmatrix} X_1 \\ X_2 \end{pmatrix}. \tag{D.35}$$

The overall connection between $A_i(-\infty)$ and X_i implied by (D.31), (D.34) and (D.35) is therefore

$$\begin{pmatrix} X_1 \\ X_2 \end{pmatrix} = \begin{pmatrix} e^{-\pi\delta}, & (1 - e^{-2\pi\delta})^{\frac{1}{2}} \exp(i\phi) \\ -(1 - e^{-2\pi\delta})^{\frac{1}{2}} \exp(-i\phi), & e^{-\pi\delta} \end{pmatrix} \begin{pmatrix} A_1(-\infty) \\ A_2(-\infty) \end{pmatrix}, \tag{D.36}$$

where

$$\begin{aligned} \phi &= -2\theta + \arg \Gamma(i\delta) + \pi/4 \\ &= -\tfrac{2}{3}(b^3/a) - 2\delta \ln (2b/a) + \delta \ln 4\delta + \arg \Gamma(i\delta) + \pi/4. \end{aligned} \tag{D.37}$$

The following identities (Abramowitz and Stegun (1965)),

$$\left.\begin{aligned} \Gamma(1 - i\delta) &= -i\delta\, \Gamma(i\delta) \\ |\Gamma(\pm i\delta)|^{-1} &= \pi^{-\frac{1}{2}}[\delta \sinh \pi\delta]^{\frac{1}{2}} \end{aligned}\right\} \tag{D.38}$$

have been used in the reduction of (D.37).

The application of similar arguments around the point $t = b/a$ leads to the analogous connection between X_i and $A_i(+\infty)$.

$$\begin{pmatrix} X_1 \\ X_2 \end{pmatrix} = \begin{pmatrix} e^{-\pi\delta}, & (1 - e^{-2\pi\delta})^{\frac{1}{2}} \exp(-i\phi) \\ -(1 - e^{-2\pi\delta})^{\frac{1}{2}} \exp(i\phi), & e^{-\pi\delta} \end{pmatrix} \begin{pmatrix} A_1(\infty) \\ A_2(\infty) \end{pmatrix}, \tag{D.39}$$

Hence by elimination of X_i between (D.36) and (D.39)

$$\begin{pmatrix} A_1(\infty) \\ A_2(\infty) \end{pmatrix} = \begin{pmatrix} e^{-2\pi\delta} + (1 - e^{-2\pi\delta})\, e^{-2i\phi}, & 2i\, e^{-\pi\delta}(1 - e^{-2\pi\delta})^{\frac{1}{2}} \sin\phi \\ 2i\, e^{-\pi\delta}(1 - e^{-2\pi\delta})^{\frac{1}{2}} \sin\phi, & e^{-2\pi\delta} + (1 - e^{-2\pi\delta})\, e^{2i\phi} \end{pmatrix} \begin{pmatrix} A_1(-\infty) \\ A_2(-\infty) \end{pmatrix} \tag{D.40}$$

This completes the necessary connection between the initial and final amplitudes in channels 1 and 2. It may be verified, as required by flux conservation, that the transformation (D.40), like (D.36) and (D.39) is of unitary form.

D.1.2. Low Energy Limit: $b^2 \ll -1, |b^3/a| \ll 1.$

As indicated in Fig. D.1(c), when $b^2 \ll -1$, $|b^3/a| \gg 1$, the branch points t_1, t_2, t_3 and t_4 lie in pairs around the points $\pm i|b|/a$. This means that only one such pair need be taken into account in passing from $t = -\infty$ to $t = +\infty$. We choose for the sake of definiteness the pair t_1, t_2 around the point $t = i|b|/a$. Furthermore it is convenient for the discussion in Section 8.5, to approximate $K^{\frac{1}{2}}(t)$ on the real axis in the form

$$K^{\frac{1}{2}}(t) = \tfrac{1}{2}[(a^2t^2 - b^2)^2 + 1]^{\frac{1}{2}} - \frac{ia^2t}{a^2t^2 - b^2}, \tag{D.41}$$

and hence to represent $B_1(t)$ by the approximations

$$B_1(t) \overset{t \gtrless 0}{\sim} A_1(\pm\infty) \exp\left\{+\tfrac{1}{2}\int_{t_1}^{t} [(a^2t^2 - b^2)^2 + 1]^{\frac{1}{2}}\, dt + \frac{i}{2}\Omega\right\}$$

$$+ \frac{A_2(\pm\infty)}{2(a^2t^2 - b^2)} \exp\left\{-\frac{i}{2}\int_{t_1}^{t} [(a^2t^2 - b^2)^2 + 1]^{\frac{1}{2}}\, dt - \frac{i}{2}\Omega\right\}, \tag{D.42}$$

where t_1 is the root of $(a^2t^2 - b^2)^2 + 1$ with $0 < \arg t < \pi/2$,

$$|\delta| = \frac{1}{8a|b|}, \tag{D.43}$$

and

$$\Omega = \int_0^{t_1} [(a^2t^2 - b^2)^2 + 1]^{\frac{1}{2}}\, dt. \tag{D.44}$$

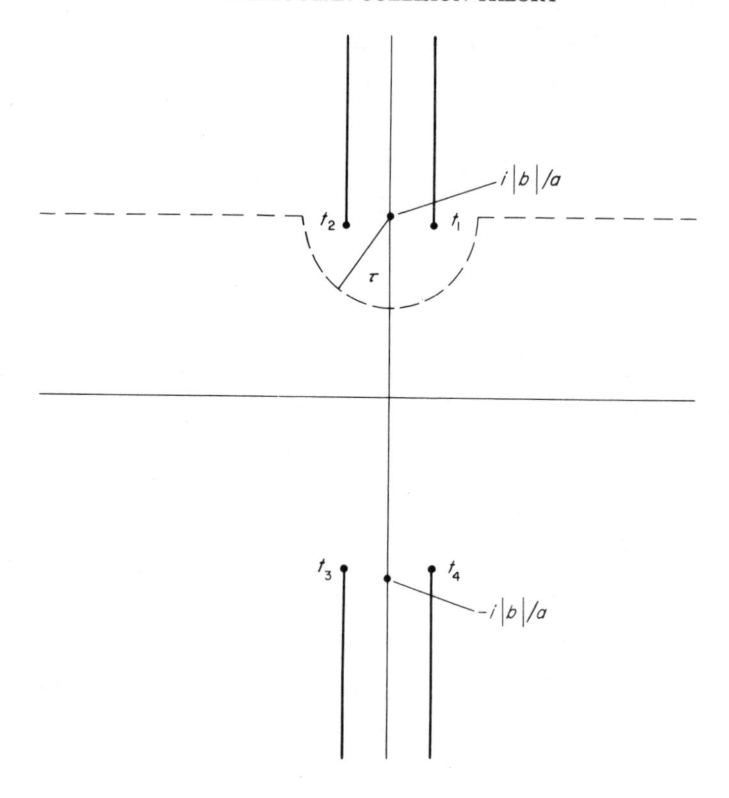

FIG. D.2. Branch cuts for the case $b^2 \ll -1$, $b^3/a \gg 1$.

With the branch cuts shown in Fig. D.2, the choice of coefficients in (D.42) ensures that

$$A_1(t) \overset{t\to\pm\infty}{\sim} A_1(\pm\infty)\left(\frac{t - b/a}{t + b/a}\right)^{|\delta|} + \frac{A_2(\pm\infty)}{2(a^2t^2 - b^2)}\left(\frac{t - b/a}{t + b/a}\right)^{-|\delta|} \exp\left[-i\left(\frac{a^2t^3}{3} - b^2t\right)\right]. \tag{D.45}$$

A connection between $A_1(-\infty)$ and $A_2(+\infty)$ is now established by following the variation in $B_1(t)$ on passing around the branch points t_1 and t_2, because in this region

$$\left.\begin{aligned} t + i|b|/a &\simeq 2i|b|/a, \\ t - i|b|/a &= \tau \end{aligned}\right\}. \tag{D.46}$$

Hence

$$\frac{i}{2}\int_{t_1}^{t} [(a^2 - b^2t^2)^2 + 1]^{\frac{1}{2}}\, dt \simeq -a|b| \int_{4|\delta|}^{\tau} [\tau^2 - |4\delta|^2]^{\frac{1}{2}}\, d\tau$$

$$\overset{\tau\to\pm\infty}{\sim} -\frac{a|b|\tau^2}{2} + \frac{|\delta|}{2} + |\delta| \ln(\tau/2|\delta|). \tag{D.47}$$

In other words,

$$B_1(t) \overset{\tau\to\pm\infty}{\sim} A_1(\pm\infty)\exp\left\{\frac{i\Omega}{2} - |\delta|\ln(2|\delta|) + \tfrac{1}{2}|\delta|\right\}\tau^{|\delta|}\exp\left(\frac{-a|b|\tau^2}{2}\right)$$

$$-iA_2(\pm\infty)\exp\left\{-\frac{i\Omega}{2} + (1+|\delta|)\ln(2|\delta|) - \tfrac{1}{2}|\delta|\right\}\tau^{-1-|\delta|}\exp\left(\frac{a|b|\tau^2}{2}\right) \tag{D.48}$$

Unfortunately the presence of the second term, which completely dominates the first as $\tau \to \pm\infty$, prevents a unique specification of $A_1(-\infty)$ (or $A_1(+\infty)$) unless $A_2(-\infty)$ (or $A_2(+\infty)$) is set equal to zero. (See for example Section C.2, and the discussion of the Stokes phenomenon by Heading (1962) or Jeffreys and Jeffreys (1956).) Hence a full connection between $A_1(\pm\infty)$ and $A_2(\pm\infty)$, of the type given by (D.40) in Section D.1.1, is inaccessible by this technique. All that can be achieved by an approximation in terms of parabolic cylinder functions is the connection $A_1(-\infty) \to A_2(+\infty)$ or $A_1(+\infty) \to A_2(-\infty)$. Fortunately these determine the necessary transition probabilities.

This connection is obtained as before, by first using (D.11) and (D.46) together with the substitution

$$z = e^{+i\pi}(4|\delta|)^{-\frac{1}{2}}\tau, \tag{D.49}$$

in order to obtain the following equation for B_1:

$$\frac{d^2B_1}{dz^2} + [\tfrac{1}{2} + |\delta| - \tfrac{1}{4}z^2]B_1 = 0. \tag{D.50}$$

Taken together with the factor $e^{i\pi}$ in the (D.49), the choice of solution

$$B_1(z) = PD_{|\delta|}(z), \tag{D.51}$$

with

$$P = A_1(-\infty)\exp\left[i\frac{\Omega}{2} + \tfrac{1}{2}|\delta| - \tfrac{1}{2}|\delta|\ln|\delta| - i\pi|\delta|\right] \tag{D.52}$$

L

ensures according to (C.40) that $B_1(t)$ reproduces (D.48) with $A_1(-\infty) = 0$ as $\tau \to e^{-i\pi}\infty$. As $\tau \to 0$ on the other hand $z \to e^{i\pi}\infty$ and

$$B_1(\tau) \sim P[\exp(i\pi|\delta| - \tfrac{1}{2}|\delta| \ln 4|\delta|)\tau^{|\delta|} \exp(-\tfrac{1}{2}a|b|\tau^2) \\ - 2(2\pi|\delta|)^{\frac{1}{2}}[\Gamma(-|\delta|)]^{-1}(4|\delta|)^{\frac{1}{2}|\delta|}\tau^{-1-|\delta|} \exp(\tfrac{1}{2}a|b|\tau^2)]. \qquad \text{(D.53)}$$

For the reason given above the coefficient of the first term in (D.53) has no strict meaning, but that of the second may be used in conjunction with (D.48) and (D.52) to determine $A_2(+\infty)$;

$$A_2(+\infty) = -i(2|\delta|/\pi)^{\frac{1}{2}} \sin(\pi|\delta|)\Gamma(|\delta|)|\delta|^{-|\delta|} \\ \times \exp[|\delta| + i\Omega - i\pi|\delta|]A_1(-\infty), \qquad \text{(D.54)}$$

after use of the identity (Abramowitz and Stegun, 1965)

$$[\Gamma(-|\delta|)]^{-1} = -\pi^{-1}|\delta| \sin \pi|\delta|\Gamma(|\delta|). \qquad \text{(D.55)}$$

The transition probability is therefore given by

$$P_{12} = |A_2(+\infty)/A_1(-\infty)|^2 = B(|\delta|) \exp[-a^{-1}\Delta(b)]$$

where (D.56)

$$B(|\delta|) = 2\pi^{-1}|\delta| \sin^2 \pi|\delta|[\Gamma(|\delta|)]^2|\delta|^{-2|\delta|} \exp[2|\delta|],$$

and

$$a^{-1}\Delta(b) = 2 \operatorname{Im} \int_0^{t_1} [1 + (a^2t^2 - b^2)^2]^{\frac{1}{2}} \, dt \\ = 2a^{-1} \operatorname{Im} \int_0^{s_1} [1 + (s^2 - b^2)^2]^{\frac{1}{2}} \, ds, \qquad \text{(D.57)}$$

s_1 being the root of the integrand with $0 < \arg s_1 < \pi/2$. It may be verified that for $|b^2| \gg 1$, $|b^3/a| \gg 1$,

$$a^{-1}\Delta(b) \simeq \frac{4}{3}\frac{|b|^3}{a}. \qquad \text{(D.58)}$$

Furthermore $B(|\delta|)$ takes the following limiting forms

$$\left.\begin{aligned} B(|\delta|) &\simeq 2\pi|\delta|, & |\delta| &\ll 0{\cdot}1 \\ &\simeq 4\sin^2|\delta|, & |\delta| &\gg 1 \end{aligned}\right\}. \qquad \text{(D.59)}$$

Numerical values of $B(|\delta|)$ for intermediate values of $|\delta|$ are given in Table D.1 at the end of this Appendix.

Finally it may be noted that for small values of $|\delta|$, the transition amplitudes derived from (D.40) and from (D.56) with $a^{-1}\Delta(b)$ and $B|\sigma|$ given by

(D.58) and (D.59) respectively, connect smoothly with the perturbation formula.

$$\begin{aligned} A_2(\infty)/A_1(-\infty) &= -\frac{i}{2}\int_{-\infty}^{\infty} \exp\left[i(a^2t^3/3 - b^2t)\right] \mathrm{d}t \\ &= -i\pi a^{-\frac{2}{3}} \,\mathrm{Ai}\,(-b^2 a^{-\frac{2}{3}}) \\ &\overset{b^2 \gg 1}{\sim} -4i\pi\delta \sin\left(\tfrac{2}{3}b^3/a + \pi/4\right) \\ &\overset{b^2 \ll -1}{\sim} -2i\pi|\delta| \exp\left(-\tfrac{2}{3}|b|^3/a\right) \end{aligned} \tag{D.60}$$

D.2 Forces with Opposite Signs: $f_1 f_2 < 0$

The case $f_1 f_2 < 0$ differs from the previous one in two respects. First the momentum representatives in (D.2) are taken in the form

$$u_i(k) = (2/|f_i|^{\frac{1}{2}}) \exp\left[\frac{i}{f_i}(\varepsilon k - k^3/3)\right] A_i(k). \tag{D.61}$$

Hence after introduction of the dimensionless parameters t, a^2 and b^2 given by (D.6), except that in this case

$$f = (f_1 |f_2|)^{\frac{1}{2}},$$

the equations for the $A_i(t)$ become,

$$\frac{\mathrm{d}A_1}{\mathrm{d}t} = -\frac{i}{2}\exp\left[i(a^2t^3/3 - b^2t)\right]A_2 \tag{D.62}$$

$$\frac{\mathrm{d}A_2}{\mathrm{d}t} = \frac{i}{2}\exp\left[-i(a^2t^3/3 - b^2t)\right]A_1$$

Equation (D.9) is therefore replaced by

$$\frac{\mathrm{d}^2 A_1}{\mathrm{d}t^2} - i(a^2t^2 - b^2)\frac{\mathrm{d}A_1}{\mathrm{d}t} - \tfrac{1}{4}A_1 = 0, \tag{D.63}$$

which reduces under the substitution

$$A_1(t) = B_1(t)\exp\left[\frac{i}{2}\left(\frac{a^2t^3}{2} - b^2t\right)\right], \tag{D.64}$$

to the form

$$\frac{\mathrm{d}^2 B_1}{\mathrm{d}t^2} + \left[-\tfrac{1}{4} + ia^2t + \tfrac{1}{4}(a^2t^2 - b^2)^2\right]B_1 = 0, \tag{D.65}$$

the important difference between (D.65) and (D.11) being the sign of the first term in the bracket.

The second important difference lies in the physical interpretation of the $A_i(k)$. Previously, with $f_1 > 0$ and $f_2 > 0$, the initial and final motion in both channels occurs in the same asymptotic region, $x \to \infty$, of coordinate space. Positive or negative momentum therefore corresponds to outgoing or incoming motion in both channels, and the transformation (D.40) between $A_i(+\infty)$ and $A_i(-\infty)$ is unitary. In the present case on the other hand if $f_1 > 0$ and $f_2 < 0$, the asymptotic regions for motion in channels 1 and 2 are $x \to +\infty$ and $x \to -\infty$ respectively. $A_1(\pm\infty)$ therefore have their previous interpretation but the significance of $A_2(\pm\infty)$ is reversed; $A_2(+\infty)$ is an incoming and $A_2(-\infty)$ is an outgoing amplitude. Although the connection between $A_i(-\infty)$ and $A_i(+\infty)$ is again obtained directly, it is the transformation between $[A_1(-\infty), A_2(+\infty)]$ and $[A_1(+\infty), A_2(-\infty)]$ which is unitary in this case. This difference is illustrated diagrammatically in Fig. D.3.

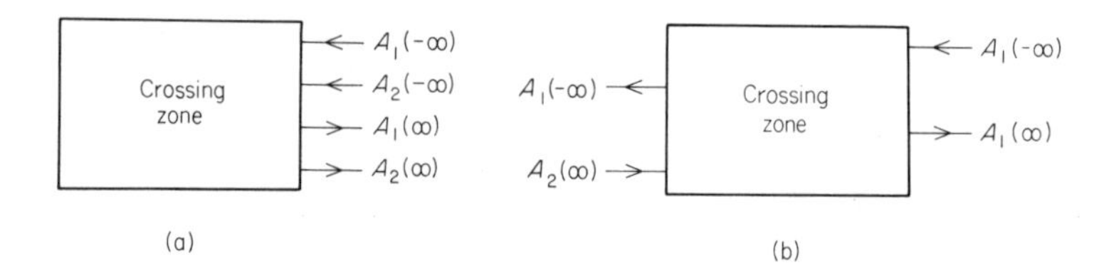

FIG. D.3. Amplitude coefficients for the cases (a) $f_1 f_2 > 0$ and (b) $f_1 f_2 < 0$.

The solution of (D.65) for the possible transition amplitudes proceeds along the previous lines, with the two sub cases $b^2 \gg 1, b^3/a \gg 1$ and $b^2 \ll -1$, $|b^3|/a \gg 1$ again being amenable to solution in terms of parabolic cylinder functions.

D.2.1 High Energy Limit: $b^2 \gg 1$, $b^3/a \gg 1$

The derivation follows, with minor variations, the arguments of (D.17)–(D.40). With $A_1(t)$ represented in the asymptotic forms

$$A_1(t) \overset{t\to\pm\infty}{\sim} A_1(\pm\infty)\left(\frac{t - b/a}{t + b/a}\right)^{i\delta} - \frac{A_2(\pm\infty)}{2(a^2t^2 - b^2)}\left(\frac{t - b/a}{t + b/a}\right)^{-i\delta} \exp\left[i\left(\frac{a^2t^3}{3} - b^2t\right)\right] \tag{D.66}$$

$$A_1(t) \overset{t\to 0}{\sim} X_1\left(\frac{b/a - t}{b/a + t}\right)^{i\delta} - \frac{X_2}{2(a^2t^2 - b^2)}\left(\frac{b/a - t}{b/a + t}\right)^{-i\delta} \exp\left[i\left(\frac{a^2t^3}{3} - b^2t\right)\right], \tag{D.67}$$

where δ is again defined by (D.18), the overall results may be written

$$\begin{pmatrix} X_1 \\ X_2 \end{pmatrix} = \begin{pmatrix} e^{\pi\delta}, & -i[e^{2\pi\delta} - 1]^{\frac{1}{2}} e^{i\tilde{\phi}} \\ i[e^{2\pi\delta} - 1]^{\frac{1}{2}} e^{-i\tilde{\phi}}, & e^{\pi\delta} \end{pmatrix} \begin{pmatrix} A_1(-\infty) \\ A_2(-\infty) \end{pmatrix} \tag{D.68}$$

$$\begin{pmatrix} A_1(\infty) \\ A_2(\infty) \end{pmatrix} = \begin{pmatrix} e^{\pi\delta}, & -i[e^{2\pi\delta} - 1]^{\frac{1}{2}} e^{-i\tilde{\phi}} \\ i[e^{2\pi\delta} - 1]^{\frac{1}{2}} e^{i\tilde{\phi}}, & e^{\pi\delta} \end{pmatrix} \begin{pmatrix} X_1 \\ X_2 \end{pmatrix}, \tag{D.69}$$

$$\begin{pmatrix} A_1(\infty) \\ A_2(\infty) \end{pmatrix} = \begin{pmatrix} e^{2\pi\delta} + [e^{2\pi\delta} - 1] e^{-2i\tilde{\phi}}, & -2i\, e^{\pi\delta}(e^{2\pi\delta} - 1)^{\frac{1}{2}} \cos\tilde{\phi} \\ +2i\, e^{\pi\delta}[e^{2\pi\delta} - 1]^{\frac{1}{2}} \cos\tilde{\phi}, & e^{2\pi\delta} + (e^{2\pi\delta} - 1) e^{2i\tilde{\phi}} \end{pmatrix} \times \begin{pmatrix} A_1(-\infty \\ A_1(-\infty) \end{pmatrix} \tag{D.70}$$

where

$$\tilde{\phi} = \tfrac{2}{3}b^3/a + \arg\Gamma(i\delta) + \delta \ln 4\delta - 2\delta \ln(2b/a) + \pi/4. \tag{D.71}$$

The corresponding unitary transformations are

$$\begin{pmatrix} X_1 \\ A_2(-\infty) \end{pmatrix} = \begin{pmatrix} e^{-\pi\delta}, & -i(1 - e^{-2\pi\delta})^{\frac{1}{2}} e^{i\tilde{\phi}} \\ -i(1 - e^{-2\pi\delta})^{\frac{1}{2}} e^{-i\tilde{\phi}}, & e^{-\pi\delta} \end{pmatrix} \begin{pmatrix} A_1(-\infty) \\ X_2 \end{pmatrix}, \tag{D.72}$$

$$\begin{pmatrix} X_1 \\ A_2(\infty) \end{pmatrix} = \begin{pmatrix} e^{-\pi\delta}, & i(1 - e^{-2\pi\delta})^{\frac{1}{2}} e^{-i\tilde{\phi}} \\ i(1 - e^{-2\pi\delta})^{\frac{1}{2}} e^{-i\tilde{\phi}}, & e^{-\pi\delta} \end{pmatrix} \begin{pmatrix} A_1(-\infty) \\ X_2 \end{pmatrix}, \tag{D.73}$$

and

$$\begin{pmatrix} A_1(\infty) \\ A_2(-\infty) \end{pmatrix} = \frac{1}{D} \begin{pmatrix} 1, & -2i\, e^{\pi\delta}(e^{2\pi\delta} - 1)^{\frac{1}{2}} \cos\tilde{\phi} \\ -2i\, e^{\pi\delta}(e^{2\pi\delta} - 1)^{\frac{1}{2}} \cos\tilde{\phi}, & 1 \end{pmatrix} \begin{pmatrix} A_1(-\infty) \\ A_2(\infty) \end{pmatrix}, \tag{D.74}$$

where

$$D = e^{2\pi\delta} + (e^{2\pi\delta} - 1) e^{2i\tilde{\phi}}. \tag{D.75}$$

Oscillations in the transmission coefficients $[A_2(-\infty)/A_1(-\infty)]$ and $[A_1(\infty)/A_2(\infty)]$ derived from (D.73) are associated with (temporary) capture between the turning points of $V_1(x)$ and $V_2(x)$, as discussed in more detail in Section 8.5. As an important special case, the transmission coefficient falls to zero when $\tilde{\phi} = (n + \frac{1}{2})\pi$.

D.2.2 Low Energy Limit: $b^2 \ll -1, |b^3/a| \gg 1$

We are concerned when $b^2 < 0$ with the possibility of penetration through the barrier imposed by the lower adiabatic term derived from $V_1(x)$ and $V_2(x)$ defined by (D.1).

Ovchinnikova (1964) finds, by the methods applied in Section D.1.2 that the transmission probability takes the form

$$T = \tilde{B}(|\delta|) \exp[-a^{-1}\tilde{\Delta}(b)], \tag{D.76}$$

where

$$\tilde{B}(|\delta|) = 2\pi(|\delta|)^{-1} [\Gamma(|\delta|)]^{-2}(|\delta|)^{2|\delta|} e^{-2|\delta|}, \tag{D.77}$$

$$\tilde{\Delta}(b) = 2 \operatorname{Im} \int_0^{s_1} [(s^2 - b^2)^2 - 1]^{\frac{1}{2}} ds, \tag{D.78}$$

s_1 being the root of the integrand with $0 < \arg s_1 < \pi/2$.

Note that the exponent in (D.75) may be identified with the modulus of the classical action required, by normal tunnelling theory (see for example Landau and Lifshitz, 1965) to penetrate from one classical turning point of the barrier, a_- say, to the other, b_-;

$$a^{-1}\tilde{\Delta}(b) = -2\left|\int_{a_-}^{b_-} k_-(x) dx\right|. \tag{D.79}$$

The pre-exponent, $B|\delta|$, therefore represents a non-adiabatic correction,

TABLE D.1

$\|\delta\|$	$B(\|\delta\|)$	$\tilde{B}(\|\delta\|)$	$\|\delta\|$	$B(\|\delta\|)$	$\tilde{B}(\|\delta\|)$
0·01	0·069	0·057	0·10	1·07	0·358
0·02	0·149	0·106	0·20	2·64	0·524
0·03	0·239	0·149	0·30	4·21	0·622
0·04	0·338	0·187	0·40	5·25	0·690
0·05	0·441	0·222	0·50	5·44	0·736
0·06	0·556	0·254	0·60	4·70	0·770
0·07	0·673	0·283	0·70	3·28	0·799
0·08	0·791	0·312	0·80	1·69	0·817
0·09	0·931	0·335	0·90	0·46	0·837
0·10	1·067	0·358	1·00	0·00	0·848

$B(|\delta|)$ and $\tilde{B}(|\delta|)$ may be approximated within 1 % outside these ranges by the functions

$$\left.\begin{aligned} B(|\delta|) &\simeq 2\pi|\delta|\,|\delta|^{-2|\delta|} \\ \tilde{B}(|\delta|) &\simeq 2\pi|\delta|\,|\delta|^{2|\delta|} \end{aligned}\right\} |\delta| < 0{\cdot}01.$$

$$\left.\begin{aligned} B(|\delta|) &\simeq 4 \sin^2 \pi|\delta| \left(1 + \frac{1}{12|\delta|}\right)^2 \\ \tilde{B}(|\delta|) &\simeq \left(1 + \frac{1}{12|\delta|}\right)^{-2} \end{aligned}\right\} |\delta| > 1{\cdot}00$$

which decreases in importance as the adiabatic coupling strength measured by $|\delta|$ increases, so that

$$\tilde{B}(|\delta|) = 1 \text{ for } |\delta| \gg 1. \tag{D.80}$$

At the opposite limit however

$$\tilde{B}(|\delta|) = 2\pi|\delta| \text{ for } |\delta| \ll 0{\cdot}1. \tag{D.81}$$

Numerical values of $\tilde{B}(|\delta|)$ and the function $B(|\delta|)$ defined by (D.56) at intermediate $|\delta|$ values are given in Table D.1.

APPENDIX E

Elements of Classical Mechanics

A brief survey of the elements of classical mechanics for use in generalized coordinate systems is given below. More extensive accounts may be found in the standard texts (see for example Goldstein, 1959). It is assumed that the system is conservative in the sense that the potential energy is independent of time.

E.1 Lagrange's Equations

Suppose initially that the system is specified by n cartesian coordinates $(x_1, x_2, \ldots, x_n)$, so that Newton's equations take the form

$$F_i = m_i\dot{v}_i = m_i\ddot{x}_i, \tag{E.1}$$

where

$$F_i = -(\partial V/\partial x_i). \tag{E.2}$$

Lagrange's equations of motion, for use in the generalized coordinate system $(q_1, q_2, \ldots, q_n)$ such that

$$x_i = x_i(q_1, q_2, \ldots, q_n), \tag{E.3}$$

then follow directly from (E.1) by means of D'Alembert's principle,

$$\sum_i (F_i - m_i\dot{v}_i)\delta x_i = 0 \tag{E.4}$$

for an arbitrary virtual displacement $(\delta x_1, \delta x_2, \ldots, \delta x_n)$. The immediate aim is an equation of equivalent form in δq_j, rather than δx_i. The first term of (E.4) gives

$$\sum_i F_i\delta x_i = -\sum_i \left(\frac{\partial V}{\partial x_i}\right)\delta x_i = -\sum_{ijk}\left(\frac{\partial V}{\partial q_j}\right)\left(\frac{\partial q_j}{\partial x_i}\right)\left(\frac{\partial x_i}{\partial q_k}\right)\delta q_k = -\sum_j \left(\frac{\partial V}{\partial q_j}\right)\delta q_j \tag{E.5}$$

and the second term is conveniently taken in two parts

$$\sum_i m_i \dot{v}_i \delta x_i = \sum_{ij} m_i \ddot{x}_i \left(\frac{\partial x_i}{\partial q_j}\right) \delta q_j$$

$$= \sum_{ij} \left[\frac{\mathrm{d}}{\mathrm{d}t}\left(m_i \dot{x}_i \frac{\partial x_i}{\partial q_j}\right) - (m_i \dot{x}_i)\left(\frac{\partial \dot{x}_i}{\partial q_j}\right)\right] \delta q_j. \tag{E.6}$$

Of these, the first reduces under the identity $(\partial x_i/\partial q_j) = (\partial \dot{x}_i/\partial \dot{q}_j)$ to

$$\sum_{ij} \frac{\mathrm{d}}{\mathrm{d}t}\left(m_i \dot{x}_i \frac{\partial \dot{x}_i}{\partial \dot{q}_j}\right) \delta q_j = \sum_j \frac{\mathrm{d}}{\mathrm{d}t}\left(\frac{\partial T}{\partial \dot{q}_j}\right) \delta q_j, \tag{E.7}$$

where T is the kinetic energy

$$T = \sum_i \tfrac{1}{2} m_i \dot{x}_i^2. \tag{E.8}$$

Similarly the second term in (E.6) becomes

$$\sum_{ij} m_i \dot{x}_i (\partial \dot{x}_i/\partial q_j) \delta q_j = \sum_i (\partial T/\partial q_j) \delta q_j. \tag{E.9}$$

Hence on combining (E.4)–(E.9)

$$\sum_j \left[\frac{\partial}{\partial q_j}(T - V) - \frac{\mathrm{d}}{\mathrm{d}t}\left(\frac{\partial T}{\partial \dot{q}_j}\right)\right] \delta q_j = 0. \tag{E.10}$$

Finally since the δq_j are arbitrary and V is by hypothesis independent of time

$$\frac{\mathrm{d}}{\mathrm{d}t}\left(\frac{\partial L}{\partial \dot{q}_j}\right) - \left(\frac{\partial L}{\partial q_j}\right) = 0, \tag{E.11}$$

where

$$L(q, \dot{q}) = T(q, \dot{q}) - V(q). \tag{E.12}$$

$L(q, \dot{q})$ is termed the Lagrangian for the system and (E.11) are the n coupled second-order equations of Lagrange.

E.2 Hamilton's Equations

Hamilton's equations employ coordinates, q, and conjugate momenta, p, in place of the coordinates and velocities $\dot{q}$ in (E.12). The Hamiltonian function is taken in the form,

$$H(p, q) = \sum_j \dot{q}_j p_j - L(q, \dot{q}), \tag{E.13}$$

with the momenta p_j defined so that $H(p, q)$ is independent of $\dot{q}_j$; in other words

$$\mathrm{d}H = \sum_j \left[\left(\frac{\partial H}{\partial q_j} \right) \mathrm{d}q_j + \left(\frac{\partial H}{\partial q_j} \right) \mathrm{d}p_j \right] \tag{E.14}$$

for an arbitrary change in the system. Such a change applied to (E.13) gives however

$$\mathrm{d}H = \sum_j \left\{ - \left(\frac{\partial L}{\partial q_j} \right) \mathrm{d}q_j + \dot{q}_j \,\mathrm{d}p_j + \left[p_j - \frac{\partial L}{\partial q_j} \right] \mathrm{d}\dot{q}_j \right\}. \tag{E.15}$$

Hence, in order to eliminate terms in $\mathrm{d}\dot{q}_j$

$$p_j = \left(\frac{\partial L}{\partial \dot{q}_j} \right). \tag{E.16}$$

This defines the conjugate momentum p_j. Hamilton's equations then follow by substituting for $(\partial L/\partial \dot{q}_j)$ in (E.11) to obtain

$$\dot{p}_j = \frac{\mathrm{d}}{\mathrm{d}t} \left(\frac{\partial L}{\partial \dot{q}_j} \right) = \frac{\partial L}{\partial q_j}, \tag{E.17}$$

so that (E.15) reduces to the form

$$\mathrm{d}H = \sum_j [-\dot{p}_j \,\mathrm{d}q_j + \dot{q}_j \,dp_j] \tag{E.18}$$

Hence by comparison with (E.14)

$$\left. \begin{aligned} \dot{q}_j &= (\partial H/\partial p_j) \\ \dot{p}_j &= -(\partial H/\partial q_j). \end{aligned} \right\} \tag{E.19}$$

These n pairs of first order equations are the equations of Hamilton. They are supplemented for a conservative system by a further equation; H which is supposed independent of time takes a constant value equal to the energy of the system

$$H(p, q) = E. \tag{E.20}$$

E.3 Canonical Transformations

Both Lagrange's and Hamilton's equations may also be obtained by Hamilton's principle, according to which any motion from $(\mathbf{q}_1, t_1)$ to $(\mathbf{q}_2, t_2)$ must follow a path for which the classical action

$$W = \int_{t_1}^{t_2} L(\mathbf{q}, \dot{\mathbf{q}}) \,\mathrm{d}t = \int_{t_1}^{t_2} (\sum_j \dot{q}_j p_j - H) \,\mathrm{d}t \tag{E.21}$$

is stationary (see Goldstein, 1959 pp. 36–38, 225–227). This principle also serves to provide a set of valid generators for a canonical transformation from one system (p, q) with Hamiltonian $H(p, q)$ to another system (P, Q) with Hamiltonian $\tilde{H}(P, Q)$, such that if

$$\left.\begin{aligned} \dot{q}_j &= (\partial H/\partial p_j), \quad \dot{p}_j = -(\partial H/\partial q_j) \\ \text{then} \qquad & \\ \dot{Q}_i &= (\partial \tilde{H}/\partial P_i), \quad \dot{P}_i = -(\partial \tilde{H}/\partial Q_i). \end{aligned}\right\} \tag{E.22}$$

The condition on such a transformation is, using (E.21),

$$\delta \int_{t_1}^{t_2} [\sum_j \dot{q}_j p_j - H(p, q)]\, \mathrm{d}t = \delta \int_{t_1}^{t_2} [\sum \dot{Q}_i P_i - \tilde{H}(P, Q)]\, \mathrm{d}t = 0, \tag{E.23}$$

where δ refers to any variation in the path with end points t_1, t_2. This means that the integrands can differ at most by the total time derivative of an arbitrary function F;

$$\sum_j \dot{q}_j p_j - H(p, q) = \sum_i \dot{Q}_i P_i - \tilde{H}(P, Q) + \mathrm{d}F/\mathrm{d}t, \tag{E.24}$$

because the integrals in (E.23) then differ only by a function of the (fixed) end-points, which is necessarily constant under any variation of the path. Since there are $2n$ independent variables in the problem, the function F may now be chosen to depend explicitly on the old and new coordinates q and Q respectively. This is termed a generator of type $F_1(q, Q)$. The significance of this choice is that it implies, by the rules for partial differentiation, that

$$\frac{\mathrm{d}F_1}{\mathrm{d}t} = \sum_j \left(\frac{\partial F_1}{\partial q_j}\right) \dot{q}_j + \sum_i \left(\frac{\partial F_1}{\partial Q_i}\right) \dot{Q}_i. \tag{E.25}$$

Hence since the hamiltonians $H(p, q)$ and $\tilde{H}(P, Q)$ must both be equal to the energy for a conservative system, the identity in (E.24) requires that

$$\left(\frac{\partial F_1}{\partial q_j}\right) = p_j, \qquad \left(\frac{\partial F_1}{\partial Q_i}\right) = P_i. \tag{E.26}$$

The first of these equations implies a constraint on the function $F_1(q, Q)$; the second determines the momenta P_i conjugate to the new coordinates Q_i.

Alternative forms of the generator are

$$F_2(q, P) = F_1(q, Q) + \sum_i P_i Q_i \tag{E.27}$$

subject to according to (E.26) to

$$p_j = (\partial F_2/\partial q_j), \qquad Q_i = (\partial F_2/\partial P_i); \tag{E.28}$$

or

$$F_3(Q, p) = F_1(q, Q) - \sum_j p_j q_j \tag{E.29}$$

for which

$$q_j = -(\partial F_3/\partial p_j), \quad P_i = -(\partial F_3/\partial Q_i); \tag{E.30}$$

or finally

$$F_4(p, P) = F_1(q, Q) + \sum_i P_i Q_i - \sum_j p_j q_j \tag{E.31}$$

in which case

$$q_j = -(\partial F_4/\partial p_j), \quad Q_i = (\partial F_4/\partial P_i). \tag{E.32}$$

In all cases

$$H(p, q) = \tilde{H}(P, Q). \tag{E.33}$$

Example: We take as an example the transformation from cartesian (x, y) to plane polar (r, θ) coordinates, and note that a combination of the form $(-xp_x - yp_y)$ will always satisfy the first of equations (E.30). This corresponds to the choice $F_1(q, Q) = 0$ in (E.29). Hence it is necessary merely to substitute for (x, y) in terms of (r, θ) to obtain

$$F_3(r, \theta; p_x, p_y) = -p_x r\cos\theta - p_y r\sin\theta. \tag{E.34}$$

It then follows from the second of (E.30) that the new conjugate momenta are

$$p_r = -\left(\frac{\partial F_3}{\partial r}\right) = p_x\cos\theta + p_y\sin\theta$$

$$p_\theta = -\left(\frac{\partial F_3}{\partial \theta}\right) = -p_x r\sin\theta + p_y r\cos\theta. \tag{E.35}$$

Hence on inverting (E.35) and substituting for p_x and p_y

$$H = \frac{1}{2m}(p_x^2 + p_y^2) + V(x, y) = \frac{1}{2m}\left(p_r^2 + \frac{1}{r^2}p_\theta^2\right) + V(r\cos\theta, r\sin\theta). \tag{E.36}$$

Note that if the potential is independent of θ, then according to Hamilton's equations, (E.19),

$$\dot{p}_\theta = -(\partial H/\partial\theta) = 0. \tag{E.37}$$

Hence p_θ is a constant of the motion.

E.4 Angle-action Variables

The above example illustrates the use of a generator when the transformed coordinates can be chosen in advance. Suppose, however, that the system is known to be amenable to a transformation such that the transformed hamiltonian is independent of one coordinate, say w_k, but that the form of w_k and the conjugate momentum†, N_k, are unknown. N_k is then necessarily a constant of the motion. We shall in fact be concerned with situations where the motion is periodic and an ambiguity in the magnitude of w_k will be removed by the requirement that w_k should increase by unity for each complete period of the motion.

For the sake of simplicity, attention is restricted to the one dimensional problem. Given a Hamiltonian $H(p, q)$ we seek a generator of the type $F_2(q, N)$, conventionally termed Hamilton's characteristic function, $W(q, N)$ such that the new momentum (or action) N should be a constant of the motion. The requirements, set by (E.28), are that

$$(\partial W/\partial q) = p$$
$$(\partial W/\partial N) = w; \tag{E.38}$$

it is also required that the transformed Hamiltonian should be independent of w.

The first step in the argument is to note that a function satisfying the first of Eqns (E.38) can in principle always be found by solving the Hamilton–Jacobi equation

$$H(p, q) = H\left(\frac{\partial W}{\partial q}, q\right) = E. \tag{E.39}$$

The second point is that the unknown action N may be determined by formally expressing the unknown angle w as a function $w(q, N)$ of q and N, and imposing the constraint that N should be a constant of the motion. Thus for motion along the classical path

$$\begin{aligned} \mathrm{d}w &= \left(\frac{\partial w}{\partial q}\right)\mathrm{d}q + \left(\frac{\partial w}{\partial N}\right)\mathrm{d}N \\ &= \left(\frac{\partial w}{\partial q}\right)\mathrm{d}q = \left(\frac{\partial^2 W}{\partial q \partial N}\right)\mathrm{d}q, \end{aligned} \tag{E.40}$$

after use of (E.38). Finally we note that w is defined to increase by unity for one complete cycle; hence, on combining (E.38) and (E.40)

† The symbol N is used in preference to the conventional J in order to underline the relation between "action" and "quantum number" in Chapter 9.

$$1 = \int_0^1 \mathrm{d}w = \frac{\partial}{\partial N}\left[\int\left(\frac{\partial W}{\partial q}\right)\mathrm{d}q\right], \tag{E.41}$$

with $(\partial W/\partial q)$ determined by solution of (E.29). It follows by integration of (E.41) that

$$N = \oint\left(\frac{\partial W}{\partial q}\right)\mathrm{d}q = \oint p\,\mathrm{d}q. \tag{E.42}$$

This equation determines N as a function of the energy.

Rearrangement of the final function yields

$$E = \tilde{H}(N), \tag{E.43}$$

so that the equations of motion become

$$\begin{aligned} \dot{N} &= \partial\tilde{H}/\partial w = 0 \\ \dot{w} &= (\partial\tilde{H}/\partial N) = \nu \end{aligned} \tag{E.44}$$

with solution

$$\begin{aligned} N &= \text{const.} \\ w &= \nu t + \text{const.} \end{aligned} \tag{E.45}$$

E.5 Linear Oscillator

We take as an example of the theory the problem of the harmonic oscillator, for which

$$H = \frac{1}{2m}p^2 + \tfrac{1}{2}kq^2 \tag{E.46}$$

Hence the Hamilton–Jacobi equation becomes

$$\frac{1}{2m}\left(\frac{\partial W}{\partial q}\right)^2 + \tfrac{1}{2}kq^2 = E \tag{E.47}$$

so that according to Eqn (E.42)

$$\begin{aligned} N = \oint\left(\frac{\partial W}{\partial q}\right)\mathrm{d}q &= 2\int_{-q_0}^{q_0}[2mE - mkq^2]^{\frac{1}{2}}\,\mathrm{d}q \\ &= 2\pi E(m/k)^{\frac{1}{2}}, \end{aligned} \tag{E.48}$$

where $q_0 = (2E/k)^{\frac{1}{2}}$. It follows that

$$\tilde{H}(N) = E = N\nu \tag{E.49}$$

where

$$\nu = \frac{1}{2\pi}\left(\frac{k}{m}\right)^{\frac{1}{2}}. \tag{E.50}$$

Finally, on combining (E.38), (E.47) and (E.48),

$$w = \frac{\partial W}{\partial N} = \frac{1}{\nu}\frac{\partial W}{\partial E} = \frac{1}{2\pi}\int \frac{dq}{(q_0^2 - q^2)^{\frac{1}{2}}}.$$

$$= \frac{1}{2\pi}\sin^{-1}(q/q_0) + \text{const.} \tag{E.51}$$

The integration constant is conveniently chosen so that

$$\begin{aligned} q &= q_0 \cos 2\pi w \\ &= (2N)^{\frac{1}{2}}(mk)^{-\frac{1}{4}}\cos 2\pi w \\ &= (2N)^{\frac{1}{2}}(mk)^{-\frac{1}{4}}\cos(\omega t + \text{const.}), \end{aligned} \tag{E.52}$$

where

$$\omega = 2\pi\nu = (k/m)^{\frac{1}{2}}.$$

References

Abramowitz, M. and Stegun, I. A. (1965). "Handbook of Mathematical Functions", Dover, London.
Alfaro, V. de and Regge, T. (1965). "Potential Scattering", North Holland, Amsterdam.
Amdur, I. (1968). *Methods of Exp. Phys.* **7A**, 341.
Amdur, I. and Jordan, J. E. (1966). *Adv. Chem. Phys.* **10**, 29.
Arthurs, A. M. and Dalgarno, A. (1960). *Proc. Roy. Soc. A***256**, 50.
Baede, A. P. M., Moutino, A. M. C., De Vries, A. E. and Los, J. (1969). *Chem. Phys. Lett.* **3**, 350.
Balint-Kurti, G. G. and Levine, R. D. (1970). *Chem. Phys. Lett.* **7**, 107.
Bandrauk, A. D. and Child, M. S. (1970). *Mol. Phys.* **19**, 95.
Bates, D. R. (1960). *Proc. Roy. Soc. A***257**, 22.
Bates, D. R. (1962). "Atomic and Molecular Processes" Chap. 4, Academic Press.
Bates, D. R. and Crothers, D. S. F. (1970). *Proc. Roy. Soc. A***315**, 465.
Bates, D. R., Johnston, H. C. and Stewart, I. (1964). *Proc. Phys. Soc.* (*Lond*) **84**, 517.
Bates, D. R. and McCarroll, R. (1962). *Adv. Phys.* **11**, 39.
Bates, D. R. and Sprevak, D. (1970). *J. Phys.* **B3**, 1483.
Bauer, E., Fisher, E. R. and Gilmore, F. R. (1969). *J. Chem. Phys.* **51**, 4173.
Bernstein, R. B. (1960). *J. Chem. Phys.* **33**, 795.
Bernstein, R. B. (1962). *J. Chem. Phys.* **37**, 1880.
Bernstein, R. B. (1963). *J. Chem. Phys.* **38**, 2599.
Bernstein, R. B. (1966a). *Adv. Chem. Phys.* **10**, 75.
Bernstein, R. B. (1966b). *Phys. Rev. Lett.* **16**, 385.
Bernstein, R. B. and Kramer, K. H. (1966). *J. Chem. Phys.* **44**, 4473.
Bernstein, R. B. and Muckerman, J. T. (1967). *Adv. Chem. Phys.* **12**, 389.
Berry, M. V. (1966). *Proc. Phys. Soc. (Lond)* **80**, 479.
Berry, M. V. (1969). *J. Phys. B.* **2**, 381.
Berry, M. V. and Mount, K. E. (1972). *Rep. Prog. Phys.* **35**, 315.
Born, M. (1960). "The Mechanics of the Atom", Bell and Sons, London.
Brink, D. M. and Satchler, G. R. (1968). "Angular Momentum", 2nd Ed. Oxford University Press, Oxford.
Buck, U. (1971). *J. Chem. Phys.* **54**, 1923.
Buck, U. and Pauly, H. (1971). *J. Chem. Phys.* **54**, 1929.
Bunker, D. L. (1970). "Molecular Beams", Ed. Ch. Schlier, Academic Press, London, New York.

Burnett, G. M. and North, A. M. (1969). "Transfer and Storage of Energy by Molecules", Wiley, New York.
Bykovskii, V., Nikitin, E. E. and Ovchinnikova, M. Ya. (1964). *Zhur. Exp. Teor. Fiz.* **47**, 750; trans. *J. Exp. Th. Phys.* **20**, 500.
Calogero, F. (1967). "Variable Phase Approach to Potential Scattering", Academic Press, London, New York.
Carrington, T. and Polyani, J. C. (1972). Reaction Kinetics, Chap. 5. "MTP International Review of Science" (Ed. J. C. Polyani).
Chan, S. K., Light, J. C. and Lin, J. L. (1968). *J. Chem. Phys.* **49**, 86.
Chen, J. C. Y. and Watson, K. M. (1968). *Phys. Rev.* **174**, 152.
Child, M. S. (1967). *Mol. Phys.* **12**, 401.
Child, M. S. (1969). *Mol. Phys.* **16**, 313.
Child, M. S. (1970). *J. Mol. Spect.* **33**, 487.
Child, M. S. (1971). *Mol. Phys.* **20**, 271.
Child, M. S. (1972). *Mol. Phys.* **23**, 469.
Child, M. S. (1973). *J. Mol. Spect.* **45**, 293.
Connor, J. N. L. (1968). *Mol. Phys.* **15**, 621.
Connor, J. N. L. (1969). *Mol. Phys.* **16**, 525.
Connor, J. N. L. (1970). *Mol. Phys.* **19**, 65.
Connor, J. N. L. and Marcus, R. A. (1971). *J. Chem. Phys.* **55**, 5636.
Connor, J. N. L. (1973a). *Mol. Phys.* **25**, 181.
Connor, J. N. L. (1973b). *Disc. Far. Soc.* **55**, 51.
Corben, H. C. and Stehle, P. (1960). "Classical Mechanics", Wiley, New York.
Coulson, C. A. and Zalewski, K. (1962). *Proc. Roy. Soc. A***268**, 437.
Cowley, L. T., Fluendy, M. A. D. and Lawley, K. P. (1969). *Trans. Far. Soc.* **65**, 2027.
Cross, R. J. (1967). *J. Chem. Phys.* **47**, 3724.
Cross, R. J. (1968). *J. Chem. Phys.* **48**, 4838.
Cross, R. J. (1969). *J. Chem. Phys.* **51**, 5163.
Crothers, D. S. F. (1971). *Adv. Phys.* **20**, 405.
Curtiss, C. F. (1970). *J. Chem. Phys.* **52**, 4832.
Curtiss, C. F. and Adler, F. T. (1952). *J. Chem. Phys.* **20**, 249.
Delos, J. B. (1972). *Phys. Rev.* **A6**, 720.
Delos, J. B. and Thorson, W. R. (1972). *Phys. Rev.* **A6**, 728.
Delos, J. B., Thorson, W. R. and Knudson, S. K. (1972). *Phys. Rev.* **A6**, 709.
Demkov, N. Yu. (1966). *Dokl. Akad. Nauk. S.S.S.R.* **166**, 1076.
Dickinson, A. S. (1970). *Mol. Phys.* **18**, 441.
Diestler, D. J. and McKoy, V. (1968). *J. Chem. Phys.* **48**, 2951.
Dirac, P. A. M. (1926). *Proc. Roy. Soc.* **A111**, 281.
Dirac, P. A. M. (1927). *Proc. Roy. Soc.* **A114**, 243.
Dirac, P. A. M. (1958). "Principles of Quantum Mechanics", 4th Ed. Oxford University Press.
Doll, J. D. and Miller, W. H. (1972). *J. Chem. Phys.* **57**, 5019
Dubrovskii, G. V. (1964). *J. Exp. Theor. Phys.* **19**, 591.
Dwight, H. B. (1961). "Tables of Integrals and Other Mathematical Data", 4th Ed., Macmillan.
Dyson, F. J. (1949). *Phys. Rev.* **75**, 486.
Eu, B. C. (1972). *J. Chem. Phys.* **57**, 2531.
Eyring, H., Walter, J. and Kimball, G. E. (1944). "Quantum Chemistry", Wiley, New York.
Feshbach, H. (1958). *Ann. Phys.* **5**, 357.

Feynmann, R. P. and Hibbs, A. R. (1965). "Quantum Mechanics and Path Integrals", McGraw-Hill, New York.
Firsov, O. B. (1953). *J. Exp. Theor. Phys.* **24**, 279.
Ford, K. W., Hill, D. L., Wakano, M. and Wheeler, J. A. (1959). *Ann. Phys.* **7**, 239.
Ford, K. W. and Wheeler, J. A. (1959a). *Ann. Phys.* **7**, 259.
Ford, K. W. and Wheeler, J. A. (1959b). *Ann. Phys.* **7**, 287.
Freed, K. F. (1972). *J. Chem. Phys.* **56**, 692.
Gersh, M. E. and Bernstein, R. B. (1969). *Chem. Phys. Lett.* **4**, 221.
Gillen, K. T., Rulis, A. M. and Bernstein, R. B. (1971). *J. Chem. Phys.* **54**, 2831.
Goldberger, M. L. and Watson, K. M. (1967). "Collision Theory", Wiley, New York.
Goldstein, H. (1959). "Classical Mechanics", Addison-Wesley, New York.
Gordon, R. G. (1969). *J. Chem. Phys.* **51**, 14.
Gray, G. C. and Van Kranendonk, J. (1966). *Canad. J. Phys.* **44**, 2411.
Gutzwiller, M. G. (1967). *J. Math. Phys.* **8**, 1979.
Gutzwiller, M. G. (1969). *J. Math. Phys.* **10**, 1004.
Gutzwiller, M. G. (1970). *J. Math. Phys.* **11**, 1791.
Gutzwiller, M. G. (1971). *J. Math. Phys.* **12**, 343.
Harms, S. H. and Wyatt, R. E. (1972). *J. Chem. Phys.* **57**, 2722.
Hartmann, H. (1968). Chemische Elementarprozesse, Springer-Verlag, Berlin, New York.
Heading, J. (1962). "Phase Integral Methods", Methuen, New York.
Heidrich, F. E., Wilson, K. R. and Rapp, D. (1971). *J. Chem. Phys.* **54**, 3885.
Heisenberg, W. (1943). *Z. Phys.* **120**, 513–673.
Helbing, R. K. B. and Rothe, E. W. (1969). *J. Chem. Phys.* **51**, 1607.
Herm, R. R. (1967). *J. Chem. Phys.* **47**, 4290.
Herschbach, D. R. (1966). *Adv. Chem. Phys.* **10**.
Herzberg, G. (1950). "Spectra of Diatomic Molecules", 2nd Ed. Van Nostrand, New York.
Herzberg, G. (1967). "Electronic Spectra of Polyatomic Molecules", Van Nostrand, New York.
Herzfeld, K. F. and Litovitz, T. A. (1959). "Absorption and Dispersion of Ultrasonic Waves", Academic Press, London, New York.
Hirschfelder, J. O., Curtiss, C. F. and Bird, R. B. (1954). "Molecular Theory of Gases and Liquids", Wiley, New York.
Jacob, M. and Wick, G. C. (1959). *Ann. Phys.* **7**, 404.
Jackson, J. M. and Mott, N. F. (1932). *Proc. Roy. Soc. A***137**, 703.
Jeffreys, H. (1923). *Proc. Lond. Math. Soc.* **23**(2), 428.
Jeffreys, H. and Jeffreys, B. S. (1956). "Methods of Mathematical Physics", 3rd Ed. Cambridge, University Press.
Johnson, B. R. and Secrest, D. (1968). *J. Chem. Phys.* **48**, 4682.
Jost, R. and Kohn, W. (1953). *Kgl. Danske Vid. Selskab. Mat-Fys. Medd.* **27**, No. 9.
Karplus, M. (1970). "Molecular Beams and Reaction Kinetics", Ed. Ch. Schlier, Academic Press, London, New York.
Keller, J. (1958). *Ann. Phys.* **4**, 180.
Kennard, E. H. (1938). "Kinetic Theory of Gases", McGraw-Hill, New York.
Kennedy, M. and Smith, F. J. (1967). *Mol. Phys.* **13**, 443.
Klein, O. (1932). *Z. Phys.* **76**, 226.
Knudson, S. W. and Thorson, W. R. (1970). *Can. J. Phys.* **48**, 313.
Kramer, K. H. and Bernstein, R. B. (1964). *J. Chem. Phys.* **40**, 200.
Kreek, H. and Marcus, R. A. (1974) (personal communication).

Landau, L. D. (1932). *Phys. Z. Sow.* **1**, 46.
Landau, L. D. and Lifshitz, E. M. (1960). "Electrodynamics of Continuous Media", §65, Addison-Wesley, New York.
Landau, L. D. and Lifshitz, E. M. (1965). "Quantum Mechanics, Non-relativistic Theory", 2nd Ed. Addison-Wesley, New York.
Langer, R. E. (1937). *Phys. Rev.* **51**, 669.
Lawley, K. P. and Ross, J. (1965). *J. Chem. Phys.* **43**, 2930.
Leaf, B. (1969). *J. Math. Phys.* **10**, 1980.
LeRoy, R. J. (1971). University of Wisconsin, Theoretical Chemistry Institute Report WIS–TCI–387.
Levine, R. D. (1968). *J. Chem. Phys.* **49**, 51.
Levine, R. D. (1969). "Quantum Mechanics of Molecular Rate Processes", Oxford University Press.
Levine, R. D. (1971). *Mol. Phys.* **22**, 497.
Levine, R. D. (1973). Theoretical Chemistry, Ed. A. D. Buckingham Chap. 7. "MTP International Review of Science".
Levine, R. D. and Balint-Kurti, G. G. (1970). *Chem. Phys. Lett.* **6**, 101.
Levine, R. D. and Bernstein, R. B. (1972). *Chem. Phys. Lett.* **15**, 1.
Levine, R. D. and Johnson, B. R. (1970a). *Chem. Phys. Lett.* **7**, 404.
Levine, R. D. and Johnson, B. R. (1970b). *Chem. Phys. Lett.* **8**, 501.
Levine, R. D. and Johnson, B. R. (1971). *Chem. Phys. Lett.* **11**, 109.
Levine, R. D. and Johnson, B. R. (1972). *Chem. Phys. Lett.* **13**, 168.
Levine, R. D., Johnson, B. R. and Bernstein, R. B. (1969). *J. Chem. Phys.* **50**, 1694.
Levinson, N. (1949). *Kgl. Danske Vid. Selskab. Mat-Fys. Medd.* **25**, No. 9.
Lippman, B. A. and Schwinger, J. (1950). *Phys. Rev.* **79**, 469.
Livingston, P. M. (1966). *J. Chem. Phys.* **45**, 601.
Magee, J. L. (1940). *J. Chem. Phys.* **8**, 687.
Magnus, W. (1954). *Commun. Pure Appl. Math.* **7**, 649.
Marcus, R. A. (1966). *J. Chem. Phys.* **45**, 4493.
Marcus, R. A. (1970). *Chem. Phys. Lett.* **7**, 525.
Marcus, R. A. (1971). *J. Chem. Phys.* **54**, 3965.
Marcus, R. A. (1972a). *J. Chem. Phys.* **56**, 311.
Marcus, R. A. (1972b). *J. Chem. Phys.* **56**, 3548.
Marcus, R. A. (1973). *Disc. Far. Soc.* **55**, 34.
Margenau, H. and Murphy, G. M. (1956). "Mathematics of Physics and Chemistry", Van Nostrand, New York.
Mason, E. A. (1957). *J. Chem. Phys.* **26**, 667.
Massey, H. S. W. and Mohr, C. B. O. (1933). *Proc. Roy. Soc.* **A141**, 434.
Massey, H. S. W. and Mohr, C. B. O. (1934). *Proc. Roy. Soc.* **A144**, 188.
Massey, H. S. W. and Smith, R. A. (1933). *Proc. Roy. Soc.* **A142**, 142.
McCullough, E. A. and Wyatt, R. E. (1971). *J. Chem. Phys.* **54**, 3578.
Mies, F. H. (1964). *J. Chem. Phys.* **40**, 523.
Miller, S. C. and Good, R. H. (1953). *Phys. Rev.* **91**, 174.
Miller, W. H. (1970a). *J. Chem. Phys.* **53**, 1949.
Miller, W. H. (1970b). *J. Chem. Phys.* **53**, 3578.
Miller, W. H. (1970c). *Chem. Phys. Lett.* **7**, 431.
Miller, W. H. (1971). *J. Chem. Phys.* **54**, 5386.
Miller, W. H. (1972). *J. Chem. Phys.* **56**, 38.
Miller, W. H. and George, T. F. (1972a). *J. Chem. Phys.* **56**, 5668.
Miller, W. H. and George, T. F. (1972b). *J. Chem. Phys.* **57**, 2458.

Miller, W. H. and George, T. F. (1973). *J. Chem. Phys.* **58**, 1343.
Mok, M. H. and Polyani, J. C. (1969). *J. Chem. Phys.* **51**, 1451.
Mok, M. H. and Polyani, J. C. (1970). *J. Chem. Phys.* **53**, 4588.
Morse, R. I. and La Brecque, R. J. (1971). *J. Chem. Phys.* **55**, 1522.
Mortensen, E. M. and Pitzer, K. S. (1962). *Chem. Soc. (Lond) Spec. Publ.* **16**, 57.
Mott, N. F. (1931). *Proc. Camb. Phil. Soc. Math. Phys. Sci.* **27**, 553.
Mott, N. F. and Massey, H. S. W. (1965). "The Theory of Atomic Collisions", 3rd Ed. Oxford University Press.
Newton, R. G. (1962). *J. Math. Phys.* **3**, 75.
Newton, R. G. (1966). "Scattering Theory of Waves and Particles", McGraw-Hill, New York.
Nikitin, E. E. (1968). "Chemische Elementarprozesse", Ed. H. Hartmann, Springer-Verlag, New York. pp. 43–77.
Olson, R. E. and Smith, F. T. (1971). *Phys. Rev.* **A3**, 1607.
Osherov, V. I. (1965). *Zhur. Exp. Teor. Fiz.* **49**, 1157.
Ovchinnikova, M. Ya. (1964). *Opt. i. Specktr.* **17**, 822; transl. **17**, 447.
Pattengill, M. D., Curtiss, C. F. and Bernstein, R. B. (1971). *J. Chem. Phys.* **54**, 2197.
Pauly, H. and Toennies, J. P. (1968). *Methods of Exp. Phys.* **7A**, 227.
Pechukas, P. (1969a). *Phys. Rev.* **181**, 166.
Pechukas, P. (1969b). *Phys. Rev.* **181**, 174.
Pechukas, P. and Light, J. C. (1966). *J. Chem. Phys.* **44**, 3897.
Percival, I. C. and Richards, D. (1970a). *J. Phys. B.* **3**, 315.
Percival, I. C. and Richards, D. (1970b). *J. Phys. B.* **3**, 1035.
Phillips, E. G. (1957). "Functions of a Complex Variable", Oliver and Boyd, London.
Polyani, J. C. and Schreiber, J. L. (1973). "Kinetics of Gas Reactions", Eds H. Eyring, W. Jost and D. Henderson Chap. 9. Academic Press, London, New York.
Polyani, J. C. and Wong, W. H. (1969). *J. Chem. Phys.* **51**, 1439.
Present, R. D. (1958). "Kinetic Theory of Gases", McGraw-Hill, New York.
Ramsay, D. S. and Child, M. S. (1971). *Mol. Phys.* **22**, 263.
Rapp, D. and Kassal, T. (1969). *Chem. Rev.* **69**, 61.
Rapp, D. and Sharp, T. E. (1963). *J. Chem. Phys.* **38**, 2641.
Rich, W. G., Bobbio, S. M., Champion, R. L. and Doverspike, L. D. (1971). *Phys. Rev.* **A4**, 2253.
Roberts, R. E. (1971a). *J. Chem. Phys.* **54**, 1224.
Roberts, R. E. (1971b). *J. Chem. Phys.* **55**, 100.
Robinson, D. W. (1963). *Helv. Phys. Acta.* **36**, 140.
Ross, J. (1966). Molecular Beams, *Adv. Chem. Phys.* **10**, Wiley Interscience, New York.
Rotenberg, M., Bivins, R., Metropolis, N. and Wooten, J. K. (1959). "The 3j and 6j Symbols", Technology Press, M.I.T., Camb. Mass.
Rydberg, J. R. (1931). *Z. Phys.* **73**, 376.
Schlier, Ch. (1970). "Molecular Beams and Reaction Kinetics", Academic Press, London, New York.
Schwartz, R. N., Slawsky, Z. I. and Herzfeld, K. F. (1951). *J. Chem. Phys.* **22**, 767.
Schutte, A., Bassi, D., Tommasini, F. and Scoles, G. (1972). *Phys. Rev. Lett.* **29**, 979.
Secrest, D. and Johnson, B. R. (1966). *J. Chem. Phys.* **45**, 4556.
Sharma, R. D. (1969). *J. Chem. Phys.* **50**, 919.
Sharma, R. D. and Brau, C. A. (1969). *J. Chem. Phys.* **50**, 924.
Siegert, A. J. F. (1939). *Phys. Rev.* **56**, 750.

Simon, A., Vander Sluis, V. H. and Biedenharn, L. C. (1954). Oak Ridge National Laboratory Report No. 1679.
Smith, F. J. (1964). *Physica* **30**, 497.
Smith, F. T. (1960). *Phys. Rev.* **118**, 349.
Smith, F. T. (1962). *J. Chem. Phys.* **36**, 248.
Smith, F. T. (1965). *J. Chem. Phys.* **42**, 2419.
Smith, F. T. (1966). *Phys. Rev.* **150**, 79.
Stueckelberg, E. C. G. (1932). *Helv. Phys. Acta.* **5**, 369.
Swan, P. (1954). *Proc. Roy. Soc. A***228**, 10.
Thorson, W. R. (1961). *J. Chem. Phys.* **34**, 1744.
Thorson, W. R. (1963). *J. Chem. Phys.* **39**, 1431.
Truhlar, D. G. and Kupperman, A. (1972). *J. Chem. Phys.* **56**, 2234.
Van Vleck, J. H. (1928). *Proc. Natl. Acad. Sci.* **14**, 178.
Van Vleck, J. H. (1951). *Rev. Mod. Phys.* **23**, 213.
Vollmer, G. (1969). *Z. Phys.* **226**, 423.
Wergeland, P. (1945). *Avh. Norske Vide. Akad. Oslo. Mat-Nat. Klasse* **9**.
Whittaker, E. T. and Watson, G. N. (1962). "Modern Analysis", 4th Ed. Cambridge University Press.
Wigner, E. (1955). *Phys. Rev.* **98**, 145.
Wong, W. H. and Marcus, R. A. (1971). *J. Chem. Phys.* **55**, 5663.
Woolley, A. M. (1971). *Mol. Phys.* **22**, 607.
Wu, S-F. and Levine, R. D. (1971). *Mol. Phys.* **22**, 881.
Wu, S-F., Johnson, R. B. and Levine, R. D. (1973). *Mol. Phys.* **25**, 609.
Wyatt, R. E. (1972). *J. Chem. Phys.* **56**, 390.
Zener, C. (1932). *Proc. Roy. Soc.* **A137**, 696.

Subject Index

Author Index

The number in italics refer to the Bibliography pages where the references are listed in full.

E

F

G

H

J

K

L

T

V

W

Z